WATER
and Its Impurities

WATER
and Its Impurities

Second Edition

By the late **Thomas R. Camp**

Founding Partner,
Camp, Dresser & McKee,
Consulting Engineers, Boston

and **Robert L. Meserve**

Associate Professor of
Civil Engineering
Northeastern University, Boston

Dowden, Hutchinson & Ross, Inc.

Stroudsburg, Pennsylvania

Library of Congress Cataloging in Publication Data

Camp, Thomas R
    Water and its impurities.

    Includes bibliographical references.
    1. Water--Pollution. 2. Water quality.
I. Meserve, Robert L., 1929-   joint author.
II. Title.
TD420.C34 1974   628.1'6   73-20219
ISBN 0-87933-112-7

Manufactured in the United States of America.

Exclusive distributor outside the United States and Canada:
John Wiley & Sons, Inc.

To the memory of
THOMAS RINGGOLD CAMP
(1895–1971)

and MARGARET EVANS CAMP
(1890–1970)

Preface to the Second Edition

Increased awareness of environmental problems in the last decade has emphasized the importance of, and the need for, such books as Camp's *Water and Its Impurities*. Too many have become "instant ecologists" and environmental "experts"; too few have taken the time and spent the effort necessary to obtain a useful background in environmental technology. Protection of the environment is everybody's responsibility, but solving major problems in environmental management must be delegated to those with adequate preparation and experience.

Dr. Camp was eminently qualified to lead others in exploring problems connected with water quality. His long and productive career included consulting practice as a sanitary engineer, many years as a professor of sanitary engineering at the Massachusetts Institute of Technology, a lifelong interest in research and the development of understanding in water and wastewater treatment, and the publication of many papers of classical stature in the hydraulics of grit removal, sedimentation, filtration, and mixing. He founded what has become the largest international consulting firm devoted strictly to environmental engineering. His work and his example have been of inestimable value to his students, associates, and clients over five decades.

Thomas R. Camp passed away in November 1971 at the age of 76. Prior to his death he invited the present writer to collaborate on a revision of this work. Although much of the original book has been altered, it still represents mainly the effort of its senior author.   The publication of a second edition is a fitting memorial to one of the world's outstanding environmental engineers.

Boston   R.L.M.
July 1973

Preface to the
First Edition

Historically, the design of works for the conveyance, control, and treatment of water and wastewater during the nineteenth and twentieth centuries has been the responsibility of civil engineers. The development of hydraulics, hydrology, and structural theory and the techniques of application of these engineering sciences to the design of water and wastewater works must be credited to the civil engineering profession.

The modern public health movement probably originated in England about the middle of the nineteenth century as a result of the intolerable sanitary conditions that developed upon the concentration of population in urban areas in the wake of the industrial revolution. A movement was started to promote better water supplies and more satisfactory methods of sewage disposal. Water supplies were purified, sewers were built, and attention was directed to methods of sewage treatment. This movement was initiated by public sanitary commissions and medical or health officers. The movement resulted in a reduction in the incidence of disease, although the causative agents of the diseases were not then known.

Beginning about 1880, a series of epoch-making discoveries in bacteriology demonstrated that specific microorganisms are responsible for many of the common diseases: anthrax, tuberculosis, cholera, typhoid fever, diphtheria, tetanus, plague, gonorrhea, pneumonia, and others. It was proved that the mere accumulation of filth was not responsible for disease, but that the diseases were transmitted from person to person through the agency of microorganisms. Some of these disease germs, notably the agents of typhoid fever and cholera, were shown to be waterborne, passing from the intestinal tract of infected persons through sewage to the drinking water of others.

About 1890, chemists, biologists, and bacteriologists became involved in problems of water quality and water pollution. A start was made to instruct some civil engineers in water chemistry and water biology and bacteriology. Thus emerged the professional sanitary engineer. The sanitary engineering profession has, over the years, been principally responsible for the protection and improvement of our water supplies, for the development of sewage and wastewater treatment and disposal methods, and for water-pollution abatement. Sanitary engineers have administered the public agencies dealing with these problems, they have designed and supervised the construction of the water and wastewater works, and they have operated these works. Today, they are involved in many other engineering activities relating to the public health and comfort.

T.R.C.

Contents

CHAPTER 1

Introduction

1.1 USES OF WATER

Water is, of course, absolutely essential to life—not only human life but all life, animal and vegetable. Indeed, it is a part of life itself, since the protoplasm of most living cells contains about 80 percent water, and any substantial reduction in this percentage is disastrous. Most of the biochemical reactions that occur in the metabolism and growth of living cells involve water, and all take place in water, which has often been referred to as the universal solvent. Yet man's assessment of the value of water is very low until he finds himself without it. Most of the surface of the earth is covered with water; this is salt water, to be sure, but most of the habitable land masses have adequate freshwater resources. Water is so plentiful, therefore, that man believes it should be free or, at most, very cheap. As a result, water supplies of good quality continue to be developed and extended on an emergency basis and, with very few exceptions, are never quite adequate for human needs. One of the principal reasons for the rarity and inadequacy of public water supplies in the underdeveloped countries is the widespread belief that water should be free.

Man uses water not only for drinking and culinary purposes but also for bathing, washing, laundering, heating, and air conditioning; for agriculture, stock raising, and gardens; for industrial processes and cooling; for water power and steam power; for fire protection; for disposal of wastes; for fishing, swimming, boating, and other recreational purposes; for fish and wildlife propagation; and for navigation.

Every activity of man involves some use of water. For many of these uses, the water must be withdrawn from a watercourse; for others, it is not withdrawn.

1

Often, the water that is withdrawn is returned to the watercourse either with pollutants or with an increase in temperature, or both, and if taken from large watercourses, this water is used over and over again by downstream communities, with further impairment in quality each time. In agriculture, a large portion of the water is lost by transpiration and evaporation. It is obvious that some cost must be involved in the handling and controlling of water for most of these purposes. Nevertheless, water by the ton is cheaper than any other commodity required by man, except air. Municipal water at 30 cents per 1,000 gallons is collected, purified, transported, and delivered to the tap for 7.2 cents per ton; this cost includes water for fire protection and for many municipal purposes. Similarly, municipal wastewaters may be collected, treated, and discharged to the receiving waters at a cost per ton not substantially higher.

1-2. WATER QUALITY AS AFFECTING USE

The quality of water suitable for man's many needs varies widely, and what is satisfactory for one purpose may not be for another. For example, a minimum of salt is required for drinking water, but seawater is used for navigation, for swimming, boating, fishing, and recreation, for condenser cooling water, and for many industrial purposes. The total salt content in a drinking water should preferably not exceed 500 parts per million (ppm), but there are many municipal water supplies which are used, without complaint, with salt contents in excess of 2,000 ppm. On the other hand, the allowable total salt content in feed water for modern ultrahigh pressure steam power plants is less than 1 ppm. For the propagation of aquatic life, water must contain dissolved oxygen, but water containing dissolved oxygen is corrosive to all metals.

Water polluted with human sewage is likely to contain disease germs and viruses and is, therefore, not safe for drinking or swimming; however, aquatic life grows well in such water, provided it contains enough dissolved oxygen and sufficiently low concentrations of toxic metals and compounds. Fish and shellfish grown in such polluted waters may not be safe for human handling and consumption, particularly if eaten raw. The most important reason for man's concern with water quality is the protection of the public health. The earliest water and sewage treatment plants were constructed in efforts to reduce the incidence of cholera in Europe and typhoid fever in the United States, both of which were endemic in the nineteenth century. These efforts were highly successful. As a consequence, the prevention of waterborne human disease remains the most important reason for water and sewage treatment, and with the successful control of cholera and typhoid in the industrial nations, attention is now being directed to viral diseases such as hepatitis.

With the expansion of the chemical and pharmaceutical industries since World

War II, hundreds of new products are being placed on the market each year, many of which are toxic or otherwise harmful to man and other life. Among the toxic organic compounds are the pesticides used to kill insects, rodents, and weeds. These substances are sprayed on farms, forest, and water, and they accumulate with repeated applications. Synthetic detergents are widely used both in home and in industry. They are very difficult and expensive to remove from wastewater and, hence, collect in our watercourses and groundwaters. They are only mildly toxic, but are objectionable because of the foam they produce. There are also many wastes that are by-products in the manufacture of chemicals and whose composition and toxicity are largely unknown.

Since World War II there has been a growing use of radioisotopes, and many nuclear reactors have been constructed for power and other purposes, all as a result of the Atoms for Peace promotional program of the Atomic Energy Commission. All these activities result in some discharge of thermal and radioactive wastes to the air or to our watercourses. This is a public health hazard of growing importance, especially where nuclear power plants are located on streams above water supply reservoirs and intakes.

The obvious approach to problems of water quality is to identify the impurity, develop methods of analyzing for its presence and concentration in water, determine the limiting concentration in water for a particular water use, and estimate whether this concentration will be exceeded for a particular case. This should be done before any decision is made that treatment is required for the removal of a particular impurity. It may well be found that the best and cheapest solution for a particular problem is to control the impurity at the source where it is manufactured or used, before it is discharged to the receiving water. Unfortunately, the numbers of these new pollutants are growing so rapidly that it has not been possible to identify them as fast as they are produced, and the development of analytical methods for determining their concentrations and assessing their toxicities is inadequate. The simplest solution for this dilemma is to require that each manufacturer identify any substance he proposes to discharge to the atmosphere or into wastewaters, or which may be so discharged by the purchaser. In addition, he should develop means for analyzing these compounds and determining their safe concentrations, if such methods are not already known.

Meanwhile, tertiary or advanced treatment of water and wastewaters is being promoted by the Environmental Protection Agency (successor to the U.S. Public Health Service), and a vigorous campaign is underway to compel the construction of wastewater treatment plants through the United States, with the avowed purpose of completing the construction program in the 1970s. Many of these treatment plants are not urgently needed, and some of them may not be needed at all if the whole problem is studied in all its complexities.

Quality standards for drinking water were first adopted by the Public Health Service in 1914. These standards, revised in 1925, 1942, 1946, and 1962, are accepted by all the states as minimum quality standards for public water supplies. Water quality standards for public bathing places have also been set by the Conference of State Sanitary Engineers and are accepted, with minor variations, by all the states. Water quality standards for shellfish growing areas have been established by the Public Health Service in cooperation with the states and the shellfish industry; these, too, are generally accepted. Water quality standards, which, in the authors' opinion, are far from adequate, have been developed for agricultural use. Attempts have been made to promote water quality standards for industry, but industrial uses increase, develop, and change so rapidly that it appears best for each user to custom-tailor the water quality specifications to his own needs. All these standards will be discussed hereinafter.

Raw water quality standards, for water to be used for drinking after treatment, were not set by the Public Health Service in the 1962 revision of *Drinking Water Standards*, because modern water treatment methods when properly used are adequate for polluted water. It was felt that raw water standards would unnecessarily prohibit the use of some waters and would discourage the development of improved methods of water treatment. Nevertheless, most state and interstate water-pollution control agencies have set raw water quality standards for water to be used for drinking after treatment. These standards are designed to force the treatment of wastewaters in the belief that treatment of both wastes and drinking water is required to produce a satisfactory drinking water. The authors believe that as a minimum all municipal sewage should be screened and disinfected before being discharged to the receiving waters, whatever the use of the receiving waters. Further treatment of wastewaters may or may not have merit, depending upon the local circumstances in each case.

The state and interstate water-pollution control agencies, since about 1940, have been in the process of classifying public waters according to their appropriate uses. In connection with this classification, water quality standards have been established. As shown in Chapter 6, these standards vary widely from state to state, and the limits on some of the items are unrealistic. There is widespread agreement among the states, supported by the Public Health Service, to require a minimum treatment for wastewaters of primary settling, whether or not such treatment is required by the water quality standard for the receiving waters. On the other hand, no bacterial standard is in effect for the lower use classifications, despite the fact that this is the only water quality standard which directly concerns the public health. Meanwhile, the discharge without treatment in storm-water overflows from combined sewers of up to about one third of all the sewage solids produced in municipal sewage together with the accompanying bacteria and viruses has been all but ignored throughout the country.

The principal objectives in water-pollution abatement are to safeguard the health of the people, to prevent fish kills, to prevent odor nuisances, and to remove unsightly sludge banks and floating material. The protection of health requires the destruction of pathogenic organisms in human sewage. The extent of treatment required to achieve the other objectives depends upon the amount of pollution and the pollution-receiving capacity of the waters. A rational analysis of the problem requires that each watercourse be studied in detail to discover the location and extent of pollution and to determine if the receiving-water quality standards are being met. If they are not being met, bacteriological and oxygen-balance studies should be made to estimate the allowable pollution loads at each point of pollution. This method of approach has, with few exceptions, not been followed throughout the country. The progress of pollution abatement has been measured in terms of the number of wastewater treatment plants that have been constructed, rather than in terms of their effectiveness in restoring the safe use of the receiving waters, at the lowest cost. The quality of the receiving waters is the only satisfactory measure of the need and the effectiveness of pollution abatement.

In all man's uses of water, he must contend with the corrosiveness of water, that is, its ability to dissolve or disintegrate the materials used by man. The annual cost resulting from the corrosion of metals is a very substantial part of our national income and amounts to billions of dollars. All solid materials, with the possible exception of living organisms, are soluble in water to some extent, and even living organisms are completely dependent on water as a solvent in their metabolic processes. Even the noble metals, gold, silver, and platinum, are slightly soluble in water. Similarly, concrete and other construction materials dissolve slightly in water. Corrosion is relative. Some materials are very much more corrosive than others. A rational attack upon the problems of corrosion requires the identification of the reactions involved, the determination of their rates, and the control of the materials or of the water so as to avoid or minimize the reactions. Considerable progress has been made in the development of corrosion-resistant metals and protective coatings, but these are costly. A better understanding of the whole problem might result in enormous savings.

1-3. REMEDIAL MEASURES

Water that evaporates from the surface of the ocean, fresh watercourses, and vegetation is carried in the air to be precipitated as rainfall or snow. The molecules of water vapor in the air are pure water, but the falling raindrops formed by their condensation are no longer pure. Raindrops are saturated with nitrogen, oxygen, and the other gases of the atmosphere, and during their fall onto land masses, they entrain dust and smoke particles and fumes. A part of the rain that falls on land surfaces runs off over the ground to the nearest open watercourses, carrying with it eroded soil, decaying vegetation, living micro-organisms, and colloidal and suspended matter of numerous varieties. Some of these materials are dissolved in the water. The remainder of the rain seeps into the soil and flows underground as groundwater. Except in very arid regions or where groundwaters are overpumped, the groundwater eventually flows out into an open watercourse. In its transit through the topsoil to become groundwater, this water dissolves carbon dioxide, which makes it more acid. In its passage through the ground and rocks, it dissolves many minerals, notably calcium, magnesium, iron, and manganese; sulfates, silicates, and chlorides. The water in our natural watercourses, even before the intervention of man, was never pure. These natural waters have always varied widely in quality, and many were originally named in recognition of their natural appearance or characteristics (e.g., Blackwater Falls, Green Bay, Cold River). Man has always had to exercise care in selecting his water supplies. This continues to be the case, and cost is the prime factor in weighing one supply against another.

It is to be noted that the water supplies for our large metropolitan communities in the Northeast—New York, Boston, Manchester, New Hampshire, and Portland, Maine—are unfiltered surface supplies. For many years these supplies received no treatment, not even chlorination, which is now universally used. These supplies come from natural lakes or man-made reservoirs on mountain streams. The waters are, of course, not as clear and colorless as those produced by modern rapid filtration, but they are satisfactory to the users and are relatively free of quality problems. By contrast, filtration is required for municipal water supplies in most other regions of the United States because the surface waters are not clear and relatively colorless. In addition, softening is required in the midwest and some other regions, and special problems are encountered in many areas.

It is to be noted, further, that the conventional processes for treatment of wastewaters—sedimentation and biochemical oxidation—also take place in the receiving waters. These processes were always present in natural waters, even before the introduction of man-made pollution, and they will continue to be present after the wastewater treatment plants are in operation. Therefore, the receiving waters, themselves, are treatment plants, and when excess organic pollution exists, these natural treatment plants are overloaded. The problem is to determine how to divide the burden of treatment between the treatment plants and the receiving waters, in the best and most economical manner. The appearance of an unsightly sewer outlet discharging sewage from a small community into a large receiving stream does not automatically require a conventional treatment plant. The test is the quality of the receiving waters after dilution of the sewage. The unsightliness may be abated by adequate dispersion of the sewage into the receiving waters. The complete removal of settleable solids from sewage, by plain settling, will not necessarily effect a substantial reduction in the rate of accretion of bottom deposits in the receiving waters, because about 75 percent of the oxidizable organic matter remains in the effluent. Much of this matter is converted in the receiving waters to bacterial bodies, which will be flocculated and settled out if the velocity is low.

Neither water treatment nor wastewater treatment can be a satisfactory remedy for pollution of our watercourses by pesticide sprays or by salt used for melting snow and ice on our highways. These materials must be controlled at their points of use, because they are damaging to land, plants, and animals, as well as to water. Similarly, excess soil erosion cannot be abated by water or waste treatment. Better land use is needed. In water quality problems, the whole environment must be examined.

It is not the purpose of this book to describe the various methods of treating water and wastewater or to promote treatment as a remedial measure. Treatment will, of course, be required in many cases, and new methods of treatment are continually being developed. The purpose of this book is to discuss the significant characteristics of water and its impurities that affect quality for various uses, that indicate whether remedial measures are required and the various alternative measures available, and that permit an evaluation of the effectiveness and cost of these alternative remedies. This method of approach is, in the opinion of the authors, somewhat novel but badly needed. Much of the material in this book is the work of others; some has been further developed under Camp, and some is new. It will be evident to the reader who is expert in this field that the subject matter has not been developed exhaustively and that extensive research is needed in many areas.

The unit process most widely used in water and wastewater treatment is sedimentation. It is effective and economical in removing a portion of the suspended particles, provided they are large enough to be removed and do not have a density equal to that of water. Plain settling of municipal sewage will remove about 60 percent of the suspended matter and about 25 percent of the oxidizable organic matter; it will also remove a considerable amount of grease and oil, some of which settles, and some of which floats and is removed from the water surface by skimming. The effectiveness of plain settling of industrial wastewater varies widely, depending upon the amount of settleable suspended solids in the water. Some industrial wastewaters, with high concentrations of organic matter, contain no suspended solids. In the treatment of water, plain settling is of little value unless the water is taken from a highly turbid stream containing large concentrations of silt and sand. In this case, plain settling is usually followed by coagulation with alum or an iron salt and by secondary settling and filtration.

If the suspended particles in water or wastewater are not large enough for effective removal by settling, they may be made so by the addition of a coagulant and by gentle stirring in flocculation basins prior to settling. This is common practice in modern water filter plants, and is sometimes used in the treatment of wastewaters. If the impurities are colloidal in size (e.g., color from vegetation) they may be flocculated so that they can be partially removed by settling. Filtration is required following flocculation and settling for effective removal of suspended matter and color in water treatment, but filters are usually not practicable in wastewater treatment because they clog too rapidly. Some dissolved minerals, notably calcium, magnesium, iron, and manganese, may be partially removed from water by settling and filtration, provided they are first precipitated by lime or lime and soda ash, and then flocculated. For effective removal by filtration, particles should be flocculant in nature so that they adhere to the surfaces of the grains of the filter medium, and they should be relatively small so that they penetrate into the filter bed and are not strained out on top. Adequate pretreatment of the water to be applied to filters is usually required. There are many variations in the types of coagulants, and many new coagulant aids and filter aids are being developed.

If more nearly complete removal of the hardness-producing metallic ions (calcium and magnesium) is required, greensand or synthetic zeolites may be used either separately or in combination with the lime–soda softening process. These solid zeolites exchange sodium ions for calcium and magnesium ions. They may also be used to remove iron and manganese ions. For effective use of the zeolites, the water must be free of compounds of the metallic ions to be removed and must be substantially free of colloidal and suspended matter; otherwise the base-exchange capacity will be drastically reduced by solid deposits on the surfaces of the zeolite grains. The zeolite will act as a filter, rather than as a base-exchange medium. Since iron and manganese precipitate readily in the presence of dissolved oxygen, the water should preferably be oxygen-free if iron or manganese is to be removed by zeolite.

For more complete removal of dissolved minerals, evaporation and condensation are available. Distillation has been available for many generations, but it is very costly (about $1.35/1,000 gal) when applied to water for general use, even though techniques for distillation have been greatly improved within the last decade. Dissolved minerals may also be removed by ion-exchange resins: hydrogen-cation exchange for metallic ions and anion exchange for the negatively charged ions. Electrodialysis and freezing are also under development for demineralization.

In both water and wastewater treatment, there are many gas transfer problems. In some cases, carbon dioxide must be removed from water; in others, it must be dissolved in water. The same is true of oxygen. The dissolved oxygen content of polluted streams is critical. When it is too low, fish kills and odor nuisances may result. The conventional remedy is to remove from wastewaters most of the decomposable organic matter that creates the oxygen demand (BOD). This is known as secondary or biological treatment of the wastes, which in combination with settling is known as complete treatment. The unit process is aeration, in which oxygen or air is added and the organic matter is rapidly oxidized or synthesized into bacterial bodies. Approximately 90 percent removal of BOD may be accomplished by complete treatment of municipal sewage. Oxygen is naturally added to the receiving water through atmospheric reaeration and through photosynthesis by green aquatic plants, but more may be added artificially by aeration or by adding sodium nitrate.

One of the oldest methods of wastewater treatment, *lagooning*, which in the recent past has been replaced by more mechanical methods, is now receiving the approval it deserves as it becomes better understood. Well-designed lagoons are now referred to as *stabilization* or *oxidation* ponds. If sufficient area is provided in shallow ponds, 95 to 98 percent removal of BOD may be accomplished. The processes that take place in stabilization ponds are those which take place in natural watercourses: bacterial oxidation of organic matter in suspension or solution, settling out of suspended solids, anaerobic decomposition of deposited solids, and aeration by the atmosphere and by photosynthesis. Photosynthesis is fostered by encouraging the growth of algae.

Taste and odor problems are usually solved by copper sulfate treatment of the supply reservoir in order to destroy algae or by the use of activated carbon, in the filter plant, to adsorb the odor-producing substances. There are many special problems in water and wastewater treatment, such as silica removal, corrosion control, adjustment of pH, disinfection, and removal of phenols, cyanides, hexavalent chromium, and other toxic substances.

Nearly all treatment processes have, as by-products, the impurities removed from water. These by-products, usually in the form of sludges or slurries, must be handled, treated, and disposed of. The cost of this disposal is in many cases greater than the cost of treatment of the water or wastewater. Treatment processes that lead to an excessive amount of such by-products should be avoided, if possible, or modified so as to minimize the quantity of by-product and the cost of its disposal.

CHAPTER 2

Physical Properties
of Pure Water

It is characteristic of liquids, including water, that the molecules and ions of which the liquids are composed are held closely together in a relatively constant volume, although they are free to move with respect to one another. Foreign molecules, ions, or particles in water are also free to move about, subject to the electric charges and the small drag forces created by the motion. Some understanding of the physical properties of pure water is an essential foundation to an adequate grasp of the characteristics that affect the quality and treatability of natural and polluted waters.

2-1. MOLECULAR STRUCTURE

A molecule of water vapor (*1*) consists of an oxygen atom and two hydrogen atoms attached by strong covalent bonds, as illustrated in Figure 2-1. From thermochemical data, the heat of formation of gaseous water from H_2 to O_2 gas is 57.8 kcal/mole and the heats of dissociation of H_2 and O_2 gas are 103.4 and 118.2 kcal/mole, respectively. Hence the bond energy for each O—H bond is $\frac{1}{2}(57.8 + 103.4 + 59.1)$ or 110.2 kcal/mole. The distances of the boundary shells in Figure 2-1 from the centers of the atoms are equal to the *van der Waals radii*. The equivalent size of the molecule is about 3.3 Å or angstrom units $(1 \text{ Å} = 10^{-8} \text{ cm} = 10^{-4} \text{ } \mu\text{m})$. In the vapor, the molecules are separated by great distances and are moving at high velocity; their translational energy is so great that when they collide the van der Waals forces are insufficient to hold them

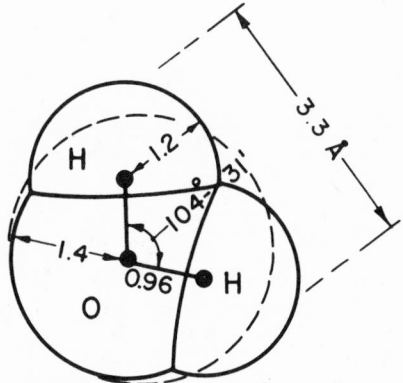

Figure 2-1
Molecule of water vapor.

together. The vapor expands and exerts pressure in conformity with the kinetic theory of gases.

In the liquid, however, the molecules are held in intimate contact with one another. Each molecule in liquid water occupies a volume of 29.7 Å³, which indicates a porosity between the molecules of about 36.7 percent. The nature of the bonding forces is not yet well understood. Some of these forces are probably of the weak van der Waals type (with bond energies of about 5 kcal/mole) and, to a considerable extent, nondirectional, since the molecules are free to move in the liquid. The magnitude of these attractive forces is nevertheless great when compared to the kinetic pressure forces, since the vapor pressure of the liquid is negligible compared to the pressure that is implied by the gas laws.

Water is known to be as associated liquid, consisting of a heterogeneous fluid structure of single molecules of H_2O, groups of such molecules, and hydrogen and hydroxyl ions, H^+ and OH^-. The concentration of H^+ or OH^- is too small to have an appreciable effect on the structure of pure liquid water, only one out of 555 million H_2O molecules being ionized at pH 7.0. The distribution of local molecular structure, however, must have great influence on the density, viscosity, and other properties of liquid water. According to Bernal and Fowler (2), liquid water consists of ice-tridymite structure (Figures 2-2 and 2-3), quartz-like structure, and close-packed ammonia-like structure.

In the ice structure, each O atom is bonded to four others by hydrogen bonds in a tetrahedral configuration (Figures 2-2 and 2-3). The H atoms in the O—H · · · O bond are no longer 0.96 Å from the O atom, as in the water vapor molecule, but may be either 0.99 or 1.77 Å away, and are free to move a distance of 0.78 Å between O atoms, according to Pauling (1). The volume per H_2O molecule in ice is 32.3 Å³. The volume per gram-molecule is 6.06 ×

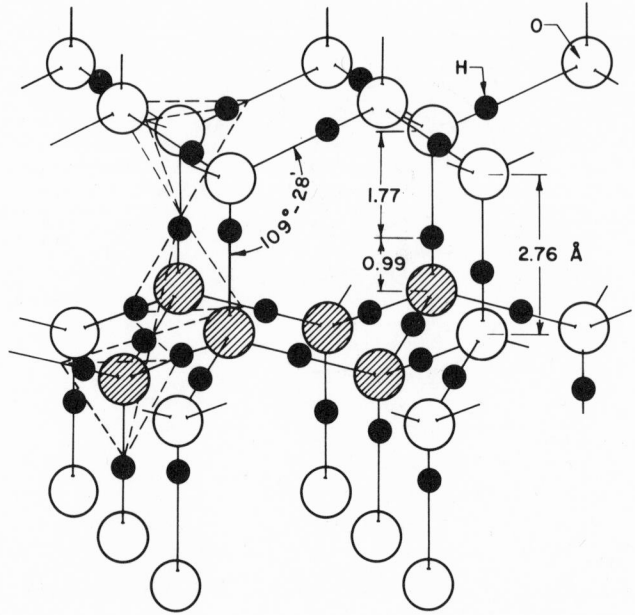

Figure 2-2
Ice crystal.

$10^{23} \times (32.3/10^{24}) = 19.56 \text{ cm}^3$, and the resulting density is $18/19.56 = 0.92$. The hydrogen bond has about the same strength as the van der Waals forces and is of great significance in the interactions of water and other associated liquids with organic compounds. Broken hydrogen bonds in water are probably re-

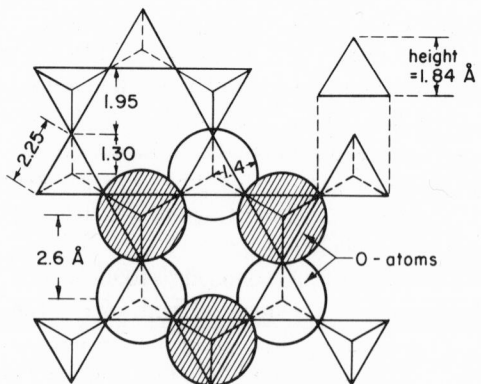

Figure 2-3
Tetrahedral configuration of ice crystal.

sponsible to a considerable extent for its high *dielectric constant* ($D = 78.55$ for water at 25°C), and thus contribute to its strength as an ionizing solvent for polar compounds.

A molecule of H_2O vapor has a volume of about 18.85 $Å^3$. A sphere of equal volume has a diameter of 3.3 Å, as shown in Figure 2-1. Assuming that liquid water is composed of such spheres, free to rotate and move about, but in contact with one another, the porosity between them would be 0.476 if they were packed rectilinearly (at the corners of cubes) and 0.258 if they were in a closest-packing arrangement. The corresponding densities at 1 atm would be 0.825 and 1.17, the arithmetic average of which is 0.9975, the density of water at about 23°C. This suggests that the packing of molecules of pure water might follow a probability curve of densities between 0.825 and 1.17 (*3*).

2-2. DENSITY AND ELASTICITY

The *mass density, ρ,* of pure liquid water at atmospheric pressure is substantially 1.0 g/cm^3 at 4°C, the temperature of maximum density (absolute density of pure water = 0.999973 g/cm^3 at 3.98°C, International Bureau of Weights and Measures, 1910). The relative density at other temperatures is shown in Table 2-1. At the ordinary pressures encountered in hydraulic practice,

Table 2-1
Relative Density of Pure Water at Atmospheric Pressure

Temperature (°C)	ρ	Temperature (°C)	ρ
0	0.99987	20	0.99823
4	1.00000	25	0.99707
10	0.99973	30	0.99567
15	0.99913	100	0.95838

water is assumed to be an incompressible fluid. It is an elastic fluid, however, and has a *modulus of elasticity* of about 300,000 psi, which indicates a volumetric decrease of about 0.000048 for each added atmosphere of pressure.

2-3. VISCOSITY

The *viscosity* of a fluid is the proportionality factor in the expression for the intensity of viscous shear at a point in the moving fluid:

$$\tau = \mu \frac{dv}{ds} \tag{2-1}$$

where τ is the *shear per unit area* of surface normal to the s direction; dv/ds is the maximum *velocity gradient* at the point, the s direction representing the direction in which the maximum occurs; and μ is the *absolute viscosity* [(force $\times$ time)/length^2]. In the CGS system, the unit of viscosity is the poise (dyne sec/cm^2), and in the English system, the unit is lb sec/ft^2.

The viscosity of pure water is a function of the temperature and is relatively independent of the pressure at the usual pressures encountered. The kinetic theory of liquids has not been developed to a stage that will permit the calculation of viscosity from density and thermal properties, as is the case for gases, but experimental measurements of viscosity are available for many liquids, including water. The viscosity of pure water at atmospheric pressure, as a function of the temperature, is presented in Figure 2-4 and Table 2-2. The *kinematic viscosity*, ν, is equal to μ/ρ.

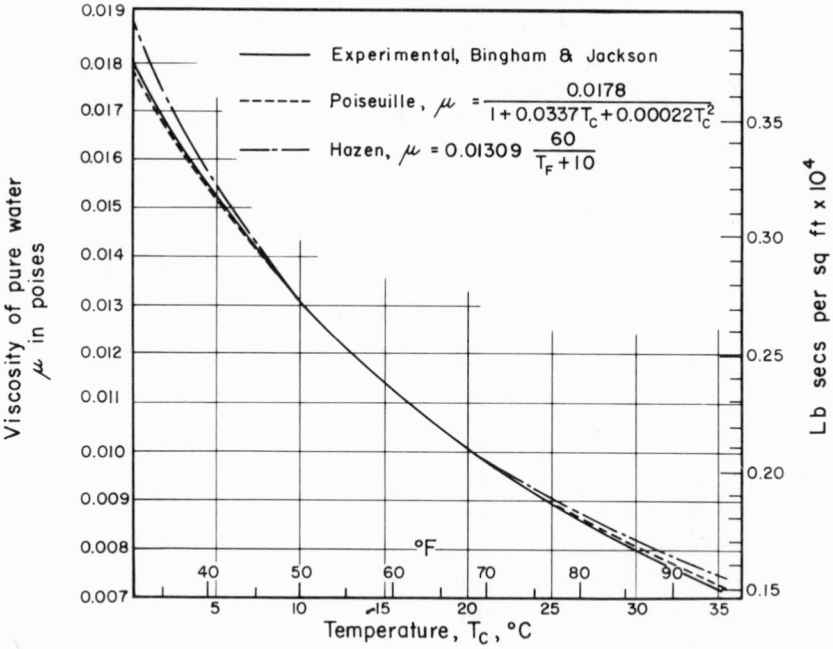

Figure 2-4
Viscosity of pure water as a function of water temperature.

The intensity of viscous shear determines the internal energy loss in problems of fluid motion. The velocity gradient, dv/ds, and the shear intensity are very important factors in flocculation, settling, and filtration.

Table 2-2
Viscosity of Water

Temperature (°C)	μ (poises)[a]	Temperature (°C)	μ (poises)	Temperature (°C)	μ (poises)
0	0.01792	35	0.00723	70	0.00406
5	0.01519	40	0.00656	75	0.00380
10	0.01308	45	0.00599	80	0.00357
15	0.01140	50	0.00549	85	0.00336
20	0.01005	55	0.00506	90	0.00317
25	0.00894	60	0.00469	95	0.00299
30	0.00801	65	0.00436	100	0.00284

[a] μ in lb sec/ft^2 = 1/478.7 μ in poises. ν in ft^2/sec = 1/929 ν in cm^2/sec.

2-4. VAPOR PRESSURE AND RELATIVE HUMIDITY

The *vapor pressure* of a liquid is the pressure of the liquid vapor in contact with the liquid at which vapor molecules condense on the liquid surface as fast as they evaporate from it. The vapor pressure is a function of the temperature for any liquid, but it increases slowly with increase in total pressure. The vapor pressure of pure water is shown in Table 2-3.

The vapor pressure of water equals the *partial pressure* of water vapor in air saturated with moisture. If the partial pressure of water vapor in contact with

Table 2-3
Vapor Pressure of Pure Water at Atmospheric Pressure

Temperature (°C)	Vapor pressure, p (atm[a])	Latent heat, ΔH (kcal/mole)	Free energy, $\Delta F°$ (kcal/mole)
0	0.006	10.73	2.77
10	0.012	10.63	2.48
15	0.0168	10.58	2.34
20	0.0231	10.53	2.20
25	0.0313	10.48	2.054
30	0.0419	10.43	1.91
40	0.0728	10.33	1.63
50	0.1217	10.23	1.35
60	0.1965	10.13	1.08
80	0.4675	9.92	0.54
100	1.0	9.70	0

[a] 1 standard atm = 760 Torr at 0°C.

liquid water is less than the vapor pressure, water will evaporate faster than the vapor condenses. The *relative humidity* of air is the ratio of the partial pressure of water vapor in the air to the vapor pressure. The vapor pressure is an equilibrium characteristic of a liquid. The relation of vapor pressure to the *latent heat* and *free energy of evaporation* will be discussed in Chapter 4. The latent heat and free energy of evaporation of water at various temperatures are shown in Table 2-3.

2-5. LATENT HEAT OF FUSION

The *latent heat of fusion* of pure water into ice at $0°C$ is 1.436 kcal/mole. The latent heat of fusion of ice from seawater at $-8.7°C$ is 0.972 kcal/mole.

2-6. SURFACE TENSION

As previously indicated, the molecules of a liquid are held together by non-directional bonding forces or attractive forces between molecules. Each molecule is presumed to be attracted to all other molecules within a certain radius beyond which the forces of attraction become negligible. Molecules that are closer to a free surface than the magnitude of this radius will be attracted to the interior of the liquid by a resultant force. The potential energy per unit of surface area, represented by this force, is known as the *surface energy*. The numerical value of the surface energy is equal to the *surface tension*, δ, of the liquid. With all liquids, the surface tension decreases with rise in temperature.

The surface tension of pure water in contact with air, at various temperatures, is shown in Table 2-4. It is pertinent to note that, although the vapor pressure of water increases more than 100-fold between its freezing and boiling points, the surface tension is decreased by only about 22 percent. Thus strong attractive forces, which hold the liquid together, are still there when water is boiling.

Table 2-4
Surface Tension of Pure Water in Contact with Air

Temperature ($°C$)	Surface tension, δ (dynes/cm)	Temperature ($°C$)	Surface tension, δ (dynes/cm)
0	75.6	50	67.91
10	74.22	60	66.18
20	72.75	70	64.4
30	71.18	80	62.6
40	69.56	100	58.9

The *interfacial tension* between water and another liquid that is immiscible with water is approximately equal to the difference between their individual surface tensions, provided each is a saturated solution of the other. The interfacial tension between two liquids also decreases with rise in temperature.

The phase rule formulated by Gibbs shows that, in the addition of a solute to a solvent, there will be marked differences between the concentration of solute in the surface and in the main body of the solution, if the solute and solvent have different surface tensions. Small amounts of a solute with low surface tension will concentrate in the surface and greatly lower the surface tension of the solution; large amounts of a solute of high surface tension will concentrate away from the surface and only slightly increase the surface tension of the solution. Although this phenomenon appears to have small effect on the vapor pressure of dilute aqueous solutions, it does affect the rate of gas transfer through the interface. It is therefore of great importance in the treatment of water and liquid wastes.

REFERENCES

1. L. Pauling, *The Nature of the Chemical Bond*, 3rd ed., Ithaca, N.Y., Cornell University Press, 1960.
2. J. D. Bernal and R. H. Fowler, "A Theory of Water and Ionic Solution," *J. Chem. Phys.*, *1*, 515 (1933).
3. T. R. Camp in C. A. Hampel and G. G. Hawley, eds., *Encyclopedia of Chemistry*, 3rd ed., New York, Van Nostrand Reinhold Company, 1973.

CHAPTER 3

Physical Properties of Impure Water

Whether the impurities in water or wastewaters are solid, liquid, or gas in their natural states, they are dispersed in three progressively finer states: suspended, colloidal, and dissolved. The state of subdivision is of the utmost importance, since it determines the methods required for the removal of an impurity. The concentration of the total impurities may be measured roughly by the *total solids* (*1*) determination, in which a sample of the unfiltered water is evaporated and the residue weighed. Volatile liquids and gases, usually a small part of the total impurities in natural water, must be determined by other means.

The impurities in natural freshwaters usually occur in concentrations which are so small that the density and the viscosity are, in effect, the same as for pure water. The impurities in natural seawater, which contains about 3.5 percent by weight of salts, increase the density about 2.5 percent, but have a negligible effect on the viscosity. The viscosity of a brine containing 20 percent sodium chloride is about 55 percent greater than the viscosity of pure water.

Impurities in municipal sewage normally occur in such low concentrations that the viscosity and density of the sewage are substantially those of pure water. The impurities in sewage sludges, heavy industrial wastewaters, and slurries, however, occur in sufficient concentrations to affect markedly both the density and viscosity. The density of a strong aqueous solution or of a sludge or slurry cannot be computed accurately from the density of the components, and should be determined experimentally if precision is required. The viscosities of sludges or slurries must also be measured, since they vary enormously. Many sludges are *thixotropic* and gel on standing. Such sludges, being *pseudoplastic*, do not have true viscosities.

3-1. SUSPENDED PARTICLES

The particles in a *suspension* are visible either to the naked eye or through an ordinary microscope, and they are sometimes large enough to be removed readily by settling (or flotation) and filtration. Such particles contribute turbidity or cloudiness to the water. The *turbidity* (1) of a water is its capacity for absorbing or scattering light, and is measured by the concentration of fine silica that produces an equivalent effect. The concentration of solids in suspension may also be measured by the *suspended solids* (1) test, in which the solids are separated by filtration through filter paper or a Gooch crucible.

There is no precise relationship between the suspended solids content and the turbidity of water, inasmuch as the latter is influenced by the size and character of the particles as well as by their concentration. The ratio of the suspended solids content to the turbidity, called the *coefficient of fineness*, is a measure of the size of particles causing turbidity; the size increases with the increasing magnitude of the coefficient.

3-2. COLLOIDAL PARTICLES

Dispersed particles that are smaller than about 1 μm and larger than individual molecules (about 10 Å) exhibit *colloidal* properties. They can be distinguished only by means of an ultramicroscope or an electron microscope. When colloidal solutions are observed in the path of an incident light beam, they appear perfectly clear; however, when examined at right angles to the beam (in a Tyndall cone), many colloids appear turbid even to the naked eye. Colloidal particles contribute little to the normal turbidity of water, but they are largely responsible for the color of natural waters. No simple means are available for measuring the concentration of colloidal particles in water. When they are responsible for color, the *color* (1) test is a rough measure of concentration.

Colloidal particles have such small mass that they have considerable velocity of translation; that is, they are subject to *diffusion* (*Brownian motion*) and move in the liquid, as gas molecules move in a vacuum. They conform approximately to the fundamental equation of the simple kinetic theory:

$$p V_m = \tfrac{1}{3} N m u^2 = RT \tag{3-1}$$

in which (in the CGS system) p is the osmotic pressure in dynes/cm^2, V_m is the volume in cubic centimeters occupied by 1 mole of the particles, N is *Avogadro's number* (the number of molecules per mole = 6.06 X 10^{23}), m is the mass of each particle in grams, u is the mean velocity of the particles in cm/sec, R is the *gas constant*, which has the value 8.315 X 10^7 ergs/mole/°K, and T is the temperature in degrees Kelvin (degrees Celsius + 273).

It is evident from Eq. (3-1) that the average kinetic energy $\frac{1}{2}mu^2$ is independent of particle size. Small particles have high velocities, whereas the velocity of suspended particles is so low that it is not observable. The velocity of particles 1 μm in size is barely discernible, being about 0.5 cm/sec from Eq. (3-1); however, the velocity of the smallest colloidal particles is extremely high, being about 10^4 cm/sec. The velocity of colloidal particles, which arises from their kinetic energy, is the cause of flocculation of colloids. Flocculation is therefore very rapid for small particles, but it ceases altogether before the particles are large enough to be visible or to be removed by settling and filtration. Visible suspended particles must be flocculated by stirring the suspension to produce velocity gradients.

The *osmotic pressure* also increases greatly as the colloidal particles become smaller. Since $V_m = M/c\rho$, where M is the molecular weight of the dispersed particles, c the concentration by mass, and ρ the density of the water, Eq. (3-1) may be written

$$p = \frac{c\rho}{M} RT \qquad\qquad (3\text{-}1a)$$

The law in this form is generally known as van't Hoff's equation for osmotic pressure of solutes. Very small colloidal particles with molecular weights of about 1,000 (size about 10 Å) will exert pressures of about 2.5 X 10^5 dynes/cm^2 (about 3.5 psi), in concentrations of 1,000 ppm. The osmotic pressure of larger particles is so small that it is difficult to observe.

Colloidal particles diffuse much more slowly than do gas molecules in a vacuum, because they collide with the water molecules as well as with themselves. The *rate of diffusion* is thus limited by the drag on the particles by the liquid. The rate of diffusion per unit area according to the law of Fick (*2*) is

$$c \frac{ds}{dt} = -D_c \frac{dc}{ds} \qquad\qquad (3\text{-}2)$$

where $c(ds/dt)$ is the quantity of particles per second diffusing through 1 cm^2 of a boundary in the direction of s, ds/dt is the diffusion velocity, $-(dc/ds)$ is the concentration gradient, and D_c is the diffusion coefficient.

For colloidal particles that are large compared with water molecules, the water may be considered as a continuous medium, and the law of Stokes can be used for the drag force. Einstein (*3*) has derived the value of the *diffusion coefficient* for this case.

$$D_c = \frac{1}{3\pi d\mu} \frac{RT}{N}$$

(3-3)

where $3\pi d\mu$ is the Stokes drag for particles of diameter d in a fluid of viscosity μ. Small colloidal particles ($d = 10$ Å $= 10^{-7}$ cm) will have a diffusion coefficient in water at 20°C of about 0.4×10^{-5} cm^2/sec and a corresponding diffusion velocity of 0.4 cm/sec, for a concentration gradient of unity and a concentration of 10 ppm. Dispersion of colloidal particles by diffusion is thus very slow.

Colloidal particles conform to the laws of settling as do larger particles, but because of their small size and the countereffect of diffusion, they settle so slowly that it is impractical to remove them by sedimentation. Their rate of removal by filters is also too slow for practical use. It is therefore necessary to build them into particles of suspended size, by flocculation, before they can be effectively removed from water.

The existence of colloidal particles in natural water is evidence of the stability of the colloidal state. Otherwise the particles would flocculate by their own Brownian motion. Coalescence of particles in a *suspensoid* colloid is prevented by the existence of an electric charge of sufficient magnitude on the surface of the particles. All interfaces are charged electrically, but the charge becomes significant compared to the mass only when the surface area per unit of volume is very large. In a stable colloid, the charges on the particles have the same sign and are of sufficient magnitude to cause the particles to repel one another.

If the colloidal particles are solid, as is the case with suspensoids, they are composed of single crystals or several crystals loosely bound together with some water molecules, and on the outside surface there are some broken bonds and free valences, which are responsible for part of the electric charge. Ions adsorbed from the water also contribute to or modify the charge; the ions are concentrated in a diffuse layer or ionic atmosphere around the particle. The material of which the particle is composed and the composition and pH of the water seem to determine the sign and magnitude of the net electric charge.

In natural waters with neutral or acid pH, materials of neutral or acid character, such as glass, silica, sulfur, and most organic particles, are negatively charged. Materials of basic character, such as alumina, Al_2O_3, and ferric oxide, Fe_2O_3, are positively charged. Usually a variation in pH or in the concentration of certain ions in the water will alter the magnitude of the charge, and the charge may pass through zero and change sign. Fe_2O_3 and $CaCO_3$ sols are negatively charged at high pH values.

Suspensoid sols may be flocculated by reducing or neutralizing the electric charges on the particles that stabilize the sols. An effective means is the addition of a salt to produce multivalent *flocculating ions* of opposite charge. According to the Schulze–Hardy rule, the flocculating value of an ion increases enormously with the valence of the ion; the relative flocculating values of trivalent, bivalent, and monovalent ions are roughly 1,000 to 30 to 1 for negative colloids. In the case of coagulation with alum or ferric sulfate in the acid region, bivalent SO_4^{2-} is much more effective than Cl^- for flocculating positively charged colloids; Al^{3+} and Fe^{3+} are more effective than Ca^{2+} and Mg^{2+}, which in turn are more effective than Na^+ for flocculating negative colloids.

Many colloidal particles in natural waters, particularly those of organic origin, are composed principally of water, and seem to be surrounded by a film of oriented water molecules, which contribute to their stability. These *emulsoid* colloids are more difficult to flocculate than suspensoids, for they require dehydration as well as reduction in the magnitude of the charge.

A colloid composed of liquid particles is called an *emulsion*. Emulsions are not prevalent in natural waters, but oils, fats, and soaps are frequently found in the emulsified form in sewage and other liquid wastes.

3-3. DISSOLVED IMPURITIES

Dissolved impurities are those which are dispersed in water as single molecules or as ions. The concentration of all the nonfluid dissolved solids may be determined by the *total dissolved solids* (*1*) test in which the residue is weighed after the water is evaporated from a sample that has first been filtered. If the residue is ignited, the *organic* or *volatile solids* (*1*) will be burned off, leaving ash, which is a rough measure of the *dissolved mineral solids* (*1*). In order to treat for removal of mineral solids, it is desirable that the composition of the solids be known from a mineral analysis (*1*) of the water in terms of ions.

The molecules of water-soluble organic and nonpolar compounds are usually held together in the solid state by van der Waals forces. These forces are readily disrupted in solution, and either electrically neutral molecules or dipole molecules are produced. Some organic molecules are so large that they exhibit all the characteristics of colloidal particles; others are very small.

Soluble polar or mineral compounds are usually held together in the solid state by ionic bonds; the acidic elements or radicals in the crystal have captured electrons from the metallic elements, with the result that acid radicals are negatively charged and the metals are positively charged. The charge on a negative ion is equal to the number of valence electrons captured, and the charge on a positive ion equals the number of valence electrons yielded.

Ions of like charge repel each other and those of unlike charge are attracted; the principal attracting or repelling force between two ions is the *coulombic force* (*4*):

$$F = \frac{z_1 z_2 \epsilon^2}{D r^2} \tag{3-4}$$

where F is the force in dynes, z_1 and z_2 are the valences of the two ions, $\epsilon =$

4.77 X 10^{-10} abs esu is the charge of an electron, r is the distance in centimeters between the ions, and D is the dielectric constant of the dispersion medium (unity for a vacuum and gases).

The bond energy corresponding to the coulombic force is Fr ergs per bond, the work done in separating the ions from the distance r to an infinite distance against the force F. For a NaCl crystal in air, for example, r is about 2.76 Å, F is 3 X 10^{-4} dyne, and the energy is 8.28 X 10^{-12} erg per bond. The bond energy per mole is 6.06 X 10^{23} X 8.28 X 10^{-12} = 5.02 X 10^{12} ergs or 120 kcal. Because of attractions of surrounding molecules and van der Waals forces, the total bond energy (also called the *lattice energy*) in the NaCl crystal is 181.3 kcal/mole, according to Pauling (4).

When a NaCl crystal is immersed in water, the high dielectric constant reduces the bond energy to 1/78.55 of its original value, or to about 2.33 kcal/mole. This is so small that the kinetic bombardment of the water molecules readily dissociates the crystal into ions; this dissociation substantially increases the value of r and leads to a further reduction in the coulombic bond energy. The bond energy appears to be transferred to *hydration* of the ions. Glasstone, Laidler, and Eyring (5) explain the abnormal mobility of hydrogen ions in electrochemical processes by the postulate that each H^+ ion is attached to a water molecule, thus giving H_3O^+, and that current is conducted mainly by the transfer of protons from one water molecule to another. Similarly, the abnormal mobility of OH^- is accounted for by the transfer of protons from water molecules to OH^- ions. If each H^+ is attached to a water molecule, the average amount of hydration of some other common ions based upon the experiments of Washburn and Millard (6) is given in Table 3-1.

Table 3-1
Hydration of Ions in Normal Solution

Ion	Moles of H_2O per ion	Ion	Moles of H_2O per ion
H^+	1	Na^+	8.5
Cs^+	4.7	Li^+	14.0
K^+	5.4	Cl^-	4

The dielectric constant of the liquid solvent, water in this case, is considered to be due to electric or *dipole moments* (7) of the liquid molecules. If a molecule has two equal charges, $+\epsilon$ and $-\epsilon$, separated by a distance r, the dipole moment η is ϵr. The value of η for water at 25°C is 1.842 X 10^{-18}. The *bond energy* between an ion and a dipole, as given by W. D. Harkins (in 8), is

$$E = \frac{z_1\, e\eta\, \cos\theta}{r^2} \tag{3-5}$$

where E is in ergs and θ is the angle of the moment with respect to the ion. The energy per H_2O molecule for Na^+ or Cl^- ions computed from Eq. (3-5) is about 13 kcal/mole; and if these ions are hydrated to the extent indicated in Table 3-1, the energy of hydration is about 165 kcal/mole of NaCl. This is an approximate accounting for the loss in crystal bond energy due to the dielectric constant of water. According to computations by Latimer (9) from heats of reaction, the energy of hydration of Na^+ is 95 kcal and of Cl^- is 89 kcal, a total of 184 kcal/mole of NaCl. These figures fully account for the transfer of lattice energy to hydration, but they indicate a greater relative hydration of Cl^- than is shown in Table 3-1.

Multivalent salts, such as $CaCO_3$, which ionizes into Ca^{2+} and CO_3^{2-} ions, will have residual coulombic bond energies in water which are several times that for NaCl, because of the valence factors z_1 and z_2 in Eq. (3-4). This residual bond energy is equal to, or greater than, the van der Waals energy. Multivalent polar compounds are, therefore, less soluble than the monovalent salts in water, and sometimes exist in water as suspended or colloidal crystals in equilibrium with the ions. Na and K salts are fully ionized in the concentrations in which they occur in natural waters.

3-4. DIFFUSION RATES

Both ions and molecules in true solutions move freely in the liquid, except that positive and negative ions must diffuse together and maintain the same local concentration of positive and negative charge. Both ions and molecules conform to the kinetic theory; they exert osmotic pressure and diffuse according to Eqs. (3-1) to (3-2). When, however, the concentration of the solute is sufficient to lower the vapor pressure of water appreciably (concentrations in excess of about 0.1 molal), the osmotic pressure will exceed the values given by Eqs. (3-1) and (3-1a). For strong solutions, the *osmotic pressure* is given by

$$p_o = \frac{RT}{V_1} \ln \frac{p_v}{p_v'} \tag{3-1b}$$

where V_1 is the volume of 1 mole of pure water, p_v is the vapor pressure of pure water, and p_v' the lowered vapor pressure.

The coefficient of diffusion for ions and small molecules is not adequately stated by Eq. (3-3), however, since the particles are of the same order of size as the water molecules, and the water cannot, therefore, be considered as a continuous medium. Glasstone, Laidler, and Eyring (5) suggest the substitution in

Eq. (3-3) of the quantity $\mu\sqrt[3]{V_m/\rho N}$ (where $V_m/\rho N$ is the volume per molecule of liquid) for the Stokes drag $3\pi\,d\mu$. The experimental value for the diffusion coefficient, D_c, may differ by a factor of 10 from that computed by Eq. (3-3), using either quantity for the drag. Some measured values for D_c are given in Table 3-2.

Table 3-2
Diffusion Coefficients in Pure Water

Solute	Molal Concentration	Temperature ($^\circ$C)	D_c (cm^2/sec)
Dissolved oxygen (O_2)	—	20	2.03×10^{-5}
Dissolved nitrogen (N_2)	—	20	1.88×10^{-5}
Dissolved helium (He)	—	25	6.3×10^{-5}
Hydrochloric acid (H^+ and Cl^-)	0.1	19	2.56×10^{-5}
Sodium chloride (Na^+ and Cl^-)	0.1	15	1.09×10^{-5}
Calcium chloride (Ca^{2+} and Cl^-)	2.0	10	0.79×10^{-5}
Chlorine (Cl_2)	0.1	12	1.41×10^{-5}
Chlorine (Cl_2)	—	16	1.26×10^{-5}
Sulfuric acid (H^+, HSO_4^-, and SO_4^{2-})	1.0	12	1.3×10^{-5}
Magnesium sulfate (Mg^{2+} and SO_4^{2-})	1.0	7	0.35×10^{-5}
Ammonia (NH_3, NH_4OH, NH_4^+, and OH^-)	1.0	15	1.78×10^{-5}
Acetic acid (CH_3COOH, H^+, and CH_3COO^-)	0.2	13.5	0.89×10^{-5}
Methyl alcohol (CH_3OH)	—	18	1.37×10^{-5}
Glycerol ($C_3H_5OH_3$)	—	20	0.83×10^{-5}
Glucose ($C_6H_{12}O_6$)	—	18	0.57×10^{-5}
Urea (NH_2CONH_2)	—	20	1.18×10^{-5}
Various dyes	—	18	$(0.17–0.58) \times 10^{-5}$
Various proteins	—	7–17	$(0.04–0.13) \times 10^{-5}$
Gamboge particles ($d = 0.5\ \mu m$)	—	20	0.85×10^{-8}

The diffusion coefficient of dissolved oxygen in pure water is reported by Davidson and Cullen (*10*) at about 2.03×10^{-5} at 20°C, as shown in Figure 3-1. Equation (3-3), the Stokes–Einstein expression for the coefficient of diffusion of colloidal particles, indicates that the value of the coefficient should vary directly with the value of T/μ. Figure 3-1 also shows values of T/μ as a function of the temperature. The dashed line in Figure 3-1, showing the diffusion coefficients for dissolved oxygen computed by means of the Stokes–Einstein relation, may be compared with the Davidson–Cullen values, shown by the solid line. Also shown in Figure 3-1 by means of a dashed line are the relative effects of temperature on D_c for dissolved oxygen, based on experiments by Adeney (*11*).

The coefficients of diffusion in pure water for both carbon dioxide and 0.05 *m* NaCl, as taken from the International Critical Tables, are shown in Figure 3-1.

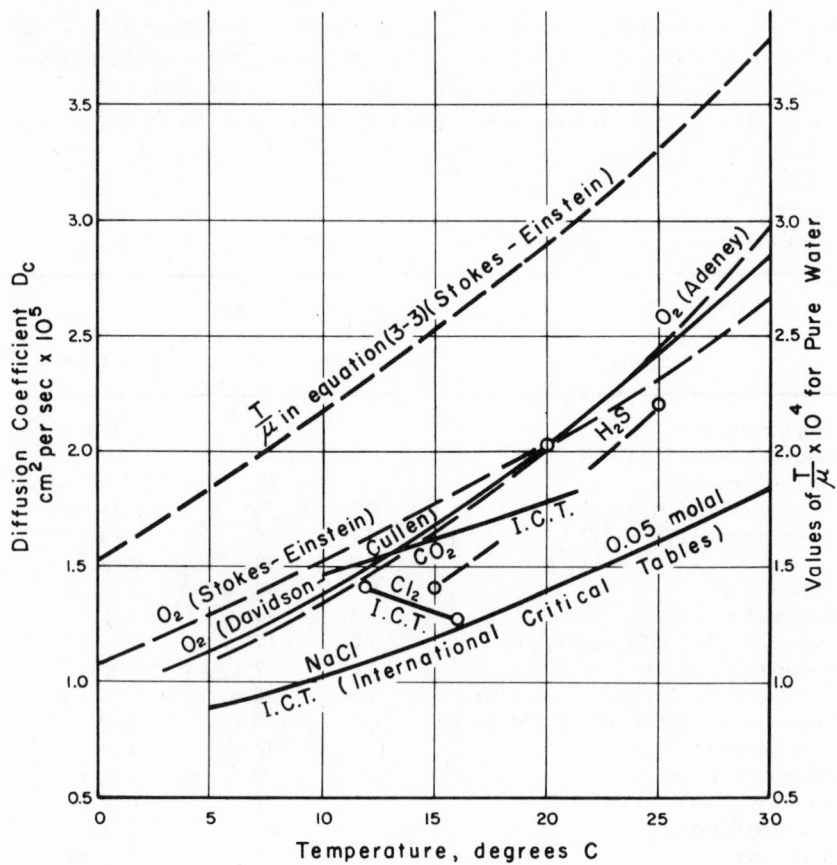

Figure 3-1
Effect of temperature on diffusion coefficients in pure water.

It will be noted that D_c for NaCl conforms almost exactly to the Stokes–Einstein temperature relation, but that D_c for CO_2 does not. The decrease in D_c for Cl_2 with increase in temperature is puzzling. The values of D_c for H_2S, as shown in Figure 3-1, were taken from two sources and are of questionable reliability for pure water.

The values of D_c in pure water for NaCl and KCl at one temperature, as taken from the *International Critical Tables*, indicate that D_c is approximately constant at low salt concentrations, but begins to increase at concentrations of about 6 percent.

There is almost a complete absence of information on diffusion coefficients for various substances in aqueous solutions and suspensions that are encountered with natural water and wastewaters. The Stokes–Einstein relation, Eq. (3-3), tends to indicate that the effect of the impurities in the water on the value of D_c, for a particular substance, is felt through the viscosity of the medium. Inasmuch as it has now been demonstrated that the rate of absorption of gases in, and extraction of gases from, aqueous solutions depends upon the diffusion coefficient of the gases in a liquid film at the gas–water interface, it is essential to have reliable values for D_c in the liquid film. Furthermore, since surface-active agents such as detergents are known to concentrate at gas–water interfaces, it is highly probable that they will affect the values of D_c for gases in the liquid film.

3-5. GAS TRANSFER INTO AND FROM WATER

In 1924, Lewis and Whitman (*12*) advanced the two-film theory of the absorption of a gas by a liquid. The theory is based on the assumption of a gas film and a liquid film located at the interface, through which the gas must pass by molecular diffusion and beyond which the concentration of the gas is uniform. In most problems relating to water and wastewater treatment, it has been found that the resistance of the gas film is negligible by comparison with that of the liquid film, and only the liquid film need be considered. The *time rate of solution* (or extraction) of the gas is as follows:

$$\frac{dc}{dt} = K_L \frac{A_s}{V} (c_s - c) \qquad (3\text{-}6)$$

and

$$\ln \frac{c_s - c_0}{c_s - c} = K_L \frac{A_s}{V} t \qquad (3\text{-}6a)$$

where c_0 and c are the concentrations of the dissolved gas in the body of the liquid at the beginning and end of time t, c_s is the concentration of the dissolved gas at the gas–liquid interface (the saturation concentration), A_s is the interfacial area including ripples or waves ($C_A A_0$, where A_0 is the quiet surface area), V is the volume of the liquid, and K_L is the *liquid film coefficient*. If the concentrations c and c_s in Eq. (3-6) are expressed in parts per million and the rate dc/dt is in ppm/hr, K_L has the dimensions of hour^{-1}. If A_s/V is in cm^2/cc, K_L has the dimensions of cm/hr. Many experimental evaluations have been made of the liquid film coefficient, K_L, for dissolving oxygen from air bubbles rising through water. For a 2.4-mm-diameter bubble rising through water at 15°C, Scouller and Watson (*13*) determined K_L to be 140 cm/hr. Adeney and Becker (*14*) found variations from 32 to 230 cm/hr. Ippen (*15*) and co-workers reported values of 7 to 295 cm/hr.

Dobbins (*16*, *17*, *18*) has presented an excellent theoretical discussion of the liquid film theory of gas transfer in which he shows that $K_L = D_c/L$ for an unbroken film, where L is the thickness of the film. With moving gas bubbles, however, shear is always present to continuously change the liquid film. Similarly, in flowing streams and in tanks or reservoirs, motion of the liquid is always present in varying degrees to change the liquid film. Dobbins presents experimental evidence to show that where the film is being continuously replaced or renewed, the value of K_L is well represented by

$$K_L = \sqrt{D_c t} \; \coth \sqrt{\frac{rL^2}{D_c}} \qquad (3\text{-}7)$$

in which r = the average rate of renewal of the liquid film.

Dobbins has shown that L and r, for a given liquid and temperature, are dependent upon the physical pattern and speed of mixing; and that for the same pattern and speed, the value of K_L is determined by the value of D_c. Using two gases in water, helium and nitrogen, which have widely different coefficients of diffusion, Dobbins (*16*) has determined the relationship of L to r at 20°C by ingenious experiments in an 8-in. diameter vertical cylinder partially filled with the water, with the gas above; turbulence is created by an oscillating lattice always positioned below the water surface. The results are shown by Figure 3-2.

Because of the difficulty of determining the true surface area A_s, Dobbins used the projected area A_0, and found it necessary to distinguish between the apparent and true values of the various parameters. If K_L', r', and L' are the apparent values of K_L, r, and L, their relations are

$$K_L = \frac{K_L'}{C_A} \qquad (3\text{-}8)$$

$$r = \frac{r'}{C_A^2} \qquad (3\text{-}9)$$

$$L = C_A L' \qquad (3\text{-}10)$$

$$rL^2 = r'(L')^2 \qquad (3\text{-}11)$$

$$rL^3 = C_A r' (L')^3 \qquad (3\text{-}12)$$

Values of K_L' were determined by Dobbins (*17*) for each gas from measured concentrations by means of Eq. (3-6a) over a wide range of speeds up to 280 rpm and four different conditions of stroke and cover over the lattice; and four curves were plotted for each gas of K_L' against rpm. For each of several different speeds and mixing conditions, L' and r' were computed by inserting the measured values of K_L' for nitrogen and helium in Eq. (3-7) and solving the resulting two equations simultaneously. It was found that a consistent relationship existed between L' and r' independent of the mixing conditions; and that

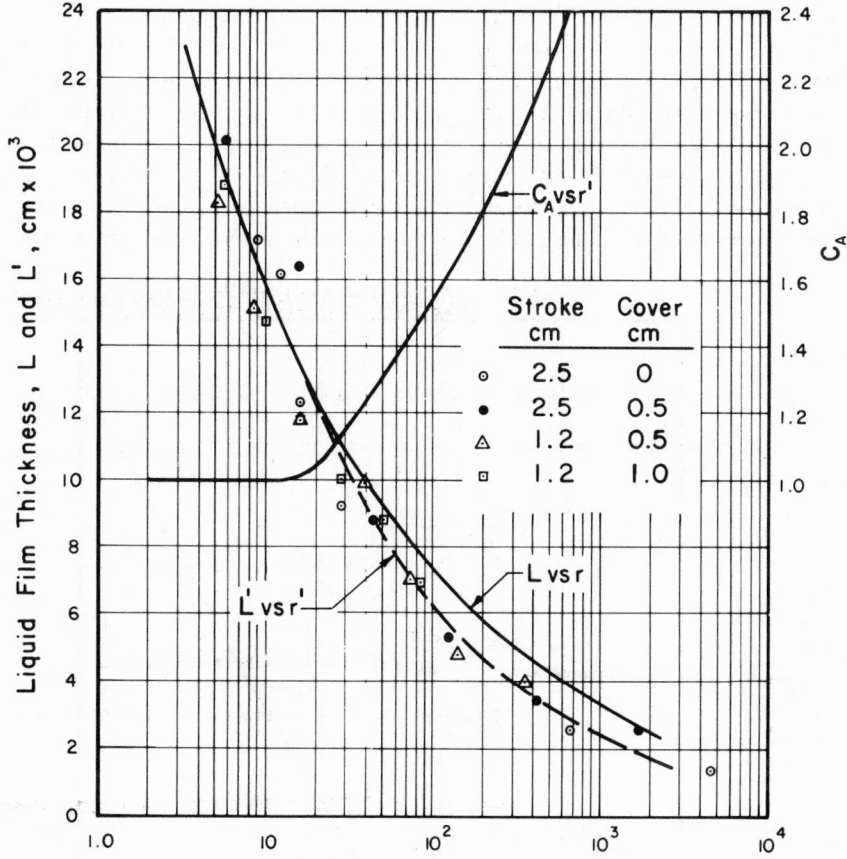

Rate of Film Renewal, r and r′ per minute at 20°C

Figure 3-2
Relationship of r and L.

$r'(L')^3$ had a constant value for small values of r' up to about 20 per minute, and a decreasing value for larger values of r' as C_A increased above 1.0. Dobbins (*17*) used values of D_c of 1.88×10^{-5} for N_2 and 4.81×10^{-5} for He, but later found a better value for He at 5.5×10^{-5}.

The authors have made independent computations of L' and r' from Dobbins's measured values of K'_L, using values of D_c of 1.88×10^{-5} for N_2 and 5.5×10^{-5} (adjusted from 6.3×10^{-5} at 25°C by ratios of T/μ) for He, and have, by means of the preceding equations, computed the values of C_A and the true values of L and r as shown on Figure 3-2. The average value of rL^3 is constant at 40×10^{-6}

cm^2/min for all values of r at 20°C. It will be noted in Figure 3-2 that, with Dobbins's experimental apparatus, C_A was about 2.2 with r' at about 500. Figure 3-3 shows the values of C_A for various ratios of wave height to length, using a sine wave. For most oxygen-balance studies of polluted streams, the value of C_A will probably be close to 1.0; but an attempt should be made to evaluate C_A when samples are collected by the field crew.

It has been observed that many impurities in water, particularly surfactants, tend to concentrate in the liquid film to reduce the surface tension but to re-

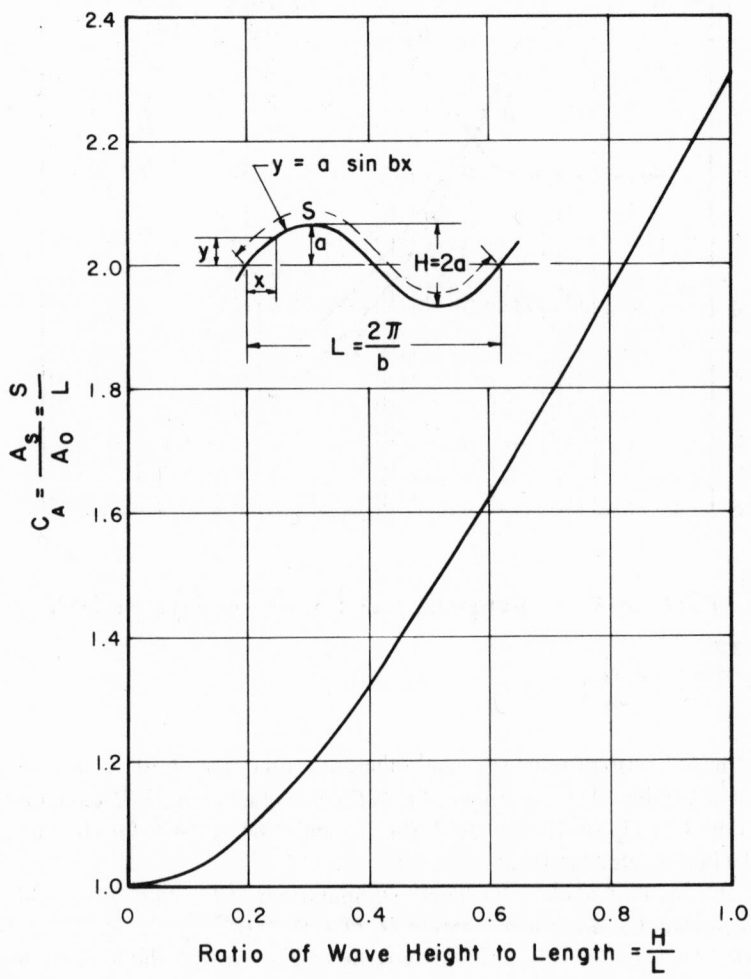

Figure 3-3
Relation of true to projected surface area with sine wave.

tard the rate of film renewal. Metzger and Dobbins (*19*) reasoned that r could not be inversely proportional to δ as previously proposed by Dobbins (*17*), but could be inversely proportional to a compressional modulus M_s (dynes/cm) resulting from local differences in surface tension.

It was reasoned that in a free surface turbulent stream the value of r in the topmost layer is proportional to $\rho LE/M_s$, and the value of L is proportional to the lower limit of the size of the eddies $(v^3/E)^{1/4}$, derived by Kolmogoroff, where E is the energy dissipated per unit mass of water, or W/ρ, where W is the mean value of the energy dissipated per unit volume of the stream. These relations lead to

$$r = \frac{C_2}{(C_1)^3} \frac{\rho v^{3/4} E^{3/4}}{M_s} \tag{3-13}$$

$$rL^2 = \frac{C_2}{C_1} \frac{\rho v^{9/4} E^{1/4}}{M_s} \tag{3-14}$$

$$rL^3 = C_1 \frac{\rho v^3}{M_s} \tag{3-15}$$

An important conclusion from Eq. (3-15) is that rL^3 is independent of the system dynamics and depends solely on the properties of the fluid in the liquid film. The experimental studies with nitrogen and helium indicate the validity of Eq. (3-15) and also provide a means for evaluating C_1/M_s for the laboratory mixer used by Dobbins. At 20°C, the value of ρv^3 is 61.3×10^{-6} g-cm^3/sec^3-min, and rL^3 was 40×10^{-6} cm^3/min, giving a value of 0.652 sec^3/g for C_1/M_s. The value of C_1/M_s for distilled water was found by Metzger and Dobbins (*19*) to be nearly constant over a temperature range of 15 to 30°C, although neither C_1 nor M_s was evaluated separately. The value of M_s is expected to increase greatly for polluted water containing surface-active impurities. The value of C_2 could not be evaluated for the laboratory mixer because E was not measured. For a natural stream, E might be estimated from the mean velocity and the slope of the energy grade line, except that both mean velocity and slope may change several times within a reach.

Assuming that the value of M_s for distilled water does not vary with the temperature, and since ρ and v do vary with temperature, Dobbins's theory automatically takes into account the effect of temperature on gas transfer. The value of r at any temperature T in terms of its value at 20°C derives directly from Eq. (3-13) as follows:

$$r_T = r_{20} \left(\frac{v_T}{v_{20}}\right)^{3/4} \left(\frac{\rho_T}{\rho_{20}}\right) \tag{3-16}$$

Table 3-3 shows the properties of pure water and the solutions for Eqs. (3-16)

Table 3-3
Effect of Temperature on K_L for Oxygen in Pure Water

	Temperature (°C)							
	0	5	10	15	20	25	30	35
ν (cm²/sec × 10²)	1.792	1.519	1.308	1.141	1.008	0.897	0.805	0.728
ρ (g/cm³)	1.0	1.0	0.9997	0.9991	0.9982	0.9971	0.9957	0.9939
$(\nu_T/\nu_{20})^{3/4}$	1.55	1.362	1.219	1.084	1.0	0.916	0.844	0.784
ρ_T/ρ_{20}	1.002	1.002	1.002	1.001	1.0	0.999	0.998	0.996
r_T/r_{20}	1.554	1.365	1.221	1.085	1.0	0.915	0.842	0.782
$\rho\nu^3$ × 10⁶ g cm³/sec³	5.76	3.51	2.24	1.485	1.023	0.725	0.521	0.383
g cm³/sec³ min r_L^3 × 10⁶ cm³/min	346	211	134.5	89.1	61.3	43.5	31.3	23.0
(He, D_c = 5.5)	225	138	87.8	58.1	40.0	28.4	20.4	15.0
D_c (cm²/sec × 10⁵) for O₂	0.93	1.12	1.38	1.68	2.037	2.42	2.85	3.30

and (3-15) at temperatures from 0 to 35°C. These solutions should be applicable to any gas in pure water for which the diffusion coefficients D_c are known. Since oxygen is of primary interest, the values of D_c for oxygen are shown in the table. The results of computations by means of Eq. (3-7) of K_L for oxygen into pure water for several values of r_{20} and temperatures from 0 to 35°C are plotted in Figure 3-4.

The effect of water temperature on K_L for oxygen has usually been expressed in terms of a factor Θ as follows:

$$K_{L(T)} = K_{L(20)}(\Theta)^{T-20} \tag{3-17}$$

Values reported for Θ have ranged from 1.015 to 1.047, according to Dobbins (17). Computed values of Θ based on Figure 3-4 show that it is not a constant, that it increases with the temperature at all values of r_{20}, and that it is greatest for low values of r_{20} and K_L. For example, for T of 5°C, Θ is 1.02 for r_{20} of 10 and 1.01 for r_{20} of 400; whereas for T of 35°C, Θ is 1.022 for r_{20} of 10 and 1.015 for r_{20} of 400.

The results of the experiments by Ippen (15) and co-workers on oxygen transfer from air and oxygen bubbles rising in tap water are shown in Figure 3-5. The experiments were conducted in a 5.5-in.-I.D. Lucite column fitted with a diffuser nozzle containing 19 equally spaced 0.115-mm-ID glass tubes for the larger bubbles and with a Venturi diffuser for the smaller bubbles. The bubble sizes obtained by the glass capillary diffuser increased from about 1.28 to 2.5 mm with increase in gas flow from about 72 to 915 cm³/min, and with the Venturi diffuser from 0.56 to 1.04 mm with increase in air flow from 35 to 340 cm³/min. The water temperature ranged from 20 to 24.5°C for the larger air bubbles, from 18 to 22.5°C for the oxygen bubbles, and from 29 to 31°C for the smaller

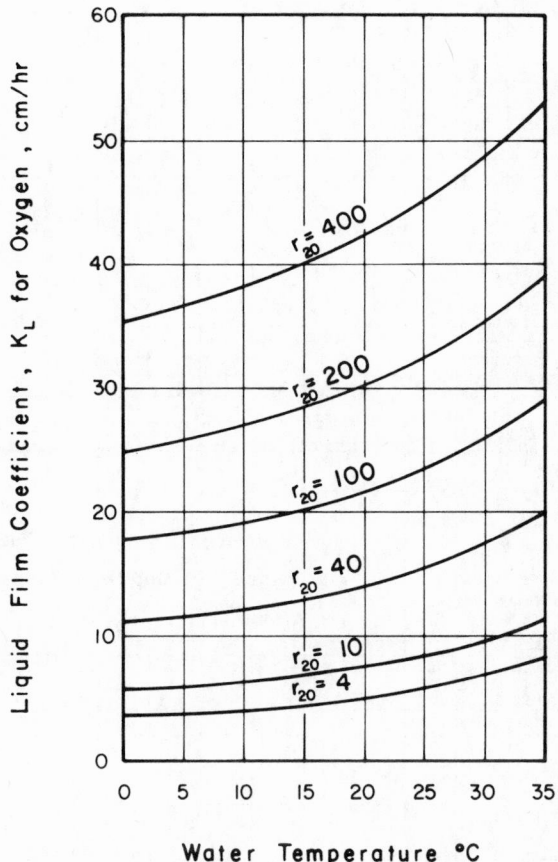

Figure 3-4
Effect of temperature on K_L for oxygen into distilled water.

air bubbles. The water column height ranged from 25 to 144 in. for the larger bubbles, and from 137.3 to 140 in. for the smaller air bubbles.

It will be noted from Figure 3-5 that the value of K_L decreases as the bubble size decreases below 1.0 mm, being about 40 for 0.6-mm bubbles and about 100 for 0.1-mm bubbles. In five of the experiments a synthetic detergent was added to the water in the concentrations (ppm) shown in Figure 3-5 at the plotted points. With the capillary diffuser, 50 ppm of detergent reduced K_L from 198 to 48 with negligible change in bubble size. With the Venturi diffuser, the added detergent resulted in a similar reduction in K_L; but the mean bubble size was also greatly reduced.

Claims have been advanced that most of the dissolved oxygen absorbed in dif-

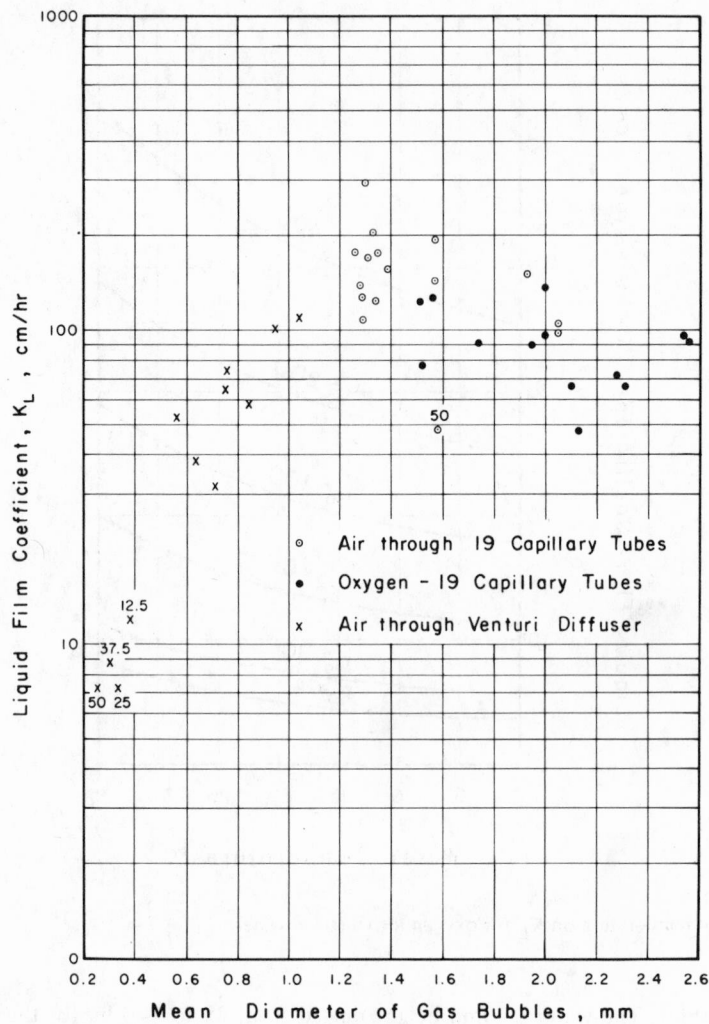

Figure 3-5
Measured values of K_L for gas bubbles.

fused air aeration comes from the atmosphere at the water surface and not from the air bubbles diffused into the water, and that the size of the air bubbles is therefore not important. To investigate this claim, Camp has estimated the water surface area per unit of volume of aeration tank for diffused air aeration at 1 ft^3/gal of sewage with a 6-hr aeration period. The aeration tank has been assumed to be 15 ft deep and 23 ft wide. The value of A_0/V in Eq. (3-6) for the water surface in this example is 0.0022 cm^2/cm^3. If the air bubbles are assumed to be 1.8 mm in diameter and to remain in the tank for 10 sec, the value of A_0/V for the air bubbles is 0.092 cm^2/cm^3. In other words, the surface area per unit of volume for the air bubbles is about 42 times as great as for the water surface. If the rate of transfer from the water surface is to be of the same order of magnitude as the rate of transfer from the rising bubbles, the rate of replacement of the liquid film, r, for the water surface must be about 1,800 times the rate of replacement of the liquid film for the rising bubbles.

Actually, the rate of replacement of the liquid film for the water surface is probably very much less than the rate for the rising air bubbles. In the experiments of Ippen and co-workers, the rate of replacement of the liquid film for the air bubbles averaged about 50 sec^{-1}. In the studies of reaeration in natural streams by O'Connor and Dobbins (20), the highest rate of renewal in the most rapidly flowing stream studied by the authors was about 1 sec^{-1}. It is reasonable to suppose that the rate of renewal of the liquid film at the water surface of an aeration tank is much greater than 1 sec^{-1} because of the turbulence produced by air bubbles breaking the surface. It is quite improbable, however, that the rate of renewal at the water surface can be as high as the rate of renewal of the liquid film on the rising air bubbles. If, for the sake of argument, it is assumed that the rate of renewal of the liquid film for the water surface is equal to that for the rising air bubbles, the conclusion is that the rate of oxygen transfer at the water surface is not more than about $\frac{1}{42}$ or 2.4 percent of the rate of transfer from the air bubbles in an average diffused air aeration tank. The bubbles, therefore, are the principal means by which gas transfer is accomplished, and the smaller the bubbles the more effective the transfer.

Experiments made under the senior author's direction on diffused aeration and carbonation of mixed municipal sewage and alkaline textile mill wastes indicated values of m, for 2.5-mm bubbles, of about 0.07 for CO_2 in flue gas containing 2.6 to 7.5 percent CO_2 and values of m for O_2 in air ranging from 0.0017 to 0.02, with an average of about 0.01. In computing these values of m, it was necessary to use values of D_c for pure water. These values are shown in Figure 3-5, and indicate that the actual value of D_c in the liquid film was very much less than that for tap water.

Laboratory and field studies (21) have been made during the past two decades on the use of monomolecular films of long-chain fatty alcohols to reduce evaporation from reservoir water surface. This gives promise as a water-conservation measure in arid and semiarid regions. As much as 90 percent reduction has been achieved in the laboratory, and 20 to 30 percent in the field where wind and wave action interfere. The alcohols that seem to offer the greatest promise are hexadecanol, $CH_3(CH_2)_{15}OH$, and octadecanol, $CH_3(CH_2)_{16}CH_2OH$. Exploratory studies by Roberts (22), using hexadecanol mixed with the soil around the roots, indicate the possibility of substantial reductions in water lost through transpiration from plants without measurable influence on plant growth. Testing with ^{14}C-hexadecanol indicated that transpiration from corn plants was reduced by the blocking action of hexadecanol molecules carried through plants and deposited at the stomate water–vapor interface.

In the use of monomolecular films for evaporation and transpiration reduction, it is important to know whether such films interfere with gas transfer across the water–air interface. Laboratory studies by Timblin (*23*) indicate that a hexadecanol film does not appreciably affect the rate of transfer of oxygen, and that it reduces the rate of transfer of carbon dioxide only slightly.

3-6. FLOCCULATION OF SUSPENSIONS

All the energy lost through the motion of a fluid is converted to heat through viscous shear. The rate of power dissipation (the work of shear per unit of volume per unit of time) is known as the *dissipation function* (*24*). The mean value of the dissipation function, W, is equal to the total power dissipation divided by the volume of the chamber or conduit containing the fluid. It has been shown (*25*) that the maximum velocity gradient at a point in a moving fluid is

$$\frac{dv}{ds} = \sqrt{\frac{W'}{\mu}} \tag{3-18}$$

where W' = the dissipation function at the point. It has also been shown (*25*) that the *root-mean-square velocity gradient* in a chamber or conduit is

$$G = \sqrt{\frac{W}{\mu}} \tag{3-19}$$

where

G = root-mean-square velocity gradient in the chamber or conduit
W = mean value of the dissipation function

The dissipation function W may be evaluated for a mixing tank (*26, 27*) or conduit of volume V rotating at speed s, or from the head loss, h_f, in the conduit in which the discharge is Q, from the following equations:

$$W = \frac{2\pi s T}{V} \tag{3-20}$$

$$W = \frac{Q g \rho h_f}{V} \tag{3-21}$$

These equations are general in that they apply to any liquid. Equations (3-18) and (3-19) are particularly useful in studies of the flocculation of suspended matter in water or wastewater. The dissipation function may be determined by actual measurement of power input or by computation from hydraulic principles. The viscosity is the viscosity of the suspension.

The rate of flocculation (*25–27*) of suspended particles, at a point, is directly proportional to the velocity gradient at that point. Similarly, the rate of flocculation of suspended matter in a conduit or chamber is directly proportional to the root-mean-square velocity gradient, G.

To remove suspended or colloidal particles from water by rapid filtration, it is necessary to use coagulating chemicals, such as alum, $Al_2(SO_4)_3 \cdot xH_2O$, or ferric sulfate, $Fe_2(SO_4)_3$, to form floc particles. The coagulating chemicals should be applied in true solutions to furnish metallic ions such as Al^{3+} or Fe^{3+}. These ions may be used as flocculating ions to flocculate color colloids at low pH, or, in the usual case, at higher optimum pH values to react within seconds with the alkalinity of the water (or added alkali if needed) to precipitate crystals such as Al_2O_3 or Fe_2O_3. The crystal bond forces are so great that when the crystals are once formed they cannot be sheared apart by the flocculation process. The free valences on the surfaces of these crystals attract clouds of ions together with water from the solution during the process of flocculation. The principal constituent of floc particles is water, which is present in fully formed floc to the extent of about 90 to 99.9 percent of the floc volume. During flocculation the suspended impurities and color particles are entrapped in the water within the floc particles. This water is probably held in the floc particle by interfacial tension at its boundary. As the floc particles grow in size by the addition of water and impurities, the total floc volume concentration also increases. For a particular velocity gradient, G, and water temperature, there is a limiting size of floc particle when flocculation is complete, and hence a limiting floc volume concentration (*28*) for each coagulant dose. This limiting concentration is proportional to the dose and increases with the water temperature.

Recent experiments (*26, 29*) indicate that the size of the aluminum or ferric oxide crystals formed is determined by the velocity gradient in the immediate vicinity of the point of application of the coagulating chemical to the water; the higher the velocity gradient, the smaller and more numerous the crystals for a particular coagulant dose. The smaller and more numerous the crystals are, the larger will be the number of free valences on the surfaces of the crystals and, in turn, the larger the floc volume concentration for a particular velocity gradient when flocculation is complete. It has been shown by recent experiments (*26*) that the floc size and floc volume concentration after flocculation is complete may be increased or decreased by subjecting the water to further flocculation at lower or higher velocity gradients, respectively. It is evident therefore, from Eq. (2-1), that the higher the velocity gradient becomes, the greater will be the shearing force. As floc particles grow in size, they become weaker and are more easily sheared apart. The magnitude of the mean shearing stresses created by the maximum velocity gradient (12,500 sec^{-1}) used in the above experiments is only 112 dynes/cm^2, which is equivalent to a shear energy of 4.8 X 10^{-8} kcal/mole of water. The bond energy of the aluminum or ferric oxide crystals is probably greater than weak van der Waals forces or hydrogen bonds (about 5 kcal/mole), which is 100 million times the shear energy. It is thus evident that the oxide crystals are not sheared apart when the floc is dispersed at high energy gradients.

Since the rate of floc formation is directly proportional to the velocity gradient, G, it should follow that the greater the magnitude of G, the less is the the time required to form the floc. Hence, for economy in the size of flocculation chambers, the velocity gradients should be made as large as practicable. The practical limit of the velocity gradient for any flocculation process is determined by the size of the floc particles required for effective settling or filtration. To form small floc particles, relatively high velocity gradients may be used (up to 500 or more for water going to filters after flocculation). For large floc particles, lower velocity gradients are required. The limiting velocity gradients required for any particular case should be determined by laboratory test. In most water and wastewater treatment problems, the limiting velocity gradient just before settling will range between 5 and 20 sec^{-1}.

3-7. SETTLING VELOCITIES OF SUSPENDED PARTICLES (*30*)

When a particle is released in a still fluid, it will, as a result of gravity, move vertically if its density differs from that of the fluid. The particle will accelerate until the frictional *drag* of the fluid approaches the value of the impelling force, after which the vertical velocity of the particle with respect to the fluid at rest will be substantially constant. The terminal velocity is known as the *settling velocity* of the particle and it is of great importance in sedimentation.

The *law for drag* was first proposed by Sir Isaac Newton (*31*) on the assumption that the drag is due to inertia only and hence is proportional to the square of the velocity. Newton's law is now usually written

$$F_D = C_D A \frac{\rho v^2}{2} \tag{3-22}$$

in which F_D is the drag force, C_D is a dimensionless number called the *drag coefficient*, A is the projected area of the body in the direction of motion [$(\pi/4)d^2$ for spheres], v is the relative velocity between the body and the fluid, $\rho v^2/2$ is the dynamic pressure, and ρ is the mass density of the fluid. Instead of being constant, as Newton assumed, the drag coefficient C_D has been found by experiment to vary widely. Experiments demonstrate, however, that when geometric similarity exists, as in the case of objects of similar shape, similarly oriented, the drag coefficient is a function of the Reynolds number, as shown in Figure 3-6 (*32, 33*).

The general equation for the settling velocity of spheres of diameter d in terms of the drag coefficient may be obtained by equating the impelling force to the drag, from which the settling velocity is

$$v = \sqrt{\frac{4}{3}\frac{g}{C_D}\frac{(\rho_1 - \rho)d}{\rho}} \tag{3-23}$$

in which g is the gravity constant and ρ_1 is the mass density of the sphere. The solution of settling velocity problems by means of Eq. (3-23) and experimental values of C_D from Figure 3-6 requires trial computations involving the Reynolds number $R = vd/\nu$, in which ν is the kinematic viscosity of the fluid. To avoid trial calculations, the curve of Figure 3-6 has been replotted in Figure 3-7 in terms of C_D/R and $C_D R^2$ as ordinates against R as abscissas (*34*). It will be noted from Eq. (3-23) that

$$\frac{C_D}{R} = \frac{4}{3}\frac{\rho_1 - \rho}{\rho^2}\frac{g\mu}{v^3} \tag{3-24a}$$

and

$$C_D R^2 = \frac{4}{3}\rho(\rho_1 - \rho)\frac{gd^3}{\mu^2} \tag{3-24b}$$

in which μ is the absolute viscosity of the fluid. For obtaining the diameter in terms of a measured settling velocity, C_D/R may be computed from Eq. (3-24a) and R can be obtained from the curve; d is then computed from the value of R taken from Figure 3-7. Similarly, for computing the settling velocity of a sphere of size d, $C_D R^2$ is calculated from Eq. (3-24b), and the corresponding value of R is taken from Figure 3-7; v is then computed from R.

The drag for a small sphere in an incompressible viscous fluid of infinite extent was first developed by Stokes (*35*), completely neglecting the inertia forces, as follows:

$$F_D = 3\pi\mu\, dv \tag{3-25}$$

Comparison of Eq. (3-25) with Eq. (3-22) reveals that, in the region of *viscous settling*, the drag coefficient is $24/R$. The corresponding settling velocity is given by *Stokes's law:*

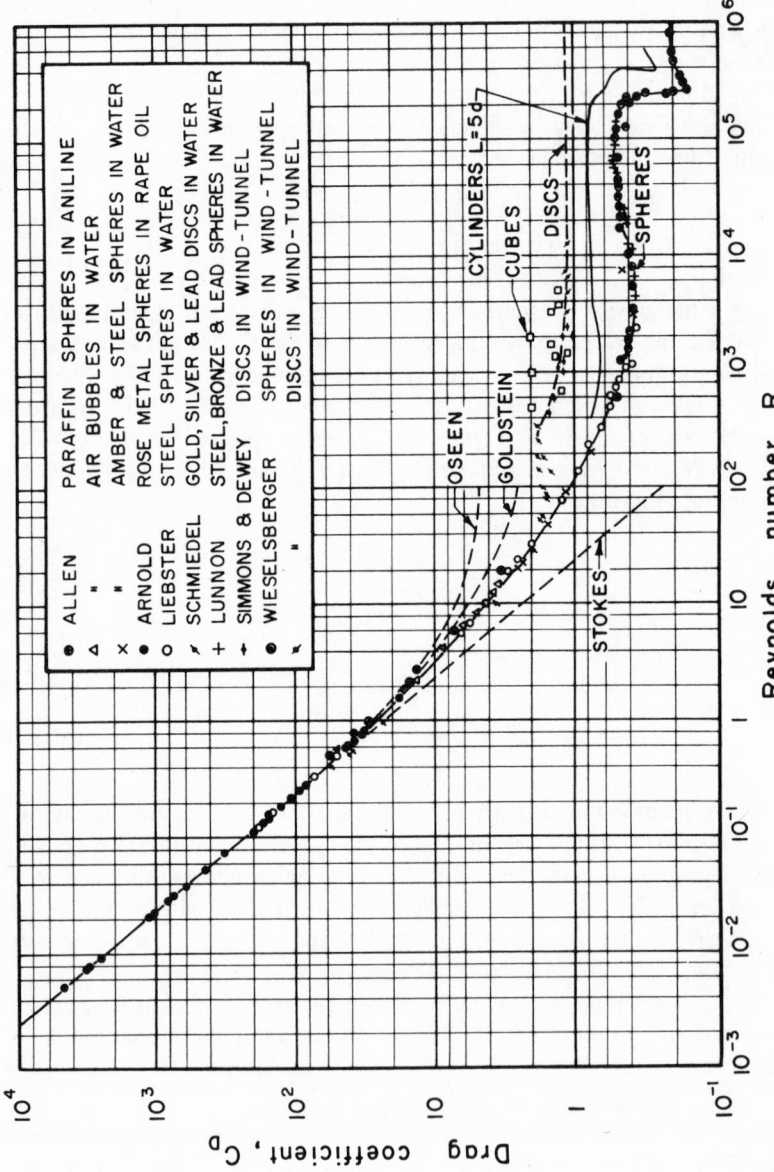

Figure 3-6
Drag coefficient as a function of the Reynolds number.

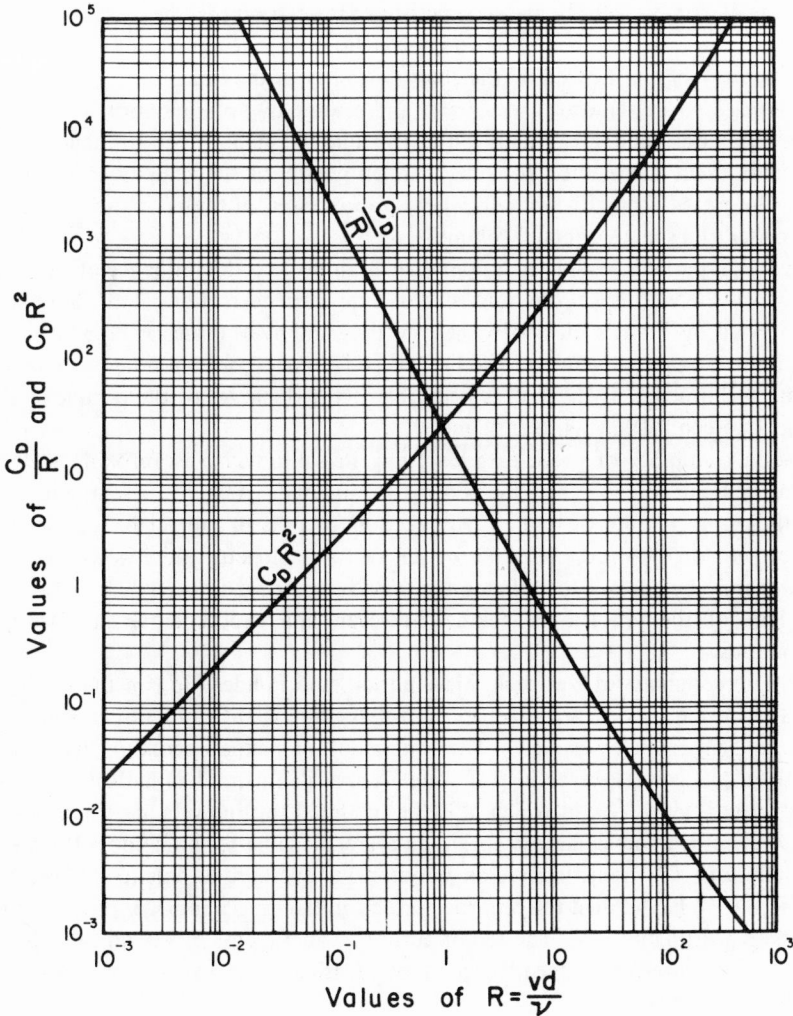

Figure 3-7
Free settling of spheres in fluids.

$$v = \frac{1}{18} \frac{g}{\mu} (\rho_1 - \rho)d^2 \qquad (3\text{-}26)$$

which has been verified experimentally by numerous investigators for fluids and spheres of various types. The data (see Figure 3-6) are in good agreement with Stokes's law for values of R from 10^{-4} to about 0.5. Although Stokes's expression for the drag force, Eq. (3-25), was developed for incompressible fluids, the effects of compressibility are so small that the relation may be used safely with gases. In fact, the coefficient of drag throughout the entire range of Figure 3-6 is not appreciably influenced by the compressibility of the fluid.

Particles to be removed from water and sewage by sedimentation are seldom truly spherical and are usually quite irregular in shape. The influence of irregularities in shape upon the drag is much greater as the velocity increases, as may be noted from Figure 3-6, where size is taken in terms of the equivalent diameter of a sphere of equal volume.

A single particle settling in a container is subject to the influence of the walls at which the velocity of the fluid is zero. Empirical correction factors have been developed by Francis (*36*) for viscous settling and by Munroe (*37*) for turbulent settling. Computed values of the Francis and Munroe correction factors indicate that the influence of the wall is negligible if the diameter of the particle is less than 1 percent of the cylinder diameter.

When a number of particles are settling in a fluid in close proximity, their velocity fields interfere and the settling is hindered. As in the case of a particle settling in a cylinder of limited cross section, there is an appreciable upward displacement of the fluid, and a correction factor analogous to the wall correction factor may be applied to estimate the *hindered settling velocity*. A theoretical analysis of the problem is lacking, however, and experimental data are not numerous.

The experiments of Kermack, McKendrick, and Ponder (*38*) on the hindered settling of red blood corpuscles are shown as curve A in Figure 3-8, in which c is the concentration by weight. The senior author's experiments on the viscous settling of Lucite spheres in still water (curve B, Figure 3-8) indicate that the correction factor v'/v, in which v' is the hindered settling velocity, increases as the Reynolds number increases. No data are available for hindered settling in a still fluid in the transition region between viscous and turbulent settling, but Gaudin (*39*) has applied the Munroe correction factor as a rough approximation, as shown in Figure 3-8. The senior author's experiments on round sand grains (curve C, Figure 3-8) suspended in a tube of rising water may also be considered as an approximation.

The rate of rise of air bubbles in still tap water, as a function of bubble size, is shown in Figure 3-9, based on measurements by Haberman and Morton (*40*) at $21°C$. The data in Figure 3-9 are equally applicable to the rise of bubbles for gases other than air. These experimenters found that the bubbles were essentially spherical in shape, with diameters less than 0.6 mm. As the size of the bubbles was increased, a change in bubble shape from spherical to ellipsoidal to cap-shaped was observed. The larger ellipsoidal and cap-shaped bubbles did not rise in a straight vertical path; their path was either helical or rectilinear with a rocking motion of the bubble. The flat shape of the larger bubbles probably accounts

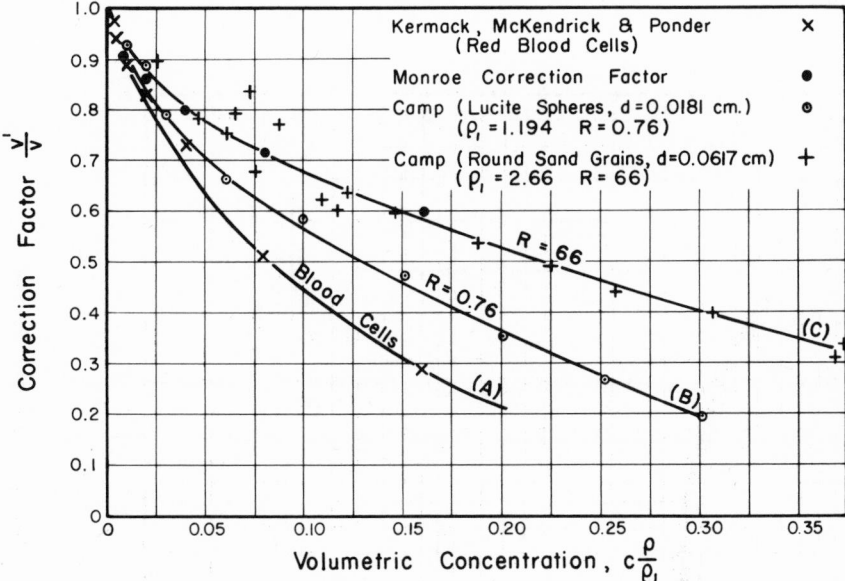

Figure 3-8
Reduction in velocity due to hindered settling.

for the relatively constant velocity for all sizes between 2 and 10 mm, as shown in Figure 3-9.

In applying the bubble velocity shown in Figure 3-9 to the design of aeration or carbonation tanks, it should be emphasized that the velocities in the figure are the relative velocities between the bubbles and the water immediately surrounding the bubbles. It should also be noted that the velocities in Figure 3-9 are the velocities for individual single bubbles. If many bubbles are rising concurrently, in close proximity to one another, the relative velocity with respect to the surrounding water will be less due to hindered settling. In estimating the velocity of rise with respect to the tank or container, the effect of both hindered settling and the motion imparted to the water by the bubbles must be considered. The latter is particularly true of spiral flow aeration tanks in which a circulating velocity of 1 to 2 ft/sec is usually imparted to the water surrounding the bubbles. Since the bubble velocities shown in Figure 3-9 are for a water temperature of 21°C, corrections should also be made for other water temperatures. With bubble sizes greater than about 1 mm, the temperature has little effect, but for smaller-sized bubbles conforming approximately to Stokes's law, temperature has a marked effect.

A study of the terminal velocities measured by Ippen et al. (*15*) indicates that

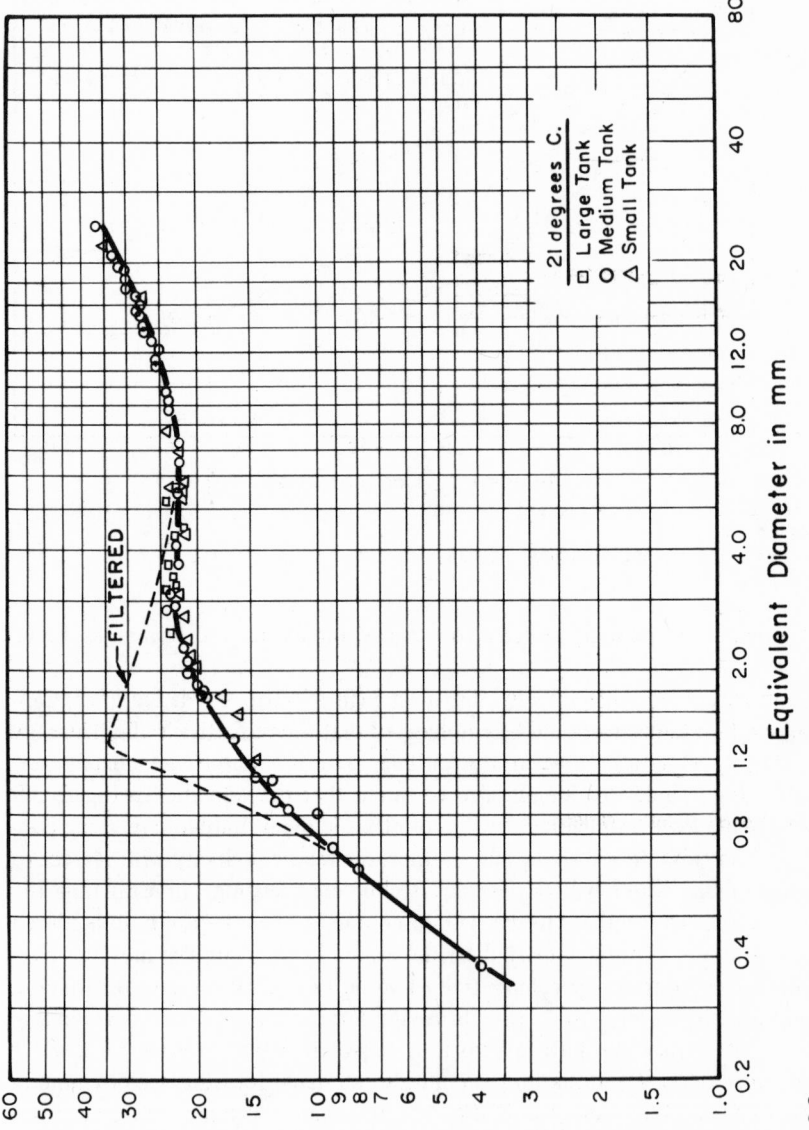

Figure 3-9
Terminal velocity of air bubbles in tap water as a function of bubble size. [After Haberman and Morton (40).]

they vary widely from those shown in Figure 3-9. Even for the same bubble sizes and water temperature, the velocities measured varied by as much as 60 percent. The difficulty of measuring bubble size and velocity, if many bubbles are rising concurrently, probably accounts for the wide discrepancy in the position of the plotted points in Figure 3-5.

3-8. RADIOACTIVE IMPURITIES

Over 100 chemically different kinds of atoms, the chemical elements, have been identified. Ninety of these are known to occur in nature, and the remainder have been made recently in the laboratory. The chemical nature of an element depends upon the number of positively charged particles, *protons*, in the nucleus, and all elements with the same number of protons are elements of the same substance and have the same chemical characteristics. The nucleus of an atom is surrounded by an atmosphere of negatively charged particles, called *electrons*, which move around the nucleus at high velocities and in regular orbits. Although the number of electrons in the orbital area of an atom is usually the same as the number of protons in its core, under certain circumstances this number can change; however, the change does not affect the number of protons, and the same kind of element prevails.

Atomic nuclei also contain *neutrons;* each neutron has approximately the same mass as a proton but no charge. The atomic mass (*41*) of a proton is 1.00758, and its mass in grams is 1.6722×10^{-24}. The atomic mass of a neutron is 1.00894, and its mass in grams is 1.6744×10^{-24}. The atomic mass of an electron is 0.0005486, and its mass in grams is 9.19×10^{-28}. Since the mass of an electron is only about $\frac{1}{1,835}$ of the mass of a proton, the mass of an element is determined primarily by the mass of the nucleus.

The *atomic number* of an atom is the number of protons in its nucleus. The chemical nature of an atom is thus indicated by its atomic number. The *mass number* of an atom is the number of protons and neutrons in the nucleus. The number of neutrons in the nucleus varies, and a particular nucleus can change its number of neutrons without altering the chemical properties of the atom. Every element has three or more *isotopes*, that is, three or more forms, each with a different number of neutrons in the nucleus. Hydrogen has three isotopes; helium has four, and the number of isotopes increases generally with the weight of the atom, with uranium having 14. There are approximately 1,100 known isotopes (*42*).

Powerful forces act upon the protons and neutrons in the nuclei of atoms, including strong attractive forces as well as potent forces tending to thrust these nuclear components apart. The magnitude of these forces and the balance between them varies from isotope to isotope, depending on the total number of particles present and the relative number of each kind. In some of the nuclei the balance is stable; in others the balance is easily disrupted. Isotopes with unstable nuclei undergo radioactive decay and are known as radioactive isotopes or radioisotopes.

Combinations of approximately equal numbers of neutrons and protons form stable nuclei in the lighter elements. Hydrogen 1 and hydrogen 2 are stable, as are carbon 12 and carbon 13. Hydrogen 3, carbon 10, carbon 11, and carbon 14 are radioactive. As the mass numbers of the successive elements increase, more neutrons than protons are required to maintain stability. The stable ratio of neutrons to protons is approximately 1.5 for elements like platinum, gold, and bismuth. In the very heavy elements above bismuth 210, all atoms are radioactive, regardless of the ratio of the neutrons to the protons. This group includes radium, uranium, and plutonium.

There is at least one known radioisotope for each of the elements. Some of the elements have many stable forms, others have but one stable form, and still others have only radioactive forms. Of the approximately 1,100 known isotopes, only about 280 are stable (*42*). The remainder are radioactive.

Despite the fact that most of the known isotopes are radioactive, the total quantity of radioactive material in existence is very small. Prior to the development of the nuclear reactor, there was available throughout the world, for medical and other uses, about 3 lb of refined radium plus limited quantities of other naturally radioactive substances. There were numbers of diagnostic, treatment, industrial, and experimental x-ray machines in hospitals, industries, and physicians' offices, and there were a few high-voltage machines. Today, however, sources of ionizing radiation are rapidly becoming more widely used. Each atomic reactor produces the radioactive equivalent of hundreds of pounds of radium. In addition, the aboveground testing of atomic bombs releases into the upper atmosphere great clouds of fission products, which encircle the earth and return in the form of radioactive fallout. With the increasing abundance of radiation in the environment, it has become essential to take account of the effects of radioactive impurities on the quality of water for drinking purposes and industrial uses.

All radioisotopes, whether natural or man-made, decay by emission of alpha, beta, or gamma radiations. *Alpha rays* are high-energy charged particles consisting of two protons and two neutrons. Upon acquiring two electrons from the environment, they become helium atoms. *Beta rays* consist of electrons or their positive counterparts called *positrons*. *Gamma rays* and *x-rays* are high-energy photons similar to light and radio waves, but of much higher frequency. They have no electric charge, and all have the velocity of light.

Radioactive decay can neither be slowed nor hurried by any practical means. For any particular radioactive atom, the decay is an all or none reaction. Among a large number of identical radioactive atoms, disintegration events occur strictly at random; so with large numbers of atoms the number that will disintegrate in any given length of time can be estimated with great accuracy. For radioactive atoms of every kind, the number decaying during a given time is proportional to the number originally present. If the interval chosen is that in which 50 percent of the atoms present will decay, then in each successive identical interval 50 percent of the remaining radioactive atoms will decay. This sort of time interval is called the *radioactive half-life* of a radioactive isotope.

The half-life of each radioisotope has a fixed value. Half-lives of the known radioisotopes range from fractions of seconds for some to millions or billions of years for others. Nitrogen 12 has a half-life of 0.013 sec. Iodine 131 has a half-life of 8.14 days. Carbon 14 has a half-life of 5,580 years. For a particular radioisotope, only one or two kinds of radiation are emitted at each disintegration. Half-lives and decay patterns are so characteristic that they can be used to help identify the various radioisotopes.

The importance of radioactive impurities in water that is to be used for drinking or industrial purposes derives from the fact that the radiation produces ions in the matter through which the fast-moving particles or photons penetrate. For this reason, high-energy radiation is called *ionizing radiation*. Ionization can occur in any kind of matter. Ionization of living matter results in tissue changes that may affect health. The degree of tissue ionization produced depends upon the composition of the tissue and on the quantity, type, and energy of the radiation.

Most particulate radiations, such as alpha and beta radiations, are completely absorbed or stopped by relatively thin layers of matter. In contrast, gamma and x-rays are only attenuated by passing through matter. Successive layers of identical material and thickness reduce these radiations by the same ratio. If a given thickness of lead reduces a given x-ray beam to one half its original intensity, a similar one twice as thick will reduce it to one half of one half or one quarter of its original intensity.

Ionization occurs when a high-velocity particle or photon strikes an atom in its path and drives out an electron from its orbit in the atom. This leaves the atom with a positive charge and a free electron in the environment, which is received by another atom to become negatively charged. The negative ion and the positive ion produced in each ionizing event are called an *ion pair*. The *specific ionization* of a particle is the number of ion pairs produced per unit of length of the particle's track in a given medium. Although specific ionization is generally stated as the number of ion pairs created per centimeter of track in air, in radiobiology it is usually given as ion pairs per micron of track in water, since water is regarded as the physical equivalent of soft tissue. Specific ionization varies directly with the magnitude of the mass and charge and inversely with the speed of travel. Alpha particles with an energy of about 1 million eV produce from 30,000 to 100,000 ion pairs per centimeter in air at $0°C$ and 760 Torr of pressure, whereas beta particles having the same energy produce only about 100 ion pairs per centimeter of track. Electromagnetic radiation, such as gamma and x-rays, produces ionization largely by secondary means. After the atom absorbs the electromagnetic radiation, it emits a charged particle. Ordinarily, the charged particle is an electron having sufficient velocity to produce ion pairs in its path. From then on the action of the electron is a repetition of the history of an ordinary ionizing particle, and the ionization produced is largely dependent upon the energy of the expelled electron.

Alpha particles have large specific ionization values. Since they create many ions per unit of path length, they dissipate their energy rapidly and penetrate only short distances. Alpha particles are normally a hazard to health only in the form of internal radiation. Beta particles are light in weight and carry single charges. Their specific ionization values are intermediate between those of alpha particles and those of gamma and x-rays. They ionize matter somewhat sparsely, dissipate their energies relatively quickly, and are moderately penetrative. Beta particles can be a health hazard either as internal or external radiation. X-rays and gamma rays have quite low specific ionization values. They ionize sparsely over long paths and are quite penetrative. As a group, these radiations constitute the chief health hazard of external radiation, although gamma rays can be a hazard also as internal radiation.

Pure water contains only about 0.01 percent of hydrogen 2 and an infinitesimal amount of radioactive hydrogen 3. The oxygen in pure water is approximately 99.76 percent oxygen 16, with the remainder consisting of stable-type oxygen 18 and oxygen 17. When natural waters are found to be radioactive, therefore, the radioactivity is due to foreign impurities, principally radium. Natural groundwaters may contain radioactive minerals that have been dissolved from the rocks through which these groundwaters pass. Natural surface waters may contain radioactive minerals, in both the dissolved and the suspended form, which originate in mining operations on the watershed. The most probable origin of widespread pollution of waters by radioactive substances in the future will be the fallout from atomic bomb tests, the wastes from radioactive isotopes used in industrial and medical institutions, and the wastes from atomic power plants. The radioactive impurities will probably occur mostly in the dissolved form, but some colloidal and suspended impurities may be present, and some concentration may take place in microorganisms and aquatic life. The concentration of radioactive impurities in water can best be measured by their ionization effect. In some cases, however, it will be necessary to determine the concentration of individual radioactive isotopes.

The quantity of radioactive material present in terms of its radioactivity is measured by the *curie*, which is the amount of radiation from approximately 1 g of pure radium. For practical purposes, the radioactivity of impurities in water is expressed in much smaller units, microcuries per milliliter (μC/ml) or micro-

microcuries per liter ($\mu\mu$C/liter). A microcurie is one millionth of a curie and a micromicrocurie is one millionth of a microcurie. Measurements (*41*) are made by means of some type of an area monitor such as the Geiger–Müller counter. These instruments count the number of radioactive discharges in a unit of time, and the count is convertible to microcuries per milliliter of water sample.

REFERENCES

1. *Standard Methods for the Examination of Water and Wastewater*, 13th ed., American Public Health Association, Washington, D.C., 1971.
2. H. R. Kruyt and H. S. van Klooster, *Colloids*, New York, John Wiley & Sons, Inc., 1930.
3. A. Einstein, *Ann. Physik* (4), *17*, 549 (1905); *19*, 371 (1906).
4. L. Pauling, *The Nature of the Chemical Bond*, 3rd ed., Ithaca, N.Y., Cornell University Press, 1960.
5. S. Glasstone, K. J. Laidler, and H. Eyring, *Theory of Rate Processes*, New York, McGraw-Hill Book Company, 1941. See also J. D. Bernal and R. H. Fowler, *J. Chem. Phys.*, *1*, 515 (1933).
6. E. W. Washburn and E. B. Millard, *J. Amer. Chem. Soc.*, *37*, 694 (1915).
7. D. A. MacInnes, *The Principles of Electrochemistry*, New York, Van Nostrand Reinhold Company, 1939.
8. F. R. Moulton, ed., *Recent Advances in Surface Chemistry and Chemical Physics*, New York, Science Press, 1939.
9. W. M. Latimer, *Oxidation Potentials*, 2nd ed., Englewood Cliffs, N.J., Prentice-Hall, Inc., 1952.
10. *Trans. Inst. Chem. Engrs. (London)*, *35*, 51 (1957).
11. W. E. Adeney and H. G. Becker, *Sci. Proc. Roy. Dublin Soc.*, *15*, 44, 609 (1919); *15*, 31, 385 (1919); *16*, 13, 143 (1920).
12. W. K. Lewis and W. C. Whitman, "Principles of Gas Adsorption," *Ind. Eng. Chem.*, *16*, 1215 (1924).
13. W. D. Scouller and W. Watson, "The Solution of Oxygen from Air Bubbles," *Surveyor*, *86*, no. 2215, 15 (1934).
14. W. E. Adeney and H. G. Becker, "The Determination of the Rate of Solution of Atmospheric Nitrogen and Oxygen by Water," *Phil. Mag.*, *38*, 317 (1919); *39*, 385 (1920).
15. A. T. Ippen, L. G. Campbell, and C. E. Carver, *Tech. Rept.* 7, May 1952, and A. T. Ippen and C. E. Carver, *Tech. Rept.* 14, April 1955, Hydrodynamics Laboratory, Massachusetts Institute of Technology, Cambridge, Mass.
16. W. E. Dobbins, "Mechanism of Gas Adsorption by Turbulent Liquids," London, Intern. Conf. Water Pollution Research, Sept. 1962.
17. W. E. Dobbins, "BOD and Oxygen Relationships in Streams," *J. San. Eng. Div., Amer. Soc. Civil Engrs.*, *90*, no. SA3, 53 (June 1964).
18. W. E. Dobbins, "BOD and Oxygen Relationships in Streams," *J. San. Eng. Div., Amer. Soc. Civil Engrs.*, *91*, no. SA5, 49 (Oct. 1965).
19. I. Metzger and W. E. Dobbins, "Role of Fluid Properties in Gas Transfer," *Env. Sci. & Tech.*, *1*, 57 (Jan. 1967).
20. D. J. O'Connor and W. E. Dobbins, "The Mechanism of Reaeration in Natural Streams," *Trans. Amer. Soc. Civil Engrs.*, *123*, 641 (1958).

21. W. W. Mansfield, "The Use of Hexadecanol for Reservoir Evaporation Control," San Antonio, Tex., Southwest Research Institute, Apr. 1956; L. O. Timblin, Jr., W. T. Moran, and W. U. Garstka, "Use of Molecular Layers for Reservoir Evaporation Reduction," *J. Amer. Water Works Assoc.*, *49*, 841 (1957); "Water-Loss Investigations, Lake Hefner," Oklahoma City, Okla., June 1959.

22. W. J. Roberts, "Reduction of Transpiration," *J. Geophys. Res.*, *66*, no. 10, 3309 (Oct. 1961).

23. *Chem. Eng. Lab. Rept. SI-10*, Denver, U. S. Bureau of Reclamation, Mar. 15, 1957.

24. G. G. Stokes, "On the Theories of Internal Friction of Fluids in Motion," *Trans. Cambridge Phil. Soc.*, *8*, 287 (1845).

25. T. R. Camp and P. C. Stein, "Velocity Gradients and Internal Work in Fluid Motion," *J. Boston Soc. Civil Engrs.*, *30*, 219 (1943).

26. T. R. Camp, "Floc Volume Concentration," *J. Amer. Water Works Assoc.*, *60*, no. 6, 656–673 (June 1968).

27. T. R. Camp, "Hydraulics of Mixing Tanks," *J. Boston Soc. Civil Engrs.*, *56*, no. 1, 1 (Jan. 1969).

28. T. R. Camp and G. F. Conklin, "Towards a Rational Jar Test for Coagulation," *J. New Eng. Water Works Assoc.*, *84*, 3, 325 (Sept. 1970).

29. L. Vrale and R. M. Jorden, "Rapid Mixing in Water Treatment," *J. Amer. Water Works Assoc.*, *63*, 52 (Jan. 1971).

30. T. R. Camp, "Sedimentation and the Design of Settling Tanks," *Trans. Amer. Soc. Civil Engrs.*, *111*, 896 (1946).

31. *Mathematical Principles of Natural Philosophy*, Isaac Newton, Book II, Secs. II and VII, trans. rev. by F. Cajori, University of California Press, Berkeley, 1966. (Original publ. London, 1686.)

32. L. Schiller, "Hydro- und Aero-Dynamik," in *Handbuch der Experimental Physik*, Bd. 4, Teil 4, Verlagsgesellschaft, Leipzig, 1932, p. 369.

33. H. Rouse, "Nomograph for the Settling Velocity of Spheres," Report of the Committee on Sedimentation for 1936–1937, Washington, D.C., Division of Geology and Geography, National Research Council, p. 57, 1937.

34. W. H. Walker, W. K. Lewis, and E. R. Gilliland, *Principles of Chemical Engineering*, 3rd ed., New York, McGraw-Hill Book Company, 1937, p. 299.

35. G. G. Stokes, "On the Theories of the Internal Friction of Fluids in Motion and of the Equilibrium and Motion of Elastic Solids," *Trans. Cambridge Phil. Soc.*, 287 (1845); also in *Mathematical and Physical Papers*, Cambridge, Cambridge University Press, Vol. 1, 1880, p. 75.

36. A. W. Francis, "Wall Effect in Falling Ball Method for Viscosity," *Physics*, *4*, 403 (1933).

37. H. S. Munroe, "The English Versus the Continental System of Jigging–Is Close Sizing Advantageous?", *Trans. Amer. Inst. Mining Met. Engrs.*, *17*, 637 (1888–1889).

38. W. O. Kermack, H. G. McKendrick, and E. Ponder, "The Stability of Suspensions: III. The Velocities of Sedimentation and Cataphoresis of Suspensions in a Viscous Fluid," *Proc. Roy. Soc. Edinburgh*, *49*, 170 (1928–1929).

39. A. M. Gaudin, *Principles of Mineral Dressing*, New York, McGraw-Hill Book Company, 1939, p. 190.

40. W. L. Haberman and R. K. Morton, "An Experimental Study of Bubbles Moving in Liquids," *Trans. Amer. Soc. Civil Engrs.*, *121*, 227 (1956).

41. "Atomic Radiation," RCA Service Company, New York, 1957.

42. "Concepts of Radiological Health," U. S. Public Health Service, Government Printing Office, Washington, D.C., Jan. 1954.

CHAPTER 4

Chemical Equilibria

The composition of a natural water as determined by an analysis is the result of its history since its precipitation as rain. All the impurities contained in a water are the result of intimate contact of the water with impurities, during which a portion are dissolved or suspended. The process of solution is a chemical reaction that proceeds as long as the water is in contact with a soluble substance or until equilibrium is reached. Subsequent contact of the water with another soluble substance may change the equilibrium conditions for a previously dissolved substance and result in its precipitation from solution either in its original form or as another compound. The composition of water is therefore a transitory condition, and the concentrations of some of the impurities may change during analysis.

Since the quality of a water depends to a considerable extent upon its content of dissolved substances, an understanding of the chemical equilibria and reaction velocities involved in aqueous solutions is an essential preliminary to a rational attack on water-treatment problems.

4-1. LAW OF MASS ACTION

Chemical reactions are due to the thermal activity of molecules or ions, and the rate at which a reaction proceeds depends upon the number of active molecules or ions available for the reaction. Since thermal activity is present in both the reactants and the products of the reaction, it follows that the reaction is proceeding in both directions at the same time, and it is only the net result that is observed. A chemical reaction (1) may be represented as follows:

$$A + B + \cdots \rightleftharpoons X + Y + \cdots \qquad (4\text{-}1)$$

in which A, B, . . . represent the reactants and X, Y, . . . the products of the reaction as it proceeds to the right, and conversely as it proceeds to the left.

If constituents A and B are dispersed as individual molecules or ions in a gas phase or solution, the velocity of the reaction to the right at any time t is proportional to the product of the numbers of active A and B molecules or ions. If constituents X and Y are also so dispersed, the velocity to the left is similarly controlled, and the reaction velocities may be stated as follows:

$$-\frac{d}{dt}\,(\alpha_A \text{ or } \alpha_B) = k_r\,\alpha_A\alpha_B \quad \text{to the right} \qquad (4\text{-}2)$$

and

$$-\frac{d}{dt}\,(\alpha_X \text{ or } \alpha_Y) = k_l\,\alpha_X\alpha_Y \quad \text{to the left} \qquad (4\text{-}2a)$$

in which α represents the number of active molecules or the *activity* of a reactant, and k is the *velocity constant* for a given temperature. Equations (4-2) and (4-2a) represent the *mass action law* for reaction velocity.

If there is only one reactant, the reaction is called a *unimolecular* reaction; if there are two reactants, it is *bimolecular*; and if three reactants, *termolecular*. In all cases, however, the velocity constant k has the dimensions t^{-1}, despite the fact that the number of particles is usually expressed as a pressure (atmospheres) or a concentration (moles/liter). This seeming anomaly arises from the fact that α represents the ratio of actual activity to unit activity, as will be demonstrated later.

In a catalyzed reaction, the catalyst appears unchanged on both sides of the reaction, but the velocity is proportional to the product of the activities of the catalyst and the reactants.

At equilibrium, the reaction velocities are equal and

$$\frac{\alpha_X\alpha_Y}{\alpha_A\alpha_B} = \frac{k_r}{k_l} = K \qquad (4\text{-}3)$$

in which K is the *equilibrium constant* of the reaction. For ionization reactions in aqueous solution, K is called the *ionization constant*. Equation (4-3) is the *mass action law* for equilibrium.

For reactions involving any gas or volatile solute A, the activity is usually taken equal to the *fugacity*, $f p_A$, where f is the *activity coefficient* and p_A is the *partial pressure* of constituent A in standard atmospheres. For reactions in aqueous solution, the activity may also be expressed as $f[A]$, where $[A]$ is the *molal concentration* (gram-molecules per liter) of constituent A. [A] is equal to $(c_A/M) \times 10^3$, where c_A is the relative concentration by weight or mass (grams of A per gram of solution) and M is the molecular weight of A. If the concentration c_A' is expressed in milligrams per liter (mg/liter), [A] is equal to $(c_A'/M) \times 10^{-3}$. In dilute solutions, parts per million by weight (ppm) is approximately equal to milligrams per liter.

The partial pressure p_A in atmospheres may be evaluated from the molal concentration [A], and vice versa, through the relation $p_A V_m = RT = 24.45$ liter-atmospheres at 25°C $(T = 298°K)$, as follows:

$$p_A = \frac{24.45}{V_m} \frac{T}{298} = 24.45 \frac{T}{298} [A] \qquad (4\text{-}4)$$

The factor $24.45 \, (T/298)$ is the volume in liters occupied by a mole of the constituent at a partial pressure of 1 atm and a temperature T. Since 1 gram-molecular weight of any gas has a volume at 25°C and 1 atm of 24.45 liters, 1 ml/liter is equal to $(M/24.45) \times 10^3$ ml/liter.

The activity coefficient f for a gas is equal to unity if the partial pressure is low enough (a few atmospheres) so that the gas follows the perfect gas law. For higher pressures, f must be evaluated from measured p-V isotherms. The activity coefficient for nonionized solutes is also unity if the concentration is low enough (about 0.1 m) so that the osmotic pressure follows the perfect gas law [Eq. (3-1) or (3-1a)].

It is customary to use molal concentrations for activities in ionic reactions in water, and f is less than unity for all but the most dilute solutions. The value of the activity coefficient for an ionic constituent may be estimated by means of the Debye–Hückel theory (2), which has the following simplified form:

$$\ln f_A = -z_A^2 \frac{\kappa \sqrt{\omega}}{1 + \beta a \sqrt{\omega}} \qquad (4\text{-}5)$$

in which z_A is the valence of any ion A, a is the distance of closest approach of the centers of two ions in centimeters (the sum of ionic radii), ω is the ionic strength, and κ and β are functions of the temperature and dielectric constant of the solvent. The *ionic strength* is defined by the relation

$$\omega = \tfrac{1}{2} ([A] \, z_A^2 + [B] \, z_B^2 + \cdots + [X] \, z_X^2 + [Y] \, z_Y^2) \qquad (4\text{-}6)$$

The values of κ and β for aqueous solutions are given in Table 4-1. Since the values of κ and β include the dielectric constant of water, the Debye–Hückel theory is limited in its application to dilute solutions in which the ionic concentration is not sufficient to change the dielectric constant appreciably (concentrations less than about 0.1 m). The value of βa may be taken as unity without

Table 4-1.
Debye–Hückel Constants for Water

Temperature (°C)	κ	β
0	0.4863	0.3243×10^8
18	0.4992	0.3272×10^8
25	0.5056	0.3286×10^8
38	0.5186	0.3314×10^8

appreciable error for dilute solutions; and if the total concentration of ions is less than 100 mg/liter of equivalent $CaCO_3$, the term including βa in the denominator of Eq. (4-5) may be omitted without introducing an error of more than 1 percent in the value of f.

4-2. HETEROGENEOUS REACTIONS

If each of the constituents of a reaction is dispersed as single molecules or ions either in the gas phase or in solution, the reaction is said to be *homogeneous* and the activities may be evaluated as described previously. If one or more of the constituents is a solid or a liquid, the reaction is called *heterogeneous*. The activities of solids and liquids are substantially constant during a reaction as long as there is any of the solid or liquid present. This results from the fact that the concentration of ions or molecules within the solid or liquid does not change during the reaction. The rate of evaporation is therefore constant, as indicated by Eqs. (4-2) and (4-2a). The net rate of evaporation is equal to the difference between this constant rate and the rate of condensation of molecules or ions on the liquid or solid. The activities of solids or liquids do not appear in the mass-action equation for equilibrium, as will be demonstrated presently. Recourse to thermodynamics will throw light on this problem and will also yield a method for the correction of equilibrium constants for changes in temperature.

The total *free energy* of a system undergoing a reaction such as represented by Eq. (4-1) is, at any instant, given by

$$F = \bar{F}_A n_A + \bar{F}_B n_B + \cdots + \bar{F}_X n_X + \bar{F}_Y n_Y \qquad (4-7)$$

where $\bar{F}_A$, $\bar{F}_B$, ... are the *partial molal free energies** and n_A, n_B, ... are the

---

*The designation *partial molal free energy*, symbol $\bar{F}$, was given by Lewis and Randall (G. N. Lewis and Merle Randall, *Thermodynamics and the Free Energy of Chemical Substances*, New York, McGraw-Hill Book Company, 1923) to the function called the *potential* μ of a constituent by Willard Gibbs. The Lewis and Randall notation will be used here.

number of moles of the constituents present at the time considered. If the reaction proceeds to the right, the increase in free energy of the system (3) is given by

$$\Delta F = -S\, dT + V\, dP - \bar{F}_A\, dn_A - \bar{F}_B\, dn_B + \cdots + \bar{F}_X\, dn_X + \bar{F}_Y\, dn_Y \qquad (4\text{-}8)$$

in which S is the *entropy*, V is the volume, and dT and dP the changes in temperture and pressure of the system; $-dn_A$ and $-dn_B$ are the decreases in the number of moles of the reactants, and $+dn_X$ and $+dn_Y$ are the increases in the number of moles of the products of the reaction.

For the special case of a reaction at constant temperature and pressure and for a change dn of 1 mole for each constituent, Eq. (4-8) becomes

$$\Delta F = -\bar{F}_A - \bar{F}_B + \cdots + \bar{F}_X + \bar{F}_Y \qquad (4\text{-}8a)$$

At equilibrium,

$$\Delta F = 0 \qquad (4\text{-}9)$$

and

$$\bar{F}_A + \bar{F}_B = \bar{F}_X + \bar{F}_Y \qquad (4\text{-}9a)$$

During a reaction at constant temperature and pressure, there is a decrease in the partial pressure of the reactants and an increase for the products. Gibbs has shown that the rate of change at constant temperature of the partial molal free energy with the partial pressure is equal to the molal volume of a constituent:

$$\left(\frac{\partial \bar{F}}{\partial p}\right)_T = V_m \qquad (4\text{-}10)$$

This equation may be integrated at constant temperature to yield the value of $\bar{F}$ at any partial pressure p in terms of some known value $\bar{F}_1$ at a partial pressure p_1:

$$\int_{\bar{F}_1}^{\bar{F}} \partial \bar{F} = \int_{p_1}^{p} V_m\, \partial p \qquad (4\text{-}11)$$

For gases and solutes, since $V_m = RT/p$, Eq. (4-11) results in the following relation:

$$\bar{F} = \bar{F}_1 + RT \ln \frac{p}{p_1} \qquad (4\text{-}12)$$

It is customary to refer free energies to the *standard free energy* $\bar{F}°$ (1 atm) at a partial pressure of 1 atm and 25°C. Hence, for gases and solutes,

$$\bar{F} = \bar{F}° \text{ (1 atm)} + RT \ln p \qquad (4\text{-}13a)$$

For ionic solutions, free energies may be referred to the *standard free energy* $\bar{F}°$ (u.a.) at unit molal activity and 25°C. From Eq. (4-4), $p = 24.45\, \alpha_m$. Integration of Eq. (4-11) with this value for p between the limits of unit activity and α_m yields

$$\bar{F} = \bar{F}° \text{ (u.a.)} + RT \ln \alpha_m \qquad (4\text{-}13b)$$

It will be noted from Eq. (4-12) that the pressure p and activity α_m in Eqs. (4-13a) and (4-13b) are dimensionless ratios, and the equations will be dimensionally consistent if RT has the same units as the free energies. It is standard practice to evaluate free energies in calories per mole, whence R is 1.985 cal/mole/°K. By convention, the standard free energy, $\bar{F}°$ (1 atm), of all pure elements is set equal to zero when the elements are in their *standard states* at 25°C and 1 atm. The standard free energy $\bar{F}°$ (1 atm) or $\Delta\bar{F}°$ (1 atm) of any substance other than the elements is numerically equal to $-\Delta\bar{F}°$ (1 atm), the *free energy of formation* of the substance from the elements. Free energies of gases, liquids, and solids are usually determined at 1 atm, but free energies of ions in aqueous solution are evaluated at unit molal activity. By convention, the free energy of hydrogen ions is set at zero. Experimental and computed values of $\bar{F}°$ for numerous substances are given by Latimer (4).

For liquids and solids that may be assumed to be incompressible, $V_m = (M/\rho) \times 10^{-3}$ liter $= 1/\alpha_m$. Since V_m is substantially constant and is not related to the pressure p through the gas laws, α_m for liquids and solids changes only slightly with pressure and is usually assumed to be constant. The changes in α_m with total pressure may be evaluated through the modulus of elasticity if precision is required. If V_m is assumed constant, the integration of Eq. (4-11) for liquids and solids yields

$$\bar{F} = \bar{F}° \text{ (1 atm)} + 0.0242\, \frac{M}{\rho} (p - 1) \qquad (4\text{-}14)$$

where $\bar{F}$ and $\bar{F}°$ are in cal/mole, M is the molecular weight, p is in atmospheres, and 0.0242 is the conversion factor 1.985 cal/82.06 cm³-atm. In this case p is the total pressure, P, on the solid or liquid. The pressure term $0.0242\,(M/\rho) \times (p - 1)$ is so small as compared to $\bar{F}°$, except for very high pressures (100 atm or more), that it may usually be neglected. Hence, the free energy of solids and liquids is nearly constant for a given temperature and

$$\bar{F} = \bar{F}° \text{ (1 atm)} = \bar{F}° \text{ (u.a.)} \qquad (4\text{-}14a)$$

If the value of $\bar{F}$ as given by Eq. (4-13a) is inserted in Eq. (4-9a) for each constituent of a *homogeneous* reaction, it is found that

$$RT \ln \frac{p_X p_Y}{p_A p_B} = \bar{F}^\circ_A \text{ (1 atm)} + \bar{F}^\circ_B \text{ (1 atm)} - \bar{F}^\circ_X \text{ (1 atm)} - \bar{F}^\circ_Y \text{ (1 atm)}$$

$$= -\Delta F^\circ \text{ (1 atm)} \tag{4-15}$$

where $-\Delta F^\circ$ (1 atm) is the *standard free energy change* per mole. Hence, for a homogeneous reaction, the equilibrium constant is

$$K = \frac{p_X p_Y}{p_A p_B} = e^{-[\Delta F^\circ \text{ (1 atm)}/RT]} \tag{4-16}$$

A similar relation exists for molal activities:

$$K = \frac{\alpha_X \alpha_Y}{\alpha_A \alpha_B} = e^{-[\Delta F^\circ \text{ (u.a.)}/RT]} \tag{4-16a}$$

Thus equilibrium constants may be computed directly from the standard free energy change of the reaction, and vice versa.

The free energy $\bar{F}$ of a dissolved gas has the same value whether stated in terms of partial pressures, Eq. (4-13a), or in terms of molal activities, Eq. (4-13b). Values of $\bar{F}^\circ$ (1 atm) and $\bar{F}^\circ$ (u.a.) may therefore be added algebraically to evaluate $\Delta \bar{F}^\circ$ for a reaction, *provided* each value of $\bar{F}^\circ$ (1 atm) is accompanied by a partial pressure p, and each value $\bar{F}^\circ$ (u.a.) is accompanied by a molal activity α_m in the term under the logarithm in Eq. (4-15). The standard free energies $\bar{F}^\circ$ (1 atm) and $\bar{F}^\circ$ (u.a.) do not have the same numerical value for a constituent. Since from Eq. (4-4) $p = 24.45 \, \alpha_m$ at 25°C, integration of Eq. (4-11) between the limits $p = 1$ and $p = 24.45$ will yield the following relation between $\bar{F}^\circ$ (1 atm) and $\bar{F}^\circ$ (u.a.):

$$\bar{F}^\circ \text{ (u.a.)} - \bar{F}^\circ \text{ (1 atm)} = RT \ln \frac{24.45}{1} = 1.985 \times 298 \times 3.192$$

or

$$\bar{F}^\circ \text{ (1 atm)} = \bar{F}^\circ \text{ (u.a.)} - 1,890 \text{ cal/mole} \tag{4-17}$$

In a *heterogeneous* reaction, such as the evaporation of a liquid or the dissociation of a solid or liquid into ions, the value of $\bar{F}_A$ for the liquid or solid as given by Eq. (4-14) or (4-14a) may be used in Eq. (4-9a) with the following result:

$$RT \ln p_X p_Y = \bar{F}_A^\circ \ (1 \text{ atm}) + 0.0242 \frac{M}{\rho} (P-1) - \bar{F}_X^\circ \ (1 \text{ atm}) - \bar{F}_Y^\circ \ (1 \text{ atm})$$

$$= -\Delta \bar{F}^\circ \ (1 \text{ atm}) + 0.0242 \frac{M}{\rho} (P-1) \tag{4-18}$$

A similar relation exists for molal activities. Hence, for a heterogeneous reaction, the equilibrium constant is

$$K = p_X p_Y = \exp \left[-\frac{\Delta F^\circ \ (1 \text{ atm}) - 0.0242 \ M/\rho \ (P-1)}{RT} \right] \tag{4-19}$$

or

$$K = f_\pm^2 \ [X][Y] = \exp \left[-\frac{\Delta F^\circ \ (\text{u.a.}) - 0.0242 \ M/\rho \ (P-1)}{RT} \right] \tag{4-19a}$$

4-3. EXAMPLES

An important application of Eq. (4-19) is in the calculation of the vapor pressure of a liquid or solid or the dissociation of a liquid or solid into components including at least one vapor. p_X and p_Y are the vapor pressures of the gaseous products. Equation (4-19a) may be applied to the ionization of a liquid. When applied to water, $K = K_w$, the *ion product constant* of water. When Eq. (4-19a) is applied to the dissociation of a solid into its ions in aqueous solution, $K = K_s$, the *solubility product* of the solid salt.

Example 4-1. Compute the vapor pressure of water at 25°C and at a pressure of (a) 1 atm; (b) 100 atm.
Solution: The reaction is

$$H_2O_{liq} \rightleftharpoons H_2O_g$$

$$\bar{F}^\circ \ (1 \text{ atm}) = -56,690 \qquad -54,636, -\Delta \bar{F}^\circ \ (1 \text{ atm}) = -2,054 \text{ cal}$$

(a) From Eq. (4-19)

$$p_X = \exp \left[-\frac{2,054 - 0.0242 \frac{18}{1} (1-1)}{1.985 \times 298} \right] = 0.0311 \text{ atm}$$

which agrees with the experimental value in Table 2-3.
(b) From Eq. (4-19)

$$p_X = \exp\left[-\frac{2{,}054 - 0.0242\frac{18}{1}(100 - 1)}{1.985 \times 298}\right] = 0.0334 \text{ atm}$$

an increase of only 7 percent for a pressure increase of 100 to 1.

Example 4-2. Compute the solubility of N_2 gas in water under a partial pressure of 1 atm and at 25°C.
Solution: The reaction is

$$N_{2g} \rightleftharpoons N_{2aq}$$

$\bar{F}°$ (1 atm) = 0 $\quad\quad$ + 2,460, $-\Delta F°$ (1 atm) = −2,460 cal

$$p_X = 24.45[N_2]_{aq} = \exp\left(-\frac{2{,}460}{1.985 \times 298}\right) = 15.62 \times 10^{-3}$$

Solubility of

$$N_2 = [N_2]_{aq} = \frac{15.62}{24.45} \times 10^{-3} = 0.639 \times 10^{-3}$$

$$c'_{N_2} = \frac{0.639 \times 10^{-3}}{10^{-3}} \times 28 = 17.9 \text{ mg/liter}$$

which agrees with the value in Table 5-2.

Example 4-3. Compute the ion product constant of water at 25°C and atmospheric pressure.
Solution: The reaction is

$$H_2O_{liq} \rightleftharpoons H^+ + OH^-$$

$\bar{F}°$ (u.a.) = −56,690 $\quad\quad$ 0 $\quad$ −37,585, $-\Delta F°$ (u.a.) = −19,105 cal/mole

From Eq. (4-19a),

$$K_w = f_\pm^2 [H^+][OH^-] = \exp\left(-\frac{19{,}105}{1.985 \times 298}\right) = 1.00 \times 10^{-14}$$

Example 4-4. Compute the solubility product of $CaCO_3$, calcite, at 25°C.

Solution: The reaction is

$$CaCO_3 \rightleftharpoons Ca^{2+} + CO_3^{2-}$$

$$\bar{F}° \text{ (u.a.)} = -269,780 \qquad -132,180 \quad -126,220$$

$$-\Delta F° \text{ (u.a.)} = -11,380 \text{ cal}$$

From Eq. (4-19a),

$$K_s = f_\pm^2 [Ca^{2+}][CO_3^{2-}] = \exp\left(-\frac{11,380}{1.985 \times 298}\right) = 4.6 \times 10^{-9}$$

The solubility product of a salt furnishes a means for determining its solubility in water.

Example 4-5. Compute the solubility of $CaCO_3$ in water at 25°C, if the composition of the solution is such that $f_\pm$ is unity and $[Ca^{2+}] = [CO_3^{2-}]$. Solution:

$$[Ca^{2+}] = \sqrt{4.6 \times 10^{-9}} = 6.8 \times 10^{-5}$$

$$c' = \frac{6.8 \times 10^{-5}}{10^{-3}} \times 100 = 6.8 \text{ mg/liter as } CaCO_3$$

The actual solubility of $CaCO_3$ in most waters is greater than 6.8 mg/liter at 25°C because the CO_3^{2-} reacts with the water to form HCO_3^- and OH^-, thus reducing $[CO_3^{2-}]$. The true solubility is a function of all the equilibria involving Ca^{2+} and CO_3^{2-}, and requires the solution of a number of simultaneous equations for its determination. The method of attack is illustrated in section 5-2.

4-4. ACIDS, ALKALIS, SALTS, AND pH VALUE

The ionization of water is one of the most important reactions in water-treatment problems. Experimental values of K_w at 1 atm and for various temperatures as quoted from Millard (1) are shown in Table 4-2.

The ion product of water may be expressed as

$$\alpha_{H^+}\alpha_{OH^-} = K_w \tag{4-20}$$

In pure water, $\alpha_{H^+} = \alpha_{OH^-} = \sqrt{K_w} = 10^{-7}$ at 25°C. Any aqueous solution in which $\alpha_{H^+} = \alpha_{OH^-}$ is said to be *neutral*. If α_{H^+} exceeds 10^{-7} when the temperature is referred to 25°C, the solution is *acid*; and if α_{H^+} is less than 10^{-7} at 25°C the solution is *alkaline*. Hence, α_{H^+} is a measure of the strength of an acid or

Table 4-2
Ion Product Constants in Water at 1 atm

Temperature (°C)	$K_w \times 10^{14}$	Temperature (°C)	$K_w \times 10^{14}$
0	0.11	40	2.91
10	0.29	50	5.48
20	0.68	60	9.65
25	1.00	80	23
30	1.47	100	52

alkaline solution. For convenience the exponent of 10, without sign, in the numerical value of α_{H^+} is used to designate the strength of an acid or alkaline solution. This figure, called the *pH value*, or simply *pH*, is defined as follows:

$$pH = \log \frac{1}{\alpha_{H^+}} \qquad (4\text{-}21)$$

Solutions with pH less than 7.0, referred to 25°C, are acid; and those with pH greater than 7.0 are alkaline.

Any electrolyte such as NaCl that ionizes in water to yield or to react with neither H^+ nor OH^- is called a *neutral salt*. Any electrolyte such as HCl or CH_3COOH that ionizes in water to yield H^+ or to combine with OH^- and decreases α_{OH^-} is an *acid*. Any electrolyte such as NaOH that ionizes in water to yield OH^-, and any electrolyte such as Na_2CO_3 that ionizes in water to furnish an ion such as CO_3^{2-}, which combines with H^+ to decrease α_{H^+}, is a *base* or *alkali*. Any substance such as HCO_3^- that may ionize under suitable conditions to yield either H^+ or OH^-, that is,

$$HCO_3^- \rightleftharpoons H^+ + CO_3^{2-}$$
$$HCO_3^- \rightleftharpoons OH^- + CO_2$$

is called an *ampholyte* or *amphoteric* substance. Such a substance is said to exhibit *buffer action*; that is, it acts as a reservoir of acidity when alkali is added and as a reservoir of alkalinity when acid is added, thus tending to maintain a constant pH in the solution containing it. Bicarbonate, HCO_3^-, is the most important buffer encountered in water treatment. Its principal sources in natural waters are dissolved CO_2 and limestone, $CaCO_3$. The equilibria involved will be treated in the discussion of alkalinity in Chapter 5.

4-5. EFFECT OF TEMPERATURE ON EQUILIBRIA

If the value of ΔF° for a reaction is known at one temperature, its value for another temperature may be computed by the van't Hoff equation (*1*, p. 545):

$$\frac{\Delta F_1^{\circ}}{T_1} - \frac{\Delta F_2^{\circ}}{T_2} = \Delta H^{\circ} \frac{T_2 - T_1}{T_2 T_1} \qquad (4\text{-}22)$$

where $-\Delta F_1^{\circ}$ and $-\Delta F_2^{\circ}$ are free energies of the reaction at temperatures T_1 and T_2, and $-\Delta H^{\circ}$ is the *heat* or *enthalpy* of the reaction. For accuracy the temperature interval $T_2 - T_1$ must be small enough so that $-\Delta H^{\circ}$ is reasonably constant. The value of $-\Delta H^{\circ}$ may be computed from the *heat contents*, H°, or the *heats of formation*, $-\Delta H^{\circ}$, of the constituents of the reaction in the same manner as for $-\Delta F^{\circ}$. Experimental values of the molal heat contents H° for numerous substances are given by Latimer (*4*).

The experimental values of H° are usually recorded at some particular temperature, 15, 18, and 25°C being the most common temperatures. To compute ΔH° for use in Eq. (4-22), the relation between the molal heat contents, H°, of each constituent and the temperature must be known. The change in molal enthalpy with temperature (*1*, pp. 106, 279) for any pure substance is

$$H_2^{\circ} - H_1^{\circ} = \int_{T_1}^{T_2} C_p \, dT \qquad (4\text{-}23)$$

where H_2° and H_1° are the molal enthalpies at temperatures T_2 and T_1, respectively, and C_p is the heat capacity at constant pressure in cal/mole/°K.

The value of C_p is a function of the temperature for most substances. The following values are sufficiently accurate for use in Eq. (4-23):

H_2O_{liquid}	$C_p = 1.00$
H_2O_{vapor} below 400°K	$= 8.36$
O_{2g}, H_{2g}, N_{2g} 300 to 2,500°K	$= 6.5 + 0.001T$
CO_{2g} and SO_{2g} 300 to 1,000°K	$= 7.0 + 0.007T$
Solid elements	$= 6.2$
Solid components	$= \Sigma C_p$ for elements

For more accurate values of C_p for numerous gases, liquids, and solids, see the review by Kelly (*5*).

Example 4-6. Compute the free energy of vaporization of water, at other temperatures, from ΔF_{25}°.

Solution: By means of Eq. (4-22) and the values of the molal latent heat of vaporization $\Delta H°$ listed in Table 2-3, the values of $\Delta F°$ listed in Table 2-3 were computed. The vapor pressure p for any temperature may then be computed from the corresponding value of $\Delta F°$ by Eq. (4-19), and it will agree with the experimental value in Table 2-3.

Equation (4-22) may be combined with Eq. (4-16) or (4-19) to give a direct relation between the value of the equilibrium constant at any temperature in terms of its known value at another temperature:

$$\frac{K_2}{K_1} = \exp\left[\frac{\Delta H°}{R}\left(\frac{T_2 - T_1}{T_2 T_1}\right)\right] \qquad (4\text{-}24)$$

This equation is identical with the Clapeyron equation (*1*, p. 106) for change in vapor pressure with temperature.

Most dissolved solids dealt with by the water engineer are ionized minerals for which the heat capacity, C_p, is unknown. It is not feasible, therefore, to use Eq. (4-23) to compute the change in molal enthalpy with temperature, nor Eq. (4-22) or (4-24) to compute the change in $\Delta F°$ or K. It is feasible, however, to measure the equilibrium concentrations of the ions in water at each desired temperature and to compute the values of K therefrom. This has been done for a few ionization reactions, but for most of the reactions of interest only the values of the equilibrium constant at 25°C are known.

REFERENCES

1. E. B. Millard, *Physical Chemistry for Colleges*, New York, McGraw-Hill Book Company, 1941.
2. D. A. MacInnes, *The Principles of Electrochemistry*, New York, Van Nostrand Reinhold Company, 1939.
3. J. W. Gibbs, *The Collected Works of J. Willard Gibbs*, Vol. 1, Essex, England, Longman Group Ltd., 1928, p. 63. See also MacInnes, *op. cit.*, p. 124.
4. W. M. Latimer, *Oxidation Potentials*, 2nd ed., Englewood Cliffs, N.J., Prentice-Hall, Inc., 1952.
5. K. K. Kelly, *U.S. Bur. Mines Bull. 371*, 1934.

CHAPTER 5

Composition of Natural and Polluted Waters and of Liquid Wastes

Most of the substances that occur in natural waters in sufficient quantity to affect the quality of the water are classified in Table 5-1. No single water may contain all these impurities, since the coexistence of some of these constituents in measurable quantities (i.e., CO_2 and CO_3^{2-}) is incompatible with chemical equilibria. Other impurities, such as lead and copper, may be introduced during treatment or conveyance, but since such constituents are not usually present in natural waters in appreciable concentration, they are omitted from Table 5-1.

Some natural waters contain selenium and arsenic in sufficient concentrations to affect the quality. All natural waters contain traces of radioactive substances, principally radium, and the concentration of radioactivity in some natural groundwaters is dangerously high. Some natural waters contain chromium, cyanides, chlorides, acids, alkalies, various metals including mercury, and organic pollutants, which are discharged in liquid wastes by mines, industries, and human habitations.

Mercury and mercury compounds may be added to natural waters from various industries and from antifouling paints containing mercury, which are used on boats; but because of their great density, mercury and its compounds usually settle to the bottom. Both mercury compounds and pesticides are difficult to detect in water because they are absorbed by algae and other aquatic life including fish. Hundreds of new exotic chemicals are

Table 5-1
Substances Occurring in Natural Waters

Origin	Suspended	Colloidal	Gases	Nonionized Solids and Dipoles	Positive Ions	Negative Ions
From mineral soils and rocks	Clay, sand, other inorganic soils	Clay, SiO_2, Fe_2O_3, Al_2O_3, MnO_2	CO_2		Ca^{2+}, Mg^{2+}, Na^+, K^+, Fe^{2+}, Mn^{2+}, Zn^{2+}	HCO_3^-, Cl^-, SO_4^{2-}, NO_3^-, CO_3^{2-}, $HSiO_3^-$, $H_2BO_3^-$, HPO_4^{2-}, $H_2PO_4^-$, OH^-, F^-
From the atmosphere			N_2, O_2, CO_2, SO_2		H^+	HCO_3^-, SO_4^{2-}
From organic decomposition	Organic soil (top soil), organic wastes	Vegetable coloring matter, organic wastes	CO_2, NH_3, H_2S, N_2, H_2S, CH_4, H_2 odorivectors	Vegetable coloring matter, organic wastes	Na^+, NH_4^+, H^+	Cl^-, HCO_3^-, NO_2^-, NO_3^-, OH^-, HS^-, organic radicals
Living organisms	Fish, algae, diatoms, and minute animals	Viruses, bacteria, algae, and diatoms				

Table 5-2
Solubility of Gases in Pure Water in Contact with the Pure Gas at a Pressure of 1 Atm

Temperature (°C)	Nitrogen,[a] 98.815% N_2 + 1.185% A		Oxygen,[a] O_2		Hydrogen,[a] H_2		Methane,[a] CH_4		Carbon Dioxide,[a] CO_2		Hydrogen Sulfide,[a] H_2S		Sulfur Dioxide,[a] SO_2		Ammonia,[b] NH_3	
	$[N_2] \times 10^3$	mg/liter	$[O_2] \times 10^3$	mg/liter	$[H_2] \times 10^3$	mg/liter	$[CH_4] \times 10^3$	mg/liter	$[CO_2]_T \times 10^3$	mg/liter$_T$[c]	$[H_2S]_T \times 10^3$	mg/liter$_T$	m_T	mg/liter$_T$	m_T	mg/liter$_T$
0	1.050	29.4	2.18	69.8	0.959	1.93	2.48	39.8	76.4	3,360	208.3	7,100	3.58	186,700	52.3	471,000
5	0.931	26.1	1.913	61.2	0.912	1.84	2.14	34.3	63.5	2,790	177.4	6,040	3.03	162,300	45.8	438,000
10	0.830	23.2	1.696	54.3	0.872	1.76	1.864	29.9	53.25	2,345	151.6	5,160	2.55	140,600	40.1	406,000
15	0.752	21.1	1.523	48.7	0.841	1.70	1.645	26.4	45.45	2,000	131.4	4,475	2.14	120,500	35.2	375,000
20	0.689	19.3	1.384	44.3	0.812	1.64	1.475	23.6	39.1	1,720	115.2	3,925	1.80	103,300	30.9	344,000
25	0.639	17.9	1.263	40.4	0.783	1.58	1.342	21.5	33.9	1,495	101.8	3,470	1.51	88,200	27.9	322,000
30	0.599	16.8	1.163	37.2	0.758	1.53	1.233	19.7	29.65	1,305	90.9	3,090	1.26	74,700	23.8	288,000
40	0.528	14.8	1.029	32.9	0.734	1.48	1.057	16.9	23.65	1,040	74.1	2,520	0.91	54,800	19.8	252,000
60	0.456	12.77	0.868	27.8	0.714	1.44	0.872	14.0	16.0	704	53.1	1,810			14.0	192,000
80	0.427	11.96	0.786	25.1	0.714	1.44	0.79	12.7			40.9	1,394			9.1	134,000
100	0.423	11.85	0.758	24.2	0.714	1.44	0.76	12.2			36.1	1,230				

[a] From Landolt-Börnstein, *Physikalisch-Chemische Tabellen*, Berlin, Springer-Verlag, 1923, Table 131, p. 763.
[b] From Landolt-Börnstein, also J. W. Mellor, *Comprehensive Treatise on Inorganic and Theoretical Chemistry*, Vol. 8, London, New York, Longmans, 1922, p. 194.
[c] T signifies total solubility.

being produced annually of which the toxicity to living organisms, including man, is yet to be determined. It is the opinion of the authors that none of these new chemicals should be released by the manufacturers until toxicological determinations have been made which show that they may be used without harm to man or the environment.

5-1. DISSOLVED GASES

The solubilities of the gases that occur in sufficient quantity to be determined by chemical analysis are shown in Table 5-2. Each figure in the table is the equilibrium concentration of the dissolved gas in pure water, including its reaction products with the water, when the water is in contact with the pure gas at a pressure of 1 atm.

The gases that do not react with water to an appreciable extent are N_2, O_2, H_2, and CH_4. Hence, the solubility of these gases in pure water at equilibrium with the atmosphere is proportional to the partial pressure of the gas in the atmosphere and follows Henry's law. Moisture-free air has the average analysis shown in Table 5-3 (1):

Table 5-3

	Percentage by Volume
Nitrogen	78.08
Oxygen	20.95
Carbon dioxide	0.03
Argon, etc.	0.94

The solubility of O_2, for example, in pure water at $20°C$ exposed to dry air at a barometric pressure of 1 standard atm is $0.2095 \times 44.3 = 9.3$ mg/liter; and if the air is saturated with moisture (see Table 2-3), the solubility is $(1 - 0.0231)9.3 = 9.1$ mg/liter.

The principal source of dissolved nitrogen and oxygen in the oceans and fresh surface waters is the air, but oxygen is also produced in water by chlorophyll-bearing aquatic plants and algae through photosynthesis. Oxygen is required in the respiration of all living plants and animals and also by all combustion processes. Nevertheless, oxygen in the atmosphere at sea level remains substantially constant with time. The presence of H_2 and CH_4 in natural waters is due primarily to the production of these gases by bacteria in anaerobic decomposition of organic matter. Buswell and Larson (2) have found CH_4 in well waters in concentrations exceeding 20 ppm and have found that methane persists even after aeration in concentrations greater than 2 ppm. Methane is also present in swamp waters and coal mine drainage. Both CH_4 and H_2 are highly combustible in the gas phase. CH_4 is not a permanent constituent of the atmosphere, and, according to Paneth (*1*), H_2 comprises less than 0.001 percent of the earth's atmosphere.

Carbon dioxide, hydrogen sulfide, sulfur dioxide, and ammonia ionize in water or react with water to produce ions. The subscript T indicates that solubilities shown in Table 5-2 are the total solubilities, including ionization products.

Hydrogen sulfide ionizes in water as follows:

$$H_2S \rightleftharpoons H^+ + HS^-$$ (5-1)

and the equilibrium relation is

$$\frac{f[H^+] \, f[HS^-]}{f[H_2S]} = 1.1 \times 10^{-7} \quad \text{at } 25°C$$ (5-2)

from which it may be seen that only about 0.1 percent of the gas is ionized in pure water at pH 4. Hence, the total solubility $[H_2S]_T$ from Table 5-2 may be used for the solubility of the nonionized gas $[H_2S]$ with an error of only 0.1 percent, and the solubility of the nonionized gas at partial pressures other than 1 atm will be proportional to the partial pressure, following Henry's law.

Hydrosulfide ionizes further in water as follows:

$$HS^- \rightleftharpoons H^+ + S^{2-}$$ (5-3)

and the equilibrium relation is

$$\frac{f[H^+] \, f[S^{2-}]}{f[HS^-]} = 1 \times 10^{-14} \quad \text{at } 25°C$$ (5-4)

from which it may be seen that the portion present as S^{2-} is negligible, even at high pH (0.1 percent at pH 11 and 1 percent at pH 12). The portions of HS^- and H_2S present at various pH values, as computed from Eq. (5-2), and the corresponding total solubilities of H_2S_T at 25°C are shown in Table 5-4.

H_2S is produced by the anaerobic decomposition of organic matter and by the dilution of waste sulfide liquors discharged from tanneries, paper mills, and textile mills. H_2S gas and the mercaptans, CH_2SH and CH_3CH_2SH, produce very disagreeable odors when released to the atmosphere, even in light concentrations; they also discolor paint. Dissolved H_2S gas is not found in waters containing even a trace of dissolved oxygen. In the presence of dissolved

Table 5-4
Percentage of Ionization and Solubility of H_2S at Various pH Values and $25°C$

pH	Percentage of H_2S	Percentage of HS^-	Solubility in Contact with 1 Atm of Pure Gas (mg/liter)
4	99.9	0.1	3,470
5	98.9	1.1	3,510
6	90.1	9.9	3,840
7	47.7	52.3	7,270
7.5	22.5	77.5	15,400
8	8.3	91.7	41,800[a]
8.5	2.80	97.20	124,000[a]
9	0.89	99.11	390,000[a]
10	0.091	99.91	—

[a]Neglecting ionic strength correction.

oxygen, H_2S is oxidized to water and free sulfur or to sulfate:

$$2H_2S + O_{2_{aq}} \rightleftharpoons 2H_2O + \underline{2S} \tag{5-5a}$$

$$H_2S_{aq} + 2O_{2_{aq}} \rightleftharpoons 2H^+ + SO_4^{2-} \tag{5-5b}$$

Since the atmosphere is normally free of H_2S, this gas may be removed readily from water containing it by aeration, which accomplishes the double function of releasing H_2S and of furnishing dissolved oxygen for the oxidation of H_2S in the water. Oxygen gas does not react with H_2S gas except at elevated temperatures.

The dissolved hydrogen sulfide gas is readily oxidized by dissolved oxygen, whereas the hydrosulfide ion is not. As the dissolved gas is oxidized, hydrosulfide ions combine with hydrogen ions of the water to form more hydrogen sulfide gas. This in turn will be oxidized by dissolved oxygen to the extent that dissolved oxygen is present or made available for the reaction. Experiments with trickling filters, treating settled sewage and tannery wastes with sulfides up to about 100 mg/liter and a pH of about 9.3, indicate substantially complete oxidation of sulfides.

The rate at which the hydrogen sulfide gas is liberated from water to the atmosphere depends upon the degree of agitation of the water. The more turbulent the mixing, the faster the gas is liberated and the more intense is its odor. Laboratory studies made under the senior author's direction indicate that the concentration of hydrogen sulfide gas in streams of moderate turbulence should probably be less than about 1.0 mg/liter to avoid odors. To avoid odors from the spray of trickling filter nozzles, the concentration should probably be less than 0.5 mg/liter.   Table 5-4 indicates that wastes having a total sulfide content of 100 mg/liter may be applied to trickling filters without odor nuisance, provided the pH is not lower than about 9.5 (0.5 percent of 100 mg/liter is 0.5 mg/liter of H_2S gas).

Sulfides may also be removed from water by oxidation with chlorine:

$$Cl_{2_{aq}} + H_2S_{aq} \rightleftharpoons 2H^+ + 2Cl^- + \underline{S} \qquad (5\text{-}6a)$$

and

$$4Cl_{2_{aq}} + H_2S_{aq} + 4H_2O \rightleftharpoons 10H^+ + 8Cl^- + SO_4^{2-} \qquad (5\text{-}6b)$$

These reactions indicate mole ratios of Cl_2 to H_2S of 1 to produce free sulfur, and of 4 to produce sulfate.   In experiments by Black and Goodson (*3*) with solutions containing 2 to 6 mg/liter of H_2S at 25°C, it was found that reactions (5-6) were substantially complete within 1 min.   The reacted mole ratios of Cl_2 to H_2S were found to range from about 1.8 at pH 9 to 3.3 at pH 5.   This indicates that the portion of H_2S oxidized to sulfate ranges from about 27 percent at pH 9 to about 77 percent at pH 5, and that more free sulfur is produced at higher pH values.

Sulfur dioxide reacts with water as follows:

$$SO_{2_{aq}} + H_2O \rightleftharpoons H^+ + HSO_3^- \qquad (5\text{-}7)$$

and the equilibrium relation is

$$\frac{f[H^+]\,f[HSO_3^-]}{f[SO_{2_{aq}}]} = 1.25 \times 10^{-2} \quad \text{at } 25°C \qquad (5\text{-}8)$$

from which it may be seen that the gas is almost completely ionized at all pH values above 4.   Neither $SO_{2_{aq}}$ nor HSO_3^- can remain in the presence of dissolved oxygen, since both are strong reducing agents. The following reaction takes place readily:

$$HSO_3^- + \tfrac{1}{2}O_{2_{aq}} \rightleftharpoons H^+ + SO_4^{2-} \qquad (5\text{-}9)$$

Sulfur dioxide enters natural waters from combustion fumes in the air and as a partially oxidized product of bacterial decomposition of organic matter. Clean air is normally free of SO_2.   Sulfurous and sulfuric acids are present in the drainage waters of coal mines.

Ammonia is a bacterial decomposition product in both aerobic and anaerobic decomposition of nitrogenous organic matter, and it is the main source of nitrogen for cell building of microorganisms. It may be discharged to natural waters in sewage along with unhydrolized urea from urine and as a waste product of industry. Nearly 80 percent of the nitrogen in human excretions is discharged in urine as urea (see Table 10-2), which is therefore the principal source of ammonia in human sewage. Urea hydrolizes slowly to NH_3 and CO_2. The gas reacts with water to ionize as follows:

$$NH_3 + H_2O \rightleftharpoons NH_4^+ + OH^- \tag{5-10}$$

and the equilibrium relation is

$$\frac{f[NH_4^+]\, f[OH^-]}{f[NH_{3_{aq}}]} = 1.81 \times 10^{-5} \quad \text{at } 25°C \tag{5-11}$$

This relation indicates that nearly all the NH_3 gas in dilute solution is ionized below pH 6 and is nonionized above pH 11, as shown in Table 5-5.

Table 5-5
Percentage of Ionization of NH_3 in Dilute Solutions
at Various pH Values and $25°$ C

pH	Percentage of NH_4^+	Percentage of NH_3
5	99.995	0.005
6	99.95	0.05
7	99.45	0.55
8	94.77	5.23
9	64.50	35.50
10	15.30	84.70
11	1.78	98.22

The solubilities given in Table 5-2 cannot be used for the solubilities of the nonionized NH_3 gas in the application of Henry's law to dilute solutions, because of the extremely high concentration of NH_3 in solution at 1 atm. According to Lewis and Randall (4), in dilute solutions of NH_3, $m/p = 56.7$ at $25°C$, where m is the molality of $NH_{3_{aq}}$ and p is the partial pressure in the gas phase. This ratio may be applied to the nonionized NH_3 with an error of less than 0.5 if the concentration is less than 1 m. Since the air is substantially free of NH_3, the presence of *free ammonia* in natural waters even in concentrations as low as 0.2 mg/liter is taken as evidence of recent organic pollution. The presence of urea is a similar indicator. The ammonia concentration in seawaters (5) ranges from 0 to 0.075 mg/liter.

Carbon dioxide, like ammonia, is a bacterial decomposition product in both aerobic and anaerobic breakdown of organic matter, but since it is a product of the respiration of all living matter and of combustion, it is more prevalent than NH_3. It is normally present in the atmosphere at about 0.03 percent by volume, but it is higher in enclosed populated spaces. Paneth (*1*) reports a concentration of about 0.064 percent in Boston subways.

CO_2 gas reacts with water to ionize in the following manner:

$$CO_2 + H_2O \rightleftharpoons H^+ + HCO_3^-$$ (5-12)

and the equilibrium relation is

$$\frac{f[H^+] f[HCO_3^-]}{f[CO_{2_{aq}}]} = 4.31 \times 10^{-7} \text{ at } 25°C$$ (5-13)

To estimate how much of the total solubility, $[CO_2]_T$ in Table 5-2, is not ionized, it is necessary to compute the $[H^+]$ of the solution. This may be done by replacing $[CO_{2_{aq}}]$ in Eq. (5-13) by $[CO_2]_T - [HCO_3^-]$, $[HCO_3^-]$ by $[H^+] - (10^{-14}/[H^+])$, and solving for $[H^+]$. The result is $[H^+] = 1.21 \times 10^{-4}$ or a pH of 3.92; and $[CO_{2_{aq}}] = 33.78 \times 10^{-3}$ or 1,490 mg/liter. The solubility of CO_2 gas in freshwater in equilibrium with the atmosphere at sea level is therefore about 0.03 percent of $1,490 = 0.447$ mg/liter, following Henry's law. Surface waters are usually supersaturated with CO_2, especially in summer when the rate of organic decomposition is high. CO_2 is produced at the expense of dissolved O_2, with the result that the CO_2 concentration near the bottom of lakes and quiet streams may be 20 ppm or higher.

The solubility of each of these gases in natural water containing mineral salts is reduced in proportion to the decrease in concentration of the solvent. The decrease in concentration is probably about equal to the concentration of water utilized in hydration of the ions of the salts. Figure 5-1 shows the effect of dissolved salts in seawater on the solubility of N_2, O_2, and CO_2, based upon the experiments of Fox (*6*). Since most freshwaters contain less than 1,000 ppm total dissolved salts, the decrease in gas solubility due to the salts is of little consequence with freshwaters.

Although the solubility of the nonionized portions of the gases in freshwater is little influenced by the presence of other solutes, the concentrations of the ions may be greatly affected. For example, if NH_3, which is basic, and CO_2, which is acid, are both present in water with a pH of 7.0, the ratio

$$\frac{f[NH_4^+]}{f[NH_{3_{aq}}]} = 181 \quad \text{and} \quad \frac{f[HCO_3^-]}{f[CO_{2_{aq}}]} = 4.16 \quad \text{at } 25°C$$

which indicates that most of both gases is in the ionized form. If such water

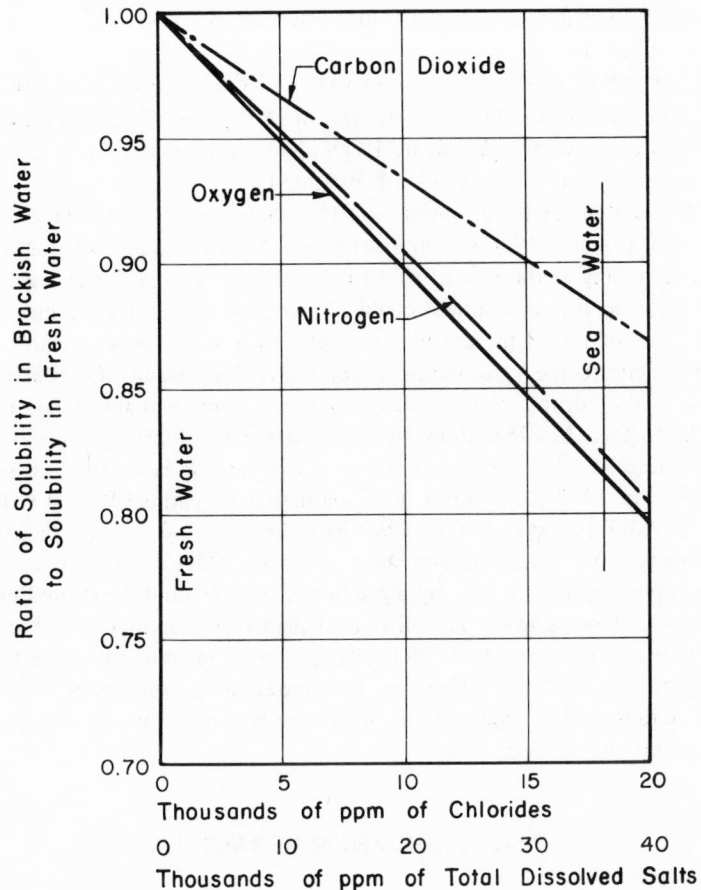

Figure 5-1
Effect of dissolved salts in seawater on solubilities of nitrogen, oxygen, and carbon dioxide.

has a free CO_2 content of 10 mg/liter, the corresponding equilibrium concentration of HCO_3^- is about 57.6 mg/liter.

At high pH values, HCO_3^- is further ionized in the following manner:

$$HCO_3^- \rightleftharpoons H^+ + CO_3^{2-} \qquad (5\text{-}14)$$

The corresponding equilibrium is

$$\frac{f[H^+]\, f[CO_3^{2-}]}{f[HCO_3^-]} = 4.69 \times 10^{-11} \quad \text{at } 25°C \qquad (5\text{-}15)$$

This relation indicates that at pH values less than 8.3 the $[CO_3^{2-}]$ is less than 1 percent of the $[HCO_3^-]$.

CO_2 gas is a less important source of $CO_{2_{aq}}$, HCO_3^-, and CO_3^{2-} in most freshwaters than are the $CaCO_3$ and $MgCO_3$ in the rocks. Since limestone and magnesite are so widely prevalent, HCO_3^- is usually present in freshwater in higher concentrations than any other constituent.

The CO_2 content of the atmosphere (7) has increased about 15 percent during the past 100 years; in 1960 it comprised about 0.032 percent of the atmosphere and was increasing at the rate of about 0.23 percent per year. A change in CO_2 concentration in the atmosphere could appreciably affect the thermal budget of the surface of the earth, and might cause a long-term change in the weather and climate due to the "greenhouse effect." The increase in CO_2 is attributed primarily to the burning of fossil fuels, which, in 1960, was about 0.41 percent of the total CO_2. In 1953, Riley (in 7) estimated the consumption of CO_2 by photosynthesis at 3.28 percent of the total per year on land and 17.8 percent in the sea. In 1957, Craig (also in 7) estimated the annual flux of CO_2 into the sea at 13.3 percent, and in 1960 Broecker (in 7) estimated this flux at 20.3 percent; both estimates are consistent with Riley's estimate of photosynthetic consumption of CO_2 by algae in the sea. It should be noted that the destruction of trees, grasses, and other land plants by man may be reducing the rate of photosynthesis on land, thus significantly contributing to the CO_2 increase in the atmosphere. There is also considerable production of CO_2 by bacterial action in the topsoils of the land masses and within the bottom muds of freshwaters and the seas.

5-2 DISSOLVED MINERALS

The mineral content of natural waters is responsible for its hardness and its alkalinity. The *hardness* of a water is defined as its content of metallic ions that react with sodium soaps to produce solid soaps and that react with negative ions, when the water is evaporated in boilers, to produce solid boiler scale. This characteristic is confined to the polyvalent metals, principally Ca and Mg, since Na and K salts are extremely soluble. Hardness is usually expressed as mg/liter of equivalent $CaCO_3$. The *alkalinity* of a water is defined as its content of negative ions or other substances which react to neutralize the hydrogen ions that are added in titration with acid. Alkalinity is due to the presence of HCO_3^-, CO_3^{2-}, OH^-, $HSiO_3^-$, $H_2BO_3^-$, HPO_4^{2-}, $H_2PO_4^-$, HS^-, and NH_3; like hardness, it is customarily expressed as mg/liter of equivalent $CaCO_3$.

Bisilicate is a product of the ionization of silica and water, as follows:

$$\underline{SiO_2} + H_2O \rightleftharpoons H^+ + HSiO_3^-$$ (5-16)

The solubility of SiO_2 (Min-U-Sil-5, a product of the Pennsylvania Glass Sand Corporation) at pH 7.6 is 9 mg/liter (8), which corresponds with a $[HSiO_3^-]$ of 1.5×10^{-4}. From thermodynamic considerations, at $25°C$ the value of $\Delta F°$ for the bisilicate ion is -232.01 and the solubility product is

$$f[H^+] \, f[HSiO_3^-] = \text{approx. } 3.75 \times 10^{-12} \quad \text{at } 25°C \qquad (5\text{-}17)$$

This value indicates that the saturation bisilicate alkalinity in terms of equivalent $CaCO_3$ is about 0.375 mg/liter at pH 6 and about 3.75 mg/liter at pH 7.

At higher pH values bisilicate ionizes as follows:

$$HSiO_3^- \rightleftharpoons H^+ + SiO_3^{2-} \qquad (5\text{-}18)$$

and the equilibrium relation is

$$\frac{f[H^+] \, f[SiO_3^{2-}]}{f[HSiO_3^-]} = \text{approx. } 2.0 \times 10^{-15} \quad \text{at } 25°C \qquad (5\text{-}19)$$

which indicates that bisilicate is only 0.02 percent ionized at pH 11 and 0.2 percent ionized at pH 12.

All silica and silicate crystals are composed of SiO_4^{4-} tetrahedra $(9, 10)$ having an Si—O distance of about 1.60 Å. In pure SiO_2 each O atom is shared by two Si atoms in a three-dimensional crystal, such as quartz. Silicates in clays and micas occur principally in two-dimensional networks similar to Figure 2-3 in which three of the four oxygen atoms of each tetrahedron are shared between two Si atoms. The network is thus an ion with the chemical formula $Si_2O_5^{2-}$ and is bonded together with positive ions to form the solid crystal. SiO_3^{2-} and $HSiO_3^-$ ions, in solution, can be accounted for as long-chain ions in which two of the oxygen atoms are shared by two Si atoms. The solution of silica or silicates probably involves the disintegration of the solid into long-chain ions. Except at high pH values, much of the silica in water is colloidal, and the crystal structure of precipitated silica is ill-defined or amorphous.

The principal sources of borate in water are borax, $Na_2B_4O_7 \cdot 10H_2O$, and solid boric acid. Both substances are quite soluble in water, yielding molecularly dispersed boric acid, $H_3BO_{3_{aq}}$, and $H_2BO_3^-$ upon solution. Boric acid ionizes as follows:

$$H_3BO_{3_{aq}} \rightleftharpoons H^+ + H_2BO_3^- \qquad (5\text{-}20)$$

and the ionization equilibrium is

$$\frac{f[H^+] \, f[H_2BO_3^-]}{f[H_3BO_3]} = 6.0 \times 10^{-10} \quad \text{at } 25°C \qquad (5\text{-}21)$$

This value indicates that less than 1 percent of the H_3BO_3 present in dilute solutions is ionized at pH values lower than 7.22 and that about 86 percent is ionized at pH 10.

The principal sources of phosphates in natural waters are probably the apatite minerals, $3Ca_3(PO_4)_2 \cdot CaF_2$ and $3Ca_3(PO_4)_2 \cdot CaCl_2$, which dissolve in water to yield $H_2PO_4^-$ and HPO_4^{2-}. PO_4^{3-} has a tetrahedral configuration like SiO_4^{4-} with a P—O distance of 1.55 Å. Polyphosphates are formed by the sharing of O atoms between two P atoms, but these compounds are less stable in solution than the corresponding silica compounds, and tend to revert to orthophosphates. Phosphates and metaphosphates are now widely used in the treatment of water for boiler use, for municipal control of red water, and for water conditioning in commercial and home laundering. Organic phosphates used as insecticides, principally phosphorothionates, and phosphates in fertilizers are washed into many surface waters.

In recent years, synthetic detergents containing phosphates have largely replaced sodium soaps used in laundering and dishwashing. Since 1963 there has developed an active public support for cleaning up the environment, including control of both air and water pollution. This public clamor has resulted in demands by some governmental agencies for the elimination of phosphates from detergents, because it was thought that phosphates are the cause of eutrophication of ponds, lakes, and the ocean by algae. Recent studies (1970) of the bottom muds of Lake Erie have shown that the concentration of solid phosphates is uniform to a depth of up to 30 ft, which indicates that there exists an inexhaustible source of phosphates to the algae that must have been the result of many decades of deposits. Laboratory experiments by Borchard (in *11*) at the University of Michigan indicate that algae settling to the bottom may ingest phosphates from the muds and store them to the extent of 20 percent of their body weight; these phosphates become available for continuous growth when the algae are distributed through the lake by turnover or turbulence. There are numerous freshwater lakes with little or no pollution but with occasional heavy blooms of algae. Green Bay, Wisconsin, was so named when it was first discovered some 300 years ago because of the prolific growths of algae in it. In view of the above and the fact that phosphates in detergents constitute a relatively small portion of the total phosphates associated with eutrophication, the removal of phosphates from detergents will not accomplish the desired end. CO_2 and HCO_3^- are perhaps the principal nutrients in the photosynthetic growth of algae; yet, as of 1973, the major substitute that has emerged for phosphates in detergents is sodium carbonate.

Orthophosphoric acid is very soluble in water, indicating a molecular dispersion as $H_3PO_4{}_{aq}$, which ionizes in the following manner:

$$H_3PO_4{}_{aq} \rightleftharpoons H^+ + H_2PO_4^- \tag{5-22}$$

The equilibrium relation is

$$\frac{f[H^+]\,f[H_2PO_4^-]}{f[H_3PO_4]} = 7.5 \times 10^{-3} \quad \text{at } 25°C \tag{5-23}$$

A second ionization takes place,

$$H_2PO_4^- \rightleftharpoons H^+ + HPO_4^{2-} \tag{5-24}$$

for which the equilibrium is

$$\frac{f[H^+]\,f[HPO_4^{2-}]}{f[H_2PO_4^-]} = 6.2 \times 10^{-8} \quad \text{at } 25°C \tag{5-25}$$

The third ionization is

$$HPO_4^{2-} \rightleftharpoons H^+ + PO_4^{3-} \tag{5-26}$$

and the equilibrium is

$$\frac{f[H^+]\,f[PO_4^{3-}]}{f[HPO_4^{2-}]} = 10^{-12} \quad \text{at } 25°C \tag{5-27}$$

These relations indicate that, in dilute solutions, less than 1 percent of the HPO_4^{2-} present is ionized at pH values less than 10; less than 1 percent of the $H_2PO_4^-$ is ionized at pH values less than 5.21; and more than 99 percent of the H_3PO_4 present is ionized at pH values greater than 4.12. It is evident that phosphate alkalinity in natural waters consists of both $H_2PO_4^-$ and HPO_4^{2-}.

The alkalinity determination is useful principally as a means for computing the bicarbonate concentration. For a reliable determination, the sample should be collected and titrated in a closed container with acid to a pH of 4 or less, and the amount of acid used, the initial pH, and the final pH should be measured accurately. The alkalinity* is given by the following relation:

$$2[Alk] = 2[CO_3^{2-}] + [HCO_3^-] + [HSiO_3^-] + [H_2BO_3^-] + 2[HPO_4^{2-}]$$
$$+ [H_2PO_4^-] + [HS^-] + [OH^-] + [NH_3] - [H^+]_f \tag{5-28}$$

in which $[Alk] = Alk \times 10^{-5}$, where Alk is the alkalinity in mg/liter of equivalent $CaCO_3$. The amount of acid used in mg/liter of equivalent $CaCO_3$ is

---

*The methods of determining alkalinity described in *Standard Methods for the Examination of Water and Wastewater* (*12*) are not sufficiently accurate for obtaining the $[HCO_3^-]$. For further discussion see refs. *13-15*.

$([\text{Alk}] + \frac{1}{2}[\text{H}^+]_f) \times 10^5$, where $[\text{H}^+]_f$ is the final hydrogen ion concentration. The value of $[\text{OH}^-]$ is determined directly from the initial pH, which also determines

$$\frac{f[\text{CO}_3^{2-}]}{f[\text{HCO}_3^-]}, \quad f[\text{HSiO}_3^-], \quad \frac{f[\text{H}_2\text{BO}_3^-]}{f[\text{H}_3\text{BO}_3]}, \quad \frac{f[\text{HPO}_4^{2-}]}{f[\text{H}_2\text{PO}_4^-]}, \quad \frac{f[\text{HS}^-]}{f[\text{H}_2\text{S}]}, \quad \frac{f[\text{NH}_4^+]}{f[\text{NH}_3]}$$

The total concentrations of silicate, borate, phosphates, sulfides, and ammonia should be titrated from separate determinations of silica, boron, phosphate, sulfides, and free ammonia. The ammonia and sulfides are usually negligible, but may be important in some industrial wastes and in water or sludge containing considerable decomposing organic matter (including urea). Alkalinity relations at 25°C are shown in Figure 5-2.

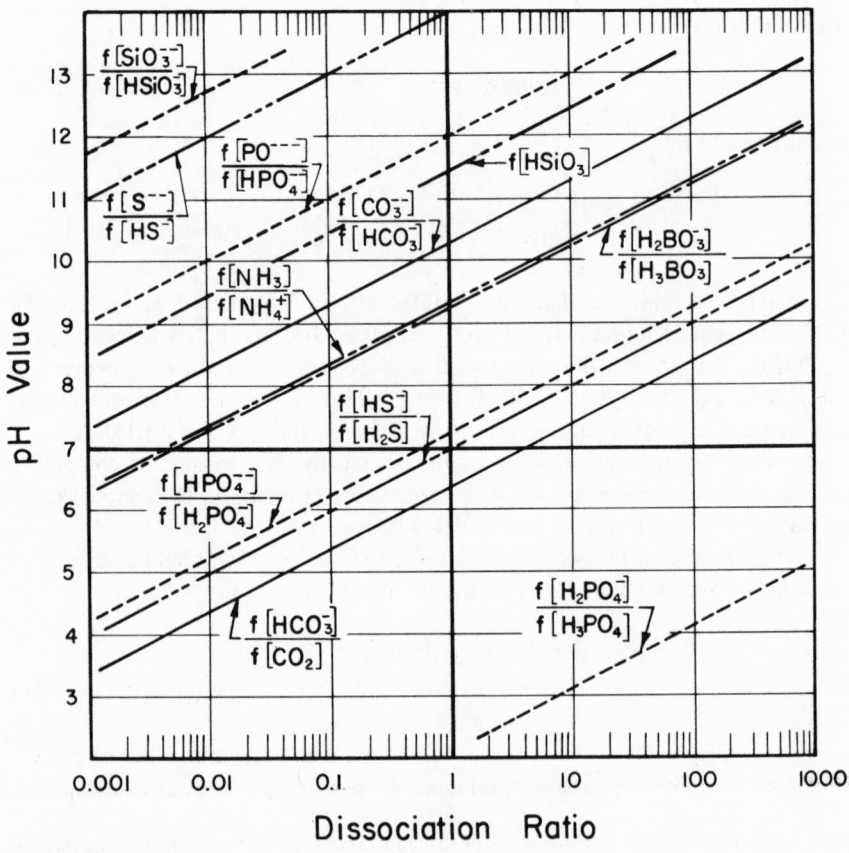

Figure 5-2
Alkalinity relations at 25°C and 1 atm.

The following substances ionize in water as indicated, and the solubility product of each is given in the following formula:

Solid $CaCO_3$ (calcite):

$$CaCO_3 \rightleftharpoons Ca^{2+} + CO_3^{2-} \tag{5-29}$$

$$f^2[Ca^{2+}][CO_3^{2-}] = 4.6 \times 10^{-9} \quad \text{at } 25°C \tag{5-30}$$

Solid $MgCO_3$:

$$MgCO_3 \rightleftharpoons Mg^{2+} + CO_3^{2-} \tag{5-31}$$

$$f^2[Mg^{2+}][CO_3^{2-}] = 4.2 \times 10^{-9} \quad \text{at } 25°C \tag{5-32}$$

Solid $MgOH_2$:

$$Mg(OH)_2 \rightleftharpoons Mg^{2+} + 2OH^- \tag{5-33}$$

$$f[Mg^{2+}]f^2[OH^-]^2 = 8.0 \times 10^{-12} \quad \text{at } 25°C \tag{5-34}$$

Solid $ZnCO_3$:

$$ZnCO_3 \rightleftharpoons Zn^{2+} + CO_3^{2-} \tag{5-35}$$

$$f^2[Zn^{2+}][CO_3^{2-}] = 2.0 \times 10^{-10} \quad \text{at } 25°C \tag{5-36}$$

Solid $Zn(OH)_2$:

$$Zn(OH)_2 \rightleftharpoons Zn^{2+} + 2OH^- \tag{5-37}$$

$$f[Zn^{2+}]f^2[OH^-]^2 = 4.5 \times 10^{-17} \quad \text{at } 25°C \tag{5-38}$$

Solid CaF_2:

$$CaF_2 \rightleftharpoons Ca^{2+} + 2F^- \tag{5-39}$$

$$f[Ca^{2+}]f^2[F^-]^2 = 1.62 \times 10^{-10} \quad \text{at } 25°C \tag{5-40}$$

Solid MgF_2:

$$MgF_2 \rightleftharpoons Mg^{2+} + 2F^- \tag{5-41}$$

$$f[Mg^{2+}]f^2[F^-]^2 = 8.3 \times 10^{-8} \quad \text{at } 25°C \tag{5-42}$$

Solid $Ca_3(PO_4)_2$ (the principal constituent of apatite minerals, bones, and teeth):

$$Ca_3(PO_4)_2 + 4H^+ \rightleftharpoons 3Ca^{2+} + 2H_2PO_4^- \qquad (5\text{-}43)$$

$$\frac{f^3[Ca^{2+}]^3 \, f^2[H_2PO_4^-]^2}{f^4[H^+]} = 8.5 \times 10^6 \quad \text{at } 25°C \qquad (5\text{-}44)$$

Solid $Cu(OH)_2$:

$$Cu(OH)_2 \rightleftharpoons Cu^{2+} + 2OH^- \qquad (5\text{-}45)$$

$$f[Cu^{2+}] \, f^2[OH^-]^2 = 1.6 \times 10^{-19} \quad \text{at } 25°C \qquad (5\text{-}46)$$

Solid Cu_2O:

$$\tfrac{1}{2}Cu_2O + \tfrac{1}{2}H_2O \rightleftharpoons Cu^+ + OH^- \qquad (5\text{-}47)$$

$$f[Cu^+] \, f[OH^-] = 1.6 \times 10^{-15} \quad \text{at } 25°C \qquad (5\text{-}48)$$

Solid $Fe(OH)_2$:

$$Fe(OH)_2 \rightleftharpoons Fe^{2+} + 2OH^- \qquad (5\text{-}49)$$

$$f[Fe^{2+}] \, f^2[OH^-]^2 = 1.8 \times 10^{-15} \quad \text{at } 25°C \qquad (5\text{-}50)$$

Solid $FeCO_3$:

$$FeCO_3 \rightleftharpoons Fe^{2+} + CO_3^{2-} \qquad (5\text{-}51)$$

$$f^2[Fe^{2+}] \, [CO_3^{2-}] = 2 \times 10^{-11} \quad \text{at } 25°C \qquad (5\text{-}52)$$

Solid $Fe(OH)_3$:

$$Fe(OH)_3 \rightleftharpoons Fe^{3+} + 3OH^- \qquad (5\text{-}53)$$

$$f[Fe^{3+}] \, f^3[OH^-]^3 = 6 \times 10^{-38} \quad \text{at } 25°C \qquad (5\text{-}54)$$

Solid $Mn(OH)_2$:

$$Mn(OH)_2 \rightleftharpoons Mn^{2+} + 2OH^- \qquad (5\text{-}55)$$

$$f[Mn^{2+}] \, f^2[OH^-]^2 = 2 \times 10^{-13} \quad \text{at } 25°C \qquad (5\text{-}56)$$

Solid $Mn(OH)_3$:

$$Mn(OH)_3 \rightleftharpoons Mn^{3+} + 3OH^- \qquad (5\text{-}57)$$

$$f[Mn^{3+}] \, f^3[OH^-]^3 = 2 \times 10^{-36} \quad \text{at } 25°C \qquad (5\text{-}58)$$

To oxidize Fe^{2+} with dissolved O_2 for removal from water as $Fe(OH)_3$:

$$Fe^{2+} + \tfrac{5}{2}H_2O + \tfrac{1}{4}O_{2\,aq} \rightleftharpoons \underline{Fe(OH)_3} + 2H^+ \qquad (5\text{-}59)$$

and the solubility product is

$$\frac{f[Fe^{2+}]\, f^{1/4}[O_{2\,aq}]^{1/4}}{f^2[H^+]^2} = 6.5 \times 10^{-3} \quad \text{at } 25^\circ C \qquad (5\text{-}60)$$

To oxidize Mn^{2+} with dissolved O_2 for removal from water as MnO_2:

$$Mn^{2+} + H_2O + \tfrac{1}{2}O_{2\,aq} \rightleftharpoons \underline{MnO_2} + 2H^+ \qquad (5\text{-}61)$$

and the solubility product is

$$\frac{f[Mn^{2+}]\, f^{1/2}[O_{2\,aq}]^{1/2}}{f^2[H^+]^2} = 3.5 \times 10^{-2} \quad \text{at } 25^\circ C \qquad (5\text{-}62)$$

These reactions require 0.143 ppm $O_2/1$ ppm Fe^{2+} and 0.29 ppm $O_2/1$ ppm Mn^{2+}. The solubilities of Fe and Mn as a function of pH at $25^\circ C$ are plotted in Figure 5-3.

The reactions and solubility products for oxidizing Fe^{2+} and Mn^{2+} with chlorine are

$$Fe^{2+} + \tfrac{5}{2}H_2O + \tfrac{1}{2}HOCl \rightleftharpoons \underline{Fe(OH)_3} + \tfrac{1}{2}Cl^- + \tfrac{5}{2}H^+ \qquad (5\text{-}63)$$

$$\frac{f[Fe^{2+}]\, f^{1/2}[HOCl]^{1/2}}{f^{1/2}[Cl^-]^{1/2} f^{5/2}[H^+]^{5/2}} = 4 \times 10^{-8} \quad \text{at } 25^\circ C \qquad (5\text{-}64)$$

$$Mn^{2+} + H_2O + HOCl \rightleftharpoons \underline{MnO_2} + Cl^- + 3H^+ \qquad (5\text{-}65)$$

$$\frac{f[Mn^{2+}]\, f[HOCl]}{f[Cl^-]\, f^3[H^+]^3} = 1.05 \times 10^{-9} \quad \text{at } 25^\circ C \qquad (5\text{-}66)$$

These reactions require 0.636 ppm $Cl_2/1$ ppm Fe^{2+} and 1.29 ppm $Cl_2/1$ ppm Mn^{2+}.

The reactions and solubility products for oxidizing Fe^{2+} and Mn^{2+} with potassium permanganate are

$$Fe^{2+} + \tfrac{7}{3}H_2O + \tfrac{1}{3}MnO_4^- \rightleftharpoons \underline{Fe(OH)_3} + \tfrac{1}{3}\underline{MnO_2} + \tfrac{5}{3}H^+ \qquad (5\text{-}67)$$

$$\frac{f[Fe^{2+}]\, f^{1/3}[MnO_4^-]^{1/3}}{f^{5/3}[H^+]^{5/3}} = 1.78 \times 10^{-11} \quad \text{at } 25^\circ C \qquad (5\text{-}68)$$

$$Mn^{2+} + \tfrac{2}{3}H_2O + \tfrac{2}{3}MnO_4^- \rightleftharpoons \tfrac{5}{3}\underline{MnO_2} + \tfrac{4}{3}H^+ \qquad (5\text{-}69)$$

$$\frac{f[Mn^{2+}]\, f^{2/3}[MnO_4^-]^{2/3}}{f^{4/3}[H^+]^{4/3}} = 2.15 \times 10^{-16} \quad \text{at } 25^\circ C \qquad (5\text{-}70)$$

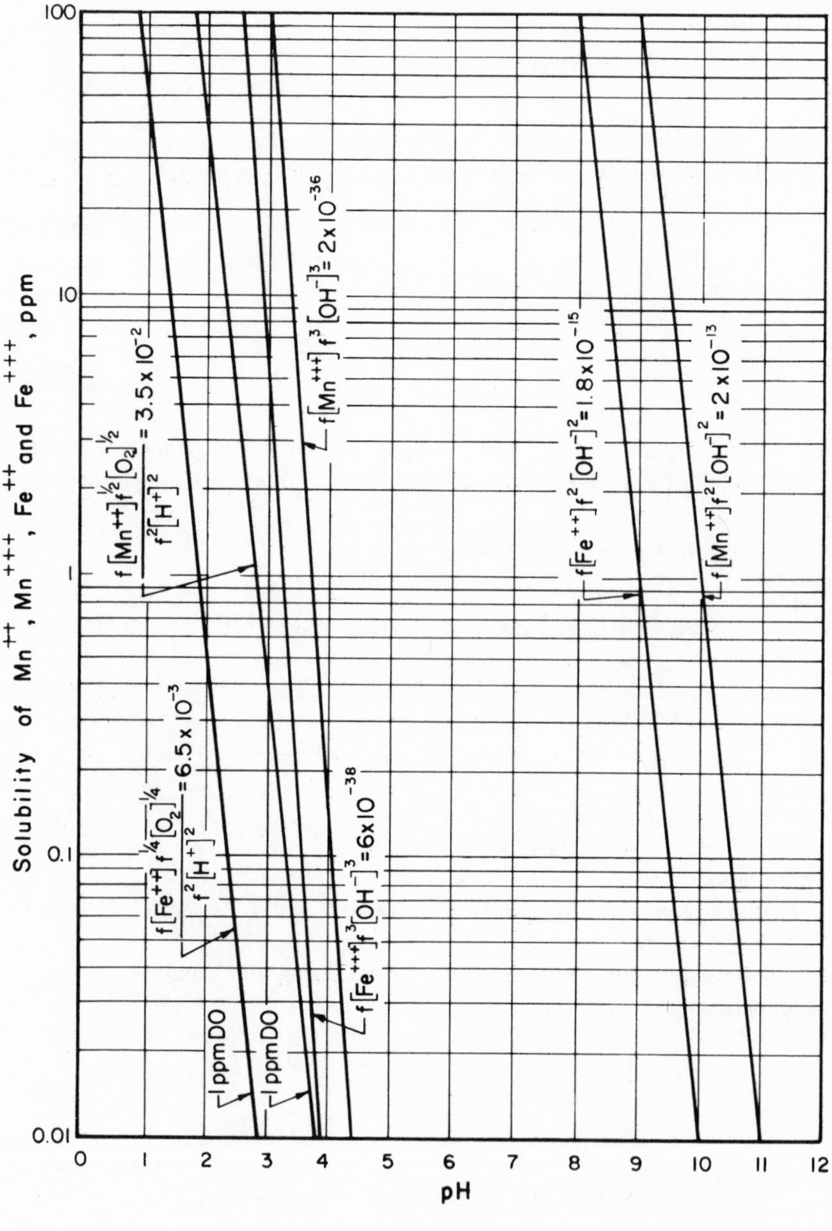

The vertical axis is labeled "Solubility of Mn^{++}, Mn^{+++}, Fe^{++} and Fe^{+++}, ppm" with values from 0.01 to 100. The horizontal axis is labeled "pH" from 0 to 12.

Curves are labeled:

$$\frac{f[Mn^{++}]f^{\frac{1}{2}}[O_2]^{\frac{1}{2}}}{f^2[H^+]^2} = 3.5 \times 10^{-2}$$

$$f[Mn^{+++}]f^3[OH^-]^3 = 2 \times 10^{-36}$$

$$f[Fe^{++}]f^2[OH^-]^2 = 1.8 \times 10^{-15}$$

$$f[Mn^{++}]f^2[OH^-]^2 = 2 \times 10^{-13}$$

$$\frac{f[Fe^{++}]f^{\frac{1}{4}}[O_2]^{\frac{1}{4}}}{f^2[H^+]^2} = 6.5 \times 10^{-3}$$

$$f[Fe^{+++}]f^3[OH^-]^3 = 6 \times 10^{-38}$$

1 ppm DO

1 ppm DO

Figure 5-3
Solubility of iron and manganese at 25° C.

These reactions require 0.944 ppm $KMnO_4$/1 ppm Fe^{2+} and 1.92 ppm $KMnO_4$/1 ppm Mn^{2+}. The solubilities with Cl_2 and $KMnO_4$ are too low to show in Figure 5-3.

In coagulation with either filter alum, $Al_2(SO_4)_3 \cdot 14H_2O$, or iron salts, the dissolved coagulant releases Al^{3+}, Fe^{2+}, or Fe^{3+}, which reacts with alkali to precipitate the hydrous oxides of the metals. These metals are added impurities, as is the Mn in $KMnO_4$, which should be removed along with other coagulated impurities in the water or wastewater by settling and/or filtration. Solid $Al(OH)_3$ ionizes in water at acid pH values as follows:

$$\underline{Al(OH)_3} \rightleftharpoons Al^{3+} + 3OH^- \tag{5-71}$$

and at higher pH values as follows:

$$\underline{Al(OH)_3} \rightleftharpoons H^+ + H_2AlO_3^- \tag{5-72}$$

The solubility products are

$$f[Al^{3+}] \, f^3[OH^-]^3 = 5 \times 10^{-33} \quad \text{at } 25°C \tag{5-73}$$

$$f^2[H_2AlO_3^-] \, [H^+] = 4 \times 10^{-13} \quad \text{at } 25°C \tag{5-74}$$

The solubilities of aluminum and iron from the hydrous oxides as a function of pH, and of Fe from $FeCO_3$ as a function of pH and carbonate and bicarbonate alkalinity are plotted in Figure 5-4.

It will be noted from Figure 5-4 that, at $25°C$, the minimum solubility of $Al(OH)_3$ is about 0.004 ppm of Al^{3+}, which occurs at about pH 5.5. This is the optimum pH for coagulation and filtration, at $25°C$, with alum to avoid after-precipitation of $Al(OH)_3$ in the water distribution system. From the figure, if the pH is less than 5 or greater than 7.25, more than 0.2 ppm of aluminum will be present in the filter effluent. For coagulation with ferrous iron, the pH should be greater than 10; but with ferric iron, any pH greater than 4 is satisfactory. Since water must be treated at temperatures ranging from $0°C$ up to about $35°C$, solubility products for $Al(OH)_3$ over this range of temperatures are badly needed to determine the optimum pH values.

Experience indicates that the optimum pH value for removal of Mn^{2+} with $KMnO_4$ is likely to be higher than the optimum pH for using alum as the coagulant. No such difficulty arises with ferric iron as the coagulant.

The effect of temperature on the values of the equilibrium constants for Eqs. (5-13) and (5-15) is given in Table 5-6.

A mineral analysis of a water may be checked by comparing the concentration of positive ionic charges with the concentration of negative charges. It is convenient for this purpose to convert the content of each ion into mg/liter of equivalent $CaCO_3$. The hardness and alkalinity are thus apparent from the analysis, and the molar concentration of each ion is simply 10^{-5} times its

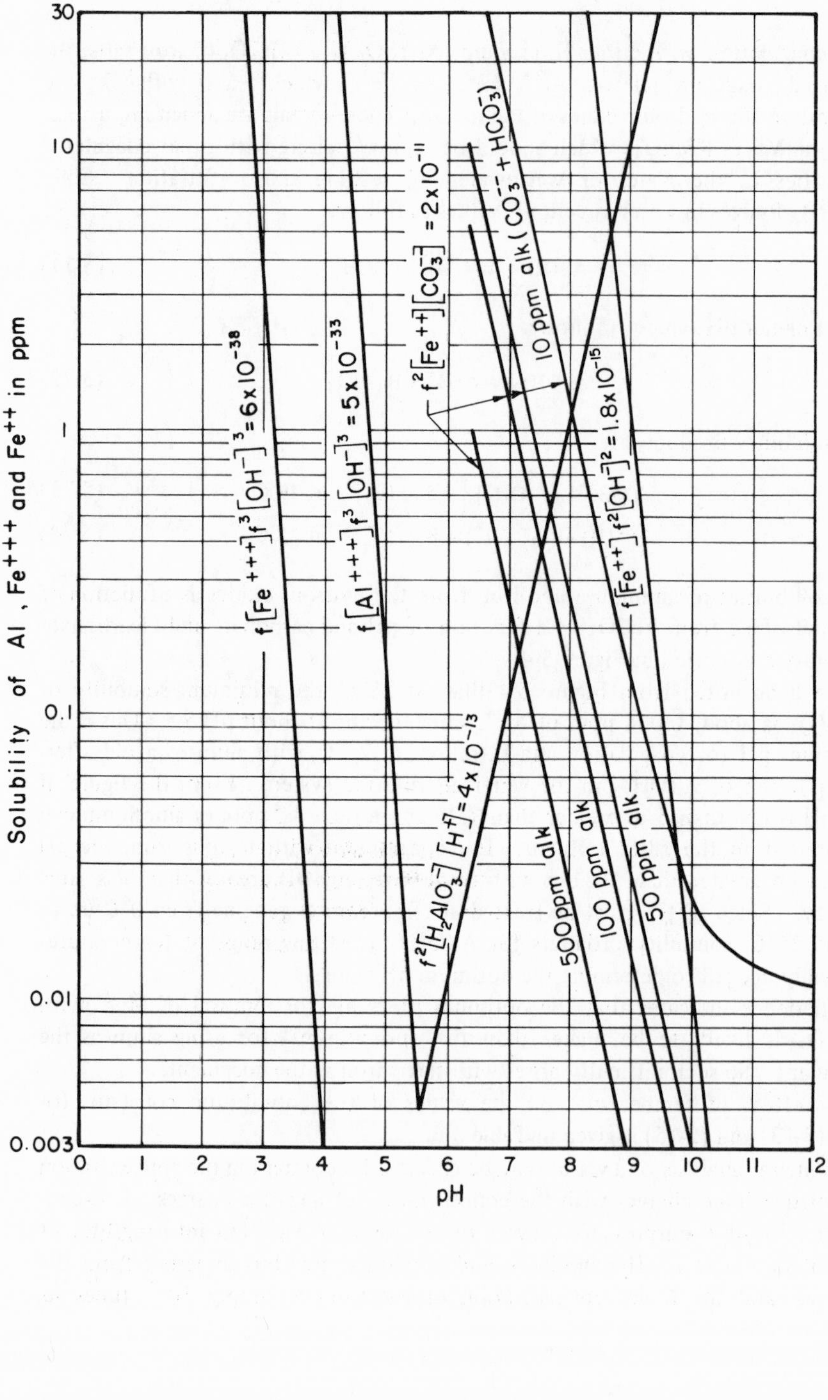

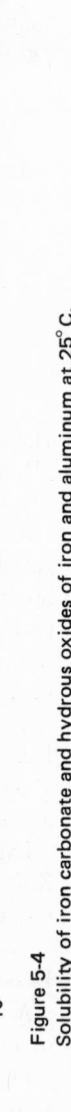

Figure 5-4
Solubility of iron carbonate and hydrous oxides of iron and aluminum at 25° C.

Table 5-6
Equilibrium Constants for Ionization of Carbonic Acid,
Bicarbonate, and Calcium Carbonate

Water	Temperature (°C)								
	0	5	10	15	20	25	30	40	50
Fresh $K_{(5-13)} \times 10^7$ (*16*)	2.61	3.0[a]	3.34	3.8[a]	4.05	4.31	4.52		
Fresh $K_{(5-15)} \times 10^{11}$ (*17*)	2.36	2.75[a]	3.24	3.72[a]	4.20	4.69	5.13	6.03	6.73
Fresh $K_{(5-29)} \times 10^9$ (*18*)		8.1	7.1	6.0	5.25	4.6		3.1	
Sea $K_{(5-13)} \times 10^7$ (*18*)		7.1				10.2			
Sea $K_{(5-15)} \times 10^{10}$ (*18*)		6.6				11.0			
Sea $K_{(5-29)} \times 10^7$ (*18*)		21.4				6.5			
Sea $K_{(5-13)} \times 10^7$ (*19*)	7.08		8.31		9.55	10.0	10.45		
Sea $K_{(5-15)} \times 10^{11}$ (*19*)	39.8		52.4		57.5	79.4	95.5		

[a]These values are from ref. *18*.

concentration in mg/liter of equivalent $CaCO_3$ for a bivalent ion, and twice this value for a monovalent ion. The checking procedure is illustrated by the following example.

Example 5-1. The analysis shown in Table 5-7 represents equilibria in a water sample at 25°C. Compute the concentrations that are not shown, balance the analysis, and determine the hardness and alkalinity of the water.

Table 5-7

Substance	Concentration (mg/liter)
Ca^{2+}	41.0
Mg^{2+}	8.5
Na^+	10.1
Fe^{2+}	0.33
CO_3^{2-}	?
HCO_3^-	128.0
SO_4^{2-}	?
Cl^-	15.0
NO_3^-	4.0
F^-	1.6
Free CO_2	?
Total silica (as SiO_2)	5.0
pH value	7.35

Solution: Table 5-8 is a convenient form in which to present the analysis.

Table 5-8

			mg/liter as $CaCO_3$	
Substance	*M*	*c'* (mg/liter)	+Ions	− Ions
Ca^{2+}	40	41.0	102.5	
Mg^{2+}	24.3	8.5	35.0	
Na^+	23	10.1	21.9	
Fe^{2+}	55.8	0.33	0.6	
CO_3^{2-}	60	☐0.17		☐0.28 (0.22)
HCO_3^-	61	128.0		105.0
$HSiO_3^-$	77.1	☐6.45		☐4.20
SO_4^{2-}	96	☐21.2		☐22.02 (22.08)
Cl^-	35.5	15.0		21.1
NO_3^-	62	4.0		3.2
F^-	19	1.6		4.2
Totals			160.0	160.0
Free CO_2	44	12.1	☐27.4	
SiO_2	60.1	5.0		
Colloidal SiO_2	60.1	☐2.48		☐4.12
pH		7.35		

$$f[H^+] = 10^{-7.35} = 4.47 \times 10^{-8}$$

From Eq. (5-17), the solubility of $HSiO_3^-$ is

$$f[HSiO_3^-] = \frac{3.75 \times 10^{-12}}{4.47 \times 10^{-8}} = 8.39 \times 10^{-5}$$

$$= \text{approx. } 4.20 \text{ mg/liter as } CaCO_3$$

$$[SiO_2]_T = \frac{5}{60.1} \times 10^{-3} = 8.32 \times 10^{-5}$$

$$= \text{approx. } 8.32 \text{ mg/liter as } CaCO_3$$

Hence, $[HSiO_3^-]$ is 8.39×10^{-5} or 4.20 mg/liter as $CaCO_3$, and the concentration of colloidal silica is $8.32 - 4.20 = 4.12$ mg/liter as $CaCO_3$.

From Eq. (5-15),

$$f[CO_3^{2-}] = 4.69 \times 10^{-11} \frac{f[HCO_3^-]}{f[H^+]}$$

$$= \text{approx. } 4.69 \times 10^{-11} \frac{2.1 \times 10^{-3}}{4.47 \times 10^{-8}}$$

$$= \text{approx. } 0.22 \times 10^{-5} \quad \text{or} \quad 0.22 \text{ ppm as } CaCO_3$$

which is entered as a trial value pending evaluation of activity coefficients. A trial value of the SO_4^{2-} concentration may now be obtained as follows:

$$c' = 160.0 - (0.22 + 105.0 + 4.20 + 21.1 + 3.2 + 4.2)$$

$$= 22.08 \text{ ppm as } CaCO_3$$

From Eq. (4-6),

$$\omega = \tfrac{1}{2}[4(102.5 + 35 + 0.6 + 0.22 + 22.08) \, 10^{-5}$$

$$+ 1(43.8 + 210 + 8.39 + 42.2 + 6.4 + 8.4) \, 10^{-5}]$$

$$= 48.04 \times 10^{-4}$$

From Eq. (4-5), for monovalent ions,

$$\log f = -1 \frac{0.5056 \sqrt{48.04 \times 10^{-4}}}{1 + \sqrt{48.04 \times 10^{-4}}} = 0.967 - 1$$

$$f = 0.93$$

For bivalent ions,

$$\log f = 4(0.967 - 1)$$

$$f = 0.74$$

Correcting concentrations for f,

$$[CO_3^{2-}] = \frac{4.69 \times 10^{-11}}{4.47 \times 10^{-8}} \times \frac{0.93 \times 2.1 \times 10^{-3}}{0.74} = 0.28 \times 10^{-5}$$

$$= 0.28 \text{ ppm as } CaCO_3$$

$$= 0.17 \text{ ppm as } CO_3^{2-}$$

The SO_4^{2-} concentration:

$$c' = 160.0 - (0.28 + 105.0 + 4.20 + 21.1 + 3.2 + 4.2)$$

$$= 22.02 \text{ ppm as } CaCO_3$$

$$= 21.2 \text{ ppm as } SO_4^{2-}$$

From Eq. (5-13),

$$[CO_2] = \frac{4.47 \times 10^{-8} \times 0.93 \times 2.1 \times 10^3}{0.74 \times 4.31 \times 10^{-7}} = 27.4 \times 10^{-5}$$

$$c' = 27.4 \times 10^{-5} \times 44 \times 10^3 = 12.1 \text{ ppm}$$

$$\text{hardness} = 102.5 + 35 + 0.6$$

$$= 138.1 \text{ ppm}$$

$$\text{alkalinity} = 0.28 + 105.0 + 4.20$$

$$= 109.48 \text{ ppm.}$$

5-3. ORGANIC IMPURITIES

The organic matter in water or wastewater is significant principally because it serves as a food supply for bacteria and other living organisms. High counts of bacteria and plankton are associated with large concentrations of organic matter. When the bacterial count is high, the probability is greater that some of the organisms may be *pathogens* (disease germs), particularly if the source of the organic matter is known to contain human or animal fecal matter. The sanitary significance of impurities is discussed in Chapter 6. Organic matter and living organisms are also detrimental to some industrial uses of water, for example, in condensers and pipes where organic growths reduce the efficiency of heat transfer and of flow. Standards of industrial water quality are discussed in Chapter 7.

The feeding processes of many microorganisms are accompanied by the production of odorivectors, which may injure the potability of the water if they are not removed. The odorivectors are usually not present in sufficient concentration to be detected by any means other than by the odors* they produce. It is not known, therefore, whether the odor-producing substance is a decomposition product of the organic matter or a secretion of the living organisms. Odors emanating from bacterial decomposition are customarily associated with the organic matter that serves as the food supply, but odors produced by plankton are usually associated with the living organism, and many plankton[†] are known to produce distinctive odors.

Colloidal and suspended organic matters can usually be removed from water by coagulation, sedimentation, and filtration. Dissolved organic matter, however, is not generally removed effectively except by biochemical means such as those used in wastewater treatment processes. Water that has a dissolved

---

*For test procedures see ref. *12*.
[†]For test procedures see refs. *12*, *20*, and *21*.

organic content of sufficient concentration to require biochemical treatment is highly polluted and should be classified as a liquid organic waste. The principles of bacterial decomposition of organic matter are discussed in Chapter 10.

Municipal sewage is the most common type of organic waste. In a municipal sewage of medium strength, the total solids content will amount to about 800 mg/liter of which about 300 mg/liter is suspended and about 500 mg/liter is colloidal and dissolved. About two thirds of the suspended solids are organic, and the remainder are mineral. About half the colloidal and dissolved solids are organic and the rest are mineral. The organic matter is about 50 percent carbohydrate, 40 percent nitrogenous matter, and 10 percent fat.

Many dissolved organic wastes that are attacked slowly by bacteria, or are not attacked at all, are now being produced by industry. Included among these wastes are tars and resins, lignins, synthetic detergents, pesticides, and many other substances produced by the petrochemical and pharmaceutical industries. Some of these substances are potentially toxic to human and aquatic life. Practical means are not yet available for identifying and measuring these substances in polluted waters, but they may be wholly or partially removed by means of activated carbon.

REFERENCES

1. F. A. Paneth, *Sci. J. Roy. Coll. Sci.*, *6*, 120 (1936).
2. A. M. Buswell and T. E. Larson, *J. Amer. Water Works Assoc.*, *29*, 1978 (1937).
3. A. P. Black and J. B. Goodson, Jr., "The Oxidation of Sulfides by Chlorine in Dilute Aqueous Solutions," *J. Amer. Water Works Assoc.*, *44*, 309 (1952).
4. G. N. Lewis and M. Randall, *Thermodynamics and the Free Energy of Chemical Substances*, New York, McGraw-Hill Book Company, 1923, p. 556.
5. D. W. Hood, in *The Encyclopedia of Oceanography*, R. W. Fairbridge, ed., New York, Van Nostrand Reinhold Company, 1966, p. 794.
6. C. J. J. Fox, *Trans. Faraday Soc.*, *5*, 68 (1909).
7. T. Takahashi, in Fairbridge, *op. cit.*, p. 170.
8. R. K. Ham and R. F. Christman, "Agglomerate Size Changes in Coagulation," *J. San. Eng. Div.*, *Amer. Soc. Civil Engrs.*, 1268 (Oct. 1970).
9. L. Pauling, *The Nature of the Chemical Bond*, 3rd ed., Ithaca, N.Y., Cornell University Press, 1960.
10. R. W. G. Wyckoff, *The Structure of Crystals*, New York, Van Nostrand Reinhold Company, 1931.
11. Excerpts from the Federal Trade Commission testimony on "Detergents, Phosphates and Eutrophication," by W. C. Krumrei of the Procter and Gamble Co., Apr. 26, 1971.
12. *Standard Methods for the Examination of Water and Wastewater*, 13th ed., New York, American Public Health Association, 1971.
13. W. F. Langelier, *J. Amer. Water Works Assoc.*, *28*, 1500 (1936).
14. F. E. DeMartini, *J. Amer. Water Works Assoc.*, *30*, 85 (1938).
15. T. E. Larson and A. M. Buswell, *J. Amer. Water Works Assoc.*, *34*, 1667 (1942).
16. T. Shedlovsky and D. A. MacInnes, *J. Amer. Chem. Soc.*, *57*, 1705 (1935).

17. H. S. Harned and S. R. Scholes, *J. Amer. Chem. Soc.*, *63*, 1706 (1941).

18. W. Stumm and J. J. Morgan, *Aquatic Chemistry*, New York, Wiley (Interscience), 1970, p. 170.

19. After Lyman (1956), quoted in *Chemical Oceanography*, Vol. 1, J. P. Riley and G. Skirrow, eds., New York, Academic Press, Inc., 1965, p. 651.

20. H. B. Ward and G. C. Whipple, *Fresh Water Biology*, New York, John Wiley & Sons, Inc., 1918.

21. G. C. Whipple, G. M. Fair, and M. C. Whipple, *Microscopy of Drinking Water*, 4th ed., New York, John Wiley & Sons, Inc., 1927.

CHAPTER 6

Sanitary Significance of Impurities in Water and Effects on Aquatic Life

6-1. WATERBORNE DISEASES

The impurities of greatest importance in water to be used for drinking or culinary purposes are the pathogenic bacteria, microorganisms, and viruses. The most serious waterborne diseases are cholera in the Old World and typhoid fever in America, both caused by specific bacteria. Other human diseases known to be transmitted through water are paratyphoid fever, bacillary dysentery, amoebic dysentery, infectious hepatitis, and other disturbances of the intestinal tract for which no specific organisms have been found. The virus of poliomyelitis has been recovered from sewage-polluted waters, and there is evidence that this disease may be waterborne in some cases. The isolation of the causative organism of any waterborne disease is impractical in routine water examination. Because most diseases known to be transmitted through water are of intestinal origin and the source of the causative organism in water is human excreta, the presence of sewage in water is evidence of the possibility of the presence of infectious organisms.

The presence of ammonia and nitrites, and of chlorides in abnormal amounts, is grounds for suspecting recent contamination of a water by sewage, because these substances are present in human sewage. It is not conclusive evidence, however, since these substances may arise from many other sources. The presence in water of the coliform group of bacilli, particularly *Escherichia coli*, whose normal habitat is the lower intestine of man and other mammals, is the best evidence of recent fecal pollution. Many natural waters are subjected to fecal contamination at many points during their course of travel toward the ocean, but are nevertheless relatively safe for drinking at other points. As a standard of safety for drinking water, complete absence of intestinal bacteria is too severe and is not feasible.

A natural water or an artificially treated water of otherwise acceptable quality, in which the total count of coliform bacteria is below a figure (about 1 per 100 ml in the United States) that experience shows to be relatively safe, may be classified as safe drinking water, provided a sanitary survey shows that the source of the supply and the water system, itself, are adequately protected against the effects of pollution. Such a standard of purity assumes that the numbers of infectious organisms of all types, if present at the source of pollution, have been reduced at least in proportion to the reduction in the count of coliform bacteria. The assumption is probably warranted in the case of most pathogenic bacteria, but it is not warranted in the case of some viruses. Viruses differ from bacteria in that they cannot reproduce except within living cells of other organisms; whereas bacteria are generally larger than 0.5 μm, viruses range from about 0.01 to 0.3 μm, and are the smallest biological form capable of producing disease in humans and in other living species. They are not removed by filtration to the same extent as bacteria, and some known viruses are more resistant to the usual doses of chlorine.

Unlike bacteria, viruses (1) are completely insensitive to antibiotics, except for the larger more complex agents of the parrot-fever group (*psittacosis*), which are generally considered not to be true viruses. Viruses multiply not by dividing, as do bacteria and larger microorganisms, but by redirecting the complex biochemical systems of invaded cells, channeling these systems into the production of virus. The cell culture technique of growing viruses in the laboratory has resulted in the discovery of many new viruses, many of which infect the gastrointestinal tract of man and animals. The plaque assay technique is used for counting viruses in terms of the plaque-forming unit (PFU). A count of the plaques in any one bottle is presumably equal to the number of PFU inoculated. A distinction must be made between infection by viruses and virus disease, because up to about 90 percent of the hosts infected show no symptoms of disease but nevertheless become carriers of the virus, which multiply greatly within the host and are excreted in enormous numbers through the gastrointestinal tract.

Outbreaks of gastroenteritis (2), of the nature of intestinal influenza, for which a virus is suspected to be the causative agent, seem to be waterborne in some cases, even when the water complies with the usual bacterial standards of purity. The 1955–1956 epidemic of infectious hepatitis in Delhi, India (3), was traced to a filtered and chlorinated water meeting the coliform standard of less than 2 per 100 ml, and there was no reported concurrent increase in typhoid and dysentery. Kelly and Sanderson (4) have found in studies of chlorination of five strains of polio and two strains of Coxsackie viruses that 0.3 ppm of free residual chlorine for at least 30 min contact is required for 99.9 percent inactivation, and that about 10 ppm of combined residual chlorine for about 60 min is required for 99.7 percent inactivation.

Two kinds of viral hepatitis (5) are known, *infectious hepatitis*, caused by virus A, and *serum hepatitis*, caused by virus B. Virus A has been demonstrated in the blood and feces and has an incubation period of 10 to 50 days; the average is 30 days. The feces and blood are infectious to others when administered orally or parenterally. Virus B has been demonstrated only in the blood and has an incubation period of 60 to 160 days. Blood and blood fractions are infectious to others. Accordingly, serum hepatitis is transmitted only by infected instruments that puncture the skin or mucous membranes or by infected blood or blood fractions used for transfusions. On the other hand, infectious hepatitis may also be transmitted through water, food, or milk. Both types may produce jaundice after about a week, but the majority of hepatitis cases in children are usually without jaundice and may be so mild as to escape notice entirely. The lack of a susceptible laboratory animal and of specific serologic tests handicaps the solution of many of the problems associated with the disease. Most of our information is derived from epidemiologic observations and from experimental studies in which human volunteers were used.

Neefe (5) and co-workers (1945, 1947), in experimental studies using human volunteers, found that coagulation, settling, and filtration of water that contained hepatitis virus A did not eliminate or inactivate the virus. Treatment of raw water, to provide a chlorine residual of 1 ppm after 30 min, did not inactivate the virus, although breakpoint chlorination did. Coagulation, settling, and filtration followed by treatment that left a 1.1 ppm total chlorine residual after 30 min effectively inactivated the virus.

The cysts of the protozoan *Endamoeba histolytica* (6), the cause of amoebic dysentery, are much more resistant to chlorination than coliform bacteria; however, they may be destroyed by superchlorination with a contact period of 30 min or more. Chang (7) has stated that even though some amoebas, in the spore stage, and nematodes are not killed by 10 ppm of residual chlorine, their ability to pass through a rapid sand filter plant is greatly reduced by exposure to chlorine at this concentration. Despite the defects of a drinking-water standard based on counts of coliform bacteria, it appears to be the best type of standard available with the present state of our knowledge, and it serves its purpose reasonably well.

Because of the importance of the coliform count as a measure of the safety of water for drinking, it has also been adopted as a measure of the safety of water for bathing and for the growing of shellfish. The bacteriological techniques for making counts of coliform bacteria have been under critical study for many years throughout the world. Prior to the introduction of the membrane-filter technique, the coliform count in a water or wastewater sample was usually made by incubation in a group of fermentation tubes where the presence or absence of organisms of the coliform group was observed for each of the tubes, but not the number in any tube. The count was only an estimate of the actual concentration. In the early years of use of the bacteriological test, it was assumed that positive tubes initially contained only one coliform bacterium. One positive 10-ml tube in a group of five would thus indicate one organism in 50 ml or 2 per 100 ml. The count estimated in this way was known as the *E. coli index*. Inasmuch as some of the positive tubes may have contained more than one bacterium, the actual number was probably greater than indicated by the *E. coli* index. The application of probability theory to the problem resulted in the development of the *most probable number* or MPN technique, which is described in standard methods (*8*). For the preceding example of one positive 10-ml tube in five, the coliform MPN is 2.2 per 100 ml. Tables of MPN counts for various combinations of sizes and numbers of fermentation tubes are presented in standard methods (*8*). For raw water both the total coliform count and the fecal coliform (*9*) count are usually made to differentiate between recent fecal pollution from warm-blooded animals and other members of the coliform group found in soil, on plants and insects, in old sewage, and in waters polluted some time in the past. Incubation is at 35° C for the total coliform count and at 43° C for the fecal coliform count.

It is becoming common practice to use the membrane-filter technique for both total and fecal coliform counts. This methodology, developed in Germany during World War II, was introduced as a tentative method in the 1955 edition of *Standard Methods For the Examination of Water and Wastewater* and became a standard method in the 1960 edition (*10*). Except for obvious limitations when the raw water is turbid or contains a high density of noncoliform organisms, the membrane-filter technique offers distinct advantages over the older bacteriological methods: (1) it is more precise than the multiple-tube fermentation test; (2) it allows the examination of more significant volumes of sample; and, (3) results are obtained in a much shorter time. It is also very useful in field determinations (*10*).

6-2. U.S. PUBLIC HEALTH SERVICE
DRINKING WATER STANDARDS

In the United States the Standards of drinking water quality established by the U.S. Public Health Service for water to be used on interstate carriers have received general acceptance and wide application in connection with nearly all public water supply. These standards, first adopted in 1914 and revised in 1925, 1942, 1946, and 1962,* specify physical, chemical, and radiological quality as well as bacterial limits. Two types of limits are used in the standards:

1. Limits which, if exceeded, shall be grounds for rejection of the supply. Substances in this category may have adverse effects on health when present in concentrations above the limit.
2. Limits which should not be exceeded whenever more suitable supplies are, or can be made, available at reasonable cost. Substances in this category, when present in concentrations above the limit, either are objectionable to an appreciable number of people or exceed the levels required by good water quality control practices.

The limits in the standards apply to the water at the free-flowing outlet of the ultimate consumer. The following are excerpts from the 1962 Drinking Water Standards (*11*):

1. DEFINITION OF TERMS

The terms used in these standards are as follows:

1.1 *Adequate protection by natural means* involves one or more of the following processes of nature that produces water consistently meeting the requirements of these Standards: dilution, storage, sedimentation, sunlight, aeration, and the associated physical and biological processes which tend to accomplish natural purification in surface waters and, in the case of ground waters, the natural purification of water by infiltration through soil and percolation through underlying material and storage below the ground water table.

1.2 *Adequate protection by treatment* means any one or any combination of the controlled processes of coagulation, sedimentation, sorption, filtration, disinfection, or other processes which produce a water consistently meeting the requirements of these Standards. This protection also includes processes which are appropriate to the source of supply; works which are of adequate capacity to meet maximum demands without creating health hazards, and which are located, designed, and constructed to eliminate or prevent pollution; and conscientious operation by well-trained and competent personnel whose qualifications are commensurate with the responsibilities of the position and acceptable to the Reporting Agency and the Certifying Authority.

1.3 *Certifying Authority* means the Surgeon General of the United States Public Health Service or his duly authorized representatives. Reference to the Certifying Authority is applicable only for those water supplies to be certified for use on carriers subject to the Public Health Service Regulations.

1.4 The *coliform group* includes all organisms considered in the coliform group as set forth in *Standard Methods for the Examination of Water*

*The 1962 Drinking Water Standards are undergoing revision by the Environmental Protection Agency at the time of publication.

and *Wastewater*, current edition, prepared and published jointly by the American Public Health Association, American Water Works Association, and Water Pollution Control Federation.

1.5   *Health hazards* mean any conditions, devices, or practices in the water supply system and its operation which create, or may create, a danger to the health and well-being of the water consumer.   An example of a health hazard is a structural defect in the water supply system, whether of location, design, or construction, which may regularly or occasionally prevent satisfactory purification of the water supply or cause it to be polluted from extraneous sources.

1.6   *Pollution*, as used in these Standards, is defined as the presence of any foreign substance (organic, inorganic, radiological, or biological) in water which tends to degrade its quality so as to constitute a hazard or impair the usefulness of the water.

1.7   *Reporting Agencies* means the respective official State health agencies or their designated representatives.

1.8   *The standard sample* for the bacteriological test shall consist of:

1.8.1   For the bacteriological fermentation tube test, five (5) standard portions of either:
   (a)   ten milliliters (10 ml)
   (b)   one hundred milliliters (100 ml)

1.8.2   For the membrane filter technique, not less than fifty milliliters (50 ml).

1.9   *Water supply system* includes the works and auxiliaries for collection, treatment, storage, and distribution of the water from the sources to the free-flowing outlet of the ultimate consumer.

2. SOURCE AND PROTECTION

2.1   The water supply should be obtained from the most desirable source which is feasible, and effort should be made to prevent or control pollution of the source.   If the source is not adequately protected by natural means, the supply shall be adequately protected by treatment.

2.2   Frequent sanitary surveys shall be made of the water supply system to locate and identify health hazards which might exist in the system.   The manner and frequency of making these surveys, and the rate at which discovered health hazards are to be removed, shall be in accordance with a program approved by the Reporting Agency and the Certifying Authority.

2.3   Approval of water supplies shall be dependent in part upon:
   (a)   Enforcement of rules and regulations to prevent development of health hazards;
   (b)   Adequate protection of the water quality throughout all parts of the system, as demonstrated by frequent surveys;
   (c)   Proper operation of the water supply system under the responsible charge of personnel whose qualifications are acceptable to the Reporting Agency and the Certifying Authority;
   (d)   Adequate capacity to meet peak demands without development of low pressures or other health hazards; and

(e) Record of laboratory examinations showing consistent compliance with the water quality requirements of these Standards.

2.4 For the purpose of application of these Standards, responsibility for the conditions in the water supply system shall be considered to be held by:

(a) The water purveyor from the source of supply to the connection to the customer's service piping, and

(b) The owner of the property served and the municipal, county, or other authority having legal jurisdiction from the point of connection to the customer's service piping to the free-flowing outlet of the ultimate consumer.

3. BACTERIOLOGICAL QUALITY

3.1 *Sampling*

3.1.1 Compliance with the bacteriological requirements of these Standards shall be based on examinations of samples collected at representative points throughout the distribution system. The frequency of sampling and the location of sampling points shall be established jointly by the Reporting Agency and the Certifying Authority after investigation by either agency, or both, of the source, method of treatment, and protection of the water concerned.

3.1.2 The minimum number of samples to be collected from the distribution system and examined each month should be in accordance with the number on the graph in Figure 6-1, for the population served by the system. For the purpose of uniformity and simplicity in application, the number determined from the graph should be in accordance with the following: for a population of 25,000 and under—to the nearest 1; 25,001 to 100,000—to the nearest 5; and over 100,000—to the nearest 10.

3.1.3 In determining the number of samples examined monthly, the following samples may be included, provided all results are assembled and available for inspection and the laboratory methods and technical competence of the laboratory personnel are approved by the Reporting Agency and the Certifying Authority:

(a) Samples examined by the Reporting Agency;

(b) Samples examined by local government laboratories;

(c) Samples examined by the water works authority;

(d) Samples examined by commercial laboratories.

3.1.4 The laboratories in which these examinations are made and the methods used in making them shall be subject to inspection at any time by the designated representatives of the Certifying Authority and the Reporting Agency. Compliance with the specified procedures and the results obtained shall be used as a basis for certification of the supply.

3.1.5 Daily samples collected following a bacteriologically unsatisfactory sample as provided in sections 3.2.1, 3.2.2, and 3.2.3 shall be considered as special samples and shall not be included in the total number of samples examined. Neither shall such special samples be used as a basis for prohibiting the supply, *provided*

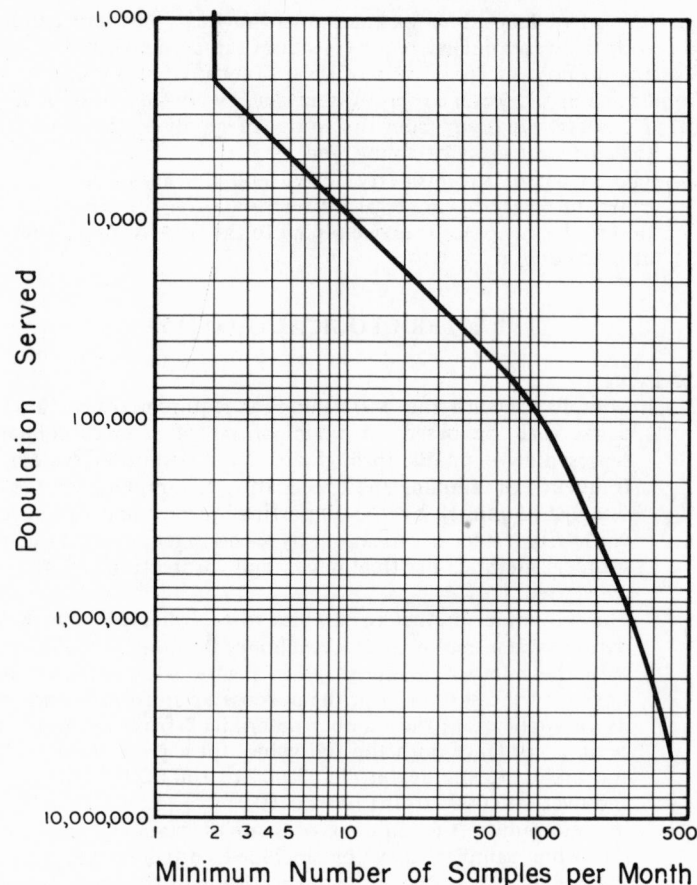

Minimum Number of Samples per Month

Figure 6-1
Relation between minimum number of samples to be collected per month and population served.

that: (1) When waters of unknown quality are being examined, simultaneous tests are made on multiple portions of a geometric series to determine a definitive coliform content; (2) Immediate and active efforts are made to locate the cause of pollution; (3) Immediate action is taken to eliminate the cause; and (4) Samples taken following such remedial action are satisfactory.

3.2 *Limits.* The presence of organisms of the coliform group as indicated by samples examined shall not exceed the following limits:

3.2.1 When 10-ml standard portions are examined, not more than 10 percent in any month shall show the presence of the coliform group. The presence of the coliform group in three or more 10-ml portions of a standard sample shall not be allowable if this occurs:

(a) In two consecutive samples;

(b) In more than one sample per month when less than 20 are examined per month; or

(c) In more than 5 percent of the samples when 20 or more are examined per month.

When organisms of the coliform group occur in three or more of the 10-ml portions of a single standard sample, daily samples from the same sampling point shall be collected promptly and examined until the results obtained from at least two consecutive samples show the water to be of satisfactory quality.

3.2.2 When 100-ml standard portions are examined, not more than 60 percent in any month shall show the presence of the coliform group. The presence of the coliform group in all five of the 100-ml portions of a standard sample shall not be allowable if this occurs:

(a) In two consecutive samples:

(b) In more than one sample per month when less than five are examined per month; or

(c) In more than 20 percent of the samples when five or more are examined per month.

When organisms of the coliform group occur in all five of the 100-ml portions of a single standard sample, daily samples from the same sampling point shall be collected promptly and examined until the results obtained from at least two consecutive samples show the water to be of satisfactory quality.

3.2.3 When the membrane filter technique is used, the arithmetic mean coliform density of all standard samples examined per month shall not exceed 3/50 ml, 4/100 ml, 7/200 ml, or 13/500 ml in:

(a) Two consecutive samples;

(b) More than one standard sample when less than 20 are examined per month; or

(c) More than 5 percent of the standard samples when 20 or more are examined per month.

When coliform colonies in a single standard sample exceed the above values, daily samples from the same sampling point shall be collected promptly and examined until the results obtained from at least two consecutive samples show the water to be of satisfactory quality.

4. PHYSICAL CHARACTERISTICS

4.1 *Sampling.* The frequency and manner of sampling shall be determined by the Reporting Agency and the Certifying Authority. Under normal circumstances samples should be collected one or more times per week from representative points in the distribution system and examined for turbidity, color, threshold odor, and taste.

4.2 *Limits.* Drinking water should contain no impurity which would cause offense to the sense of sight, taste, or smell. Under general use, the following limits should not be exceeded:

Turbidity   − 5 units
Color   − 15 units
Threshold odor number   − 3

5. CHEMICAL CHARACTERISTICS

5.1 *Sampling*

    5.1.1 The frequency and manner of sampling shall be determined by the Reporting Agency and the Certifying Authority. Under normal circumstances, analyses for substances listed below need be made only semiannually. If, however, there is some presumption of unfitness because of the presence of undesirable elements, compounds, or materials, periodic determinations for the suspected toxicant or material should be made more frequently and an exhaustive sanitary survey should be made to determine the source of the pollution. Where the concentration of a substance is not expected to increase in processing and distribution, available and acceptable source water analyses performed in accordance with standard methods may be used as evidence of compliance with these Standards.

    5.1.2 Where experience, examination, and available evidence indicate that particular substances are consistently absent from a water supply or below levels of concern, semiannual examinations for those substances may be omitted when approved by the Reporting Agency and the Certifying Authority.

    5.1.3 The burden of analysis may be reduced in many cases by using data from acceptable sources. Judgment concerning the quality of water supply and the need for performing specific local analyses may depend in part on information produced by such agencies as: (1) the U.S. Geological Survey, which determines chemical quality of surface and ground waters of the United States and publishes these data in "Water Supply Papers" and other reports, and (2) the U.S. Public Health Service which determines water quality related to pollution (or the absence of pollution) in the principal rivers of the Nation and publishes these data annually in "National Water Quality Network." Data on pollution of waters as measured by carbon chloroform extracts (CCE) may be found in the latter publication.

5.2 *Limits.* Drinking water shall not contain impurities in concentrations which may be hazardous to the health of the consumers. It should not be excessively corrosive to the water supply system. Substances used in its treatment shall not remain in the water in concentration greater than required by good practice. Substances which may have deleterious physiological effect, or for which physiological effects are not known, shall not be introduced into the system in a manner which would permit them to reach the consumer.

    5.2.1 The following chemical substances should not be present in a water supply in excess of the listed concentrations where, in the judgment of the Reporting Agency and the Certifying Authority, other more suitable supplies are or can be made available:

Substance	Concentration (mg/liter)
Alkyl benzene sulfonate (ABS)	0.5
Arsenic (As)	0.01
Chloride (Cl)	250.0
Copper (Cu)	1.0
Carbon chloroform extract (CCE)	0.2
Cyanide (CN)	0.01
Fluoride (F)	(See 5.2.3)
Iron (Fe)	0.3
Manganese (Mn)	0.05
Nitrate[a] (NO_3)	45.0
Phenols	0.001
Sulfate (SO_4)	250.0
Total dissolved solids	500.0
Zinc (Zn)	5.0

[a]In areas in which the nitrate content of water is known to be in excess of the listed concentration, the public should be warned of the potential dangers of using the water for infant feeding.

5.2.2 The presence of the following substances in excess of the concentrations listed shall constitute grounds for rejection of the supply:

Substance	Concentration (mg/liter)
Arsenic (As)	0.05
Barium (Ba)	1.0
Cadmium (Cd)	0.01
Chromium (Cr^{+6})	0.05
Cyanide (CN)	0.2
Fluoride (F)	(See 5.2.3)
Lead (Pb)	0.05
Selenium (Se)	0.01
Silver (Ag)	0.05

5.2.3 *Fluoride.* When fluoride is naturally present in drinking water, the concentration should not average more than the appropriate upper limit in Table I. Presence of fluoride in average concentrations greater than two times the optimum values in Table I shall constitute grounds for rejection of the supply.

Where fluoridation (supplementation of fluoride in drinking water) is practiced, the average fluoride concentration shall be kept within the upper and lower control limits in Table I.

In addition to the sampling required by paragraph 5.1, fluoridated and defluoridated supplies shall be sampled with sufficient frequency to determine that the desired fluoride concentration is maintained.

Table I

| | Recommended Control Limits (fluoride concentrations, mg/liter) | | |
Annual Average of Maximum Daily Air Temperatures[a]	Lower	Optimum	Upper
50.0–53.7	0.9	1.2	1.7
53.8–58.3	0.8	1.1	1.5
58.4–63.8	0.8	1.0	1.3
63.9–70.6	0.7	0.9	1.2
70.7–79.2	0.7	0.8	1.0
79.3–90.5	0.6	0.7	0.8

[a]Based on temperature data obtained for a minimum of five years.

6. RADIOACTIVITY

6.1 *Sampling*

    6.1.1 The frequency of sampling and analysis for radioactivity shall be determined by the Reporting Agency and the Certifying Authority after consideration of the likelihood of significant amounts being present. Where concentrations of Ra^{226} or Sr^{90} may vary considerably, quarterly samples composited over a period of three months are recommended. Samples for determination of gross activity should be taken and analyzed more frequently.

    6.1.2 As indicated in paragraph 5.1, data from acceptable sources may be used to indicate compliance with these requirements.

6.2 *Limits.*

    6.2.1 The effects of human radiation exposure are viewed as harmful and any unnecessary exposure to ionizing radiation should be avoided. Approval of water supplies containing radioactive materials shall be based upon the judgment that the radioactivity intake from such water supplies when added to that from all other sources is not likely to result in an intake greater than the radiation protection guidance recommended by the Federal Radiation Council (September 13, 1961) and approved by the President. Water supplies shall be approved without further consideration of other sources of radioactivity intake of radium 226 and strontium 90 when the water contains these substances in amounts not exceeding 3 and 10 $\mu\mu C$/liter, respectively. When these concentrations are exceeded, a water supply shall be approved by the Certifying Authority if surveillance of total intakes of radioactivity from all sources indicates that such intakes are within the limits recommended by the Federal Radiation Council for control action (20 $\mu\mu C$/day of radium 226 and 200 $\mu\mu C$/day of strontium 90).

    6.2.2 In the known absence of strontium 90 and alpha emitters (absence is taken here to mean a negligibly small fraction of the above specific limits, where the limit for unidentified alpha emitters is taken as the listed limit for radium 226), the water supply is acceptable when the gross beta concentrations do not exceed 1,000 $\mu\mu$C/liter. Gross beta concentrations in excess of 1,000 $\mu\mu$C/liter shall be grounds for rejection of supply except when more complete analyses indicate that concentration of nuclides are not likely to cause exposures greater than the Radiation Protection Guides as approved by the President on recommendation of the Federal Radiation Council.

7. RECOMMENDED ANALYTICAL METHODS

7.1 Analytical methods to determine compliance with the requirements of these Standards shall be those specified in *Standard Methods for the Examination of Water and Wastewater*, Am. Public Health Assoc., current edition and those specified as follows:

7.1.1 Barium—*Methods for the Collection and Analysis of Water Samples, Water Supply Paper No. 1454*, Rainwater, F. H., and Thatcher, L. L., U.S. Geological Survey, Washington, D.C.

7.1.2 Carbon Chloroform Extract (CCE)—*Manual for Recovery and Identification of Organic Chemicals in Water*, Middleton, F. M., Rosen, A. A., and Burttschell, R. H., Robert A. Taft Sanitary Engineering Center, Public Health Service, Cincinnati, Ohio.

7.1.3 Radioactivity—*Laboratory Manual of Methodology, Radionuclide Analysis of Environmental Samples*, Technical Report R59-6, Robert A. Taft Sanitary Engineering Center, Public Health Service, Cincinnati, Ohio; and *Methods of Radiochemical Analysis, Technical Report No. 173*, Report of the Joint WHO-FAO Committee, 1959, World Health Organization.

7.1.4 Selenium—*Suggested Modified Method for Colorimetric Determination of Selenium in Natural Water*, Magin, G. B., Thatcher, L. L., Rettig, S., and Levine, H. J., Am. Water Works Assoc. 52, 1199 (1960).

7.2 *Organisms of the Coliform Group*—All of the details of techniques in the determination of bacteria of this group, including the selection and preparation of apparatus and media, the collection and handling of samples and the intervals and conditions of storage allowable between collection and examination of the water sample, shall be in accordance with *Standard Methods for the Examination of Water and Wastewater*, current edition, and the procedures shall be those specified therein for:

7.2.1 The Membrane Filter Technique, Standard Test, or

7.2.2 The Completed Test, or

7.2.3 The Confirmed Test, procedure with brilliant green lactose bile broth,* or

7.2.4 The Confirmed Test, procedure with Endo or eosin methylene blue agar plates.*

*The Confirmed Test is allowed, provided the value of this test to determine the sanitary quality of the specific water supply being examined is established beyond reasonable doubt by comparisons with Completed Tests performed on the same water supply.

6-3. SANITARY AND TOXIC SIGNIFICANCE
OF 1962 STANDARDS

The 1962 Public Health Service Drinking Water Standards are accompanied by an appendix describing the rationale employed in determining the various limits. In establishing limits for toxic substances, intake from food and air was considered.

The average daily excrement of fecal matter (*12-14*) from a mixed population in the United States amounts to about 20 g dry weight, of which 4 to 5 g consists of bacterial bodies, about half of which are alive in fresh fecal matter. The total number of live bacteria in fresh feces is estimated at about 2×10^{12} per person per day, of which about 10% or 200 billion are coliform bacteria. The number of polio viruses in the stools of patients is less than 1 percent, and the number of viruses of infectious hepatitis in the stools of patients is less than 0.1 percent of the number of coliform bacteria. Domestic sewage normally contains about 10,000 times as many coliforms as pathogenic viruses.

The coliform count in normal fresh domestic sewage during dry weather averages about 100 million per 100 ml. To meet the drinking-water standard of 1 per 100 ml by dilution alone, the dilution ratio must be about 100 million to 1, or 5 to 10 bgd (billion gallons per day) per person. In practice this enormous reduction is accomplished by sewage treatment, sewage chlorination, dilution, and bacterial die-off in the receiving stream, water treatment, water chlorination, and combinations thereof.

Studies were made by Gilcreas and Kelly (*15*) on the relative rates of kill and die-off of coliform bacteria and intestinal viruses by storage in both fresh and salt water, by chlorination, by sewage treatment, and by ultraviolet irradiation. The intestinal viruses used in the tests were Bact. *coli* B bacteriophage, Coxsackie virus, and Theiler virus. The studies indicated that the relative survival of coliform bacteria and viruses in freshwater was roughly the same, except that viruses survived for much longer periods at low temperature. In salt water coliform bacteria died much more rapidly than the viruses. Chlorination was found to be effective on both coliform bacteria and viruses, but subsequent studies by Kelly and Sanderson (*4*) indicated that stronger doses and longer contact periods are required for viruses. Ultraviolet light was found to be more effective on viruses than on coliform bacteria. Sewage treatment by primary settling and trickling filters at one plant showed about 95 percent reduction of coliforms, about 80 percent reduction of phage, and nearly 100 percent reduction of Coxsackie viruses.

Experiments (*16*) were conducted under the senior author's direction in October 1959 on the chlorination of comminuted raw sewage from the Nut Island sewage-treatment plant in metropolitan Boston with chlorine doses ranging from 20 to 38 ppm. These experiments were conducted during a period when the sewage contained very little storm water. The chlorine demand ranged from about 4 to 23 ppm. The chlorine dose necessary to maintain a 10-ppm combined chlorine residual for 1 hr, the requirement for virus inactivation as found by Kelly and Sanderson, ranged from 12 to 34 ppm and averaged 25 ppm. The coliform counts in the raw sewage in terms of the MPN ranged from 43 to 250 million per 100 ml and averaged about 100 million. Most of the tests were made in a small chlorine contact chamber, 2 ft^2 in plan, and with a 2 ft depth of sewage. The chamber was operated on a fill-and-draw basis, with contact periods measured from the time of introducing the chlorine. Samples were collected for bacterial examination of the unchlorinated sewage and after contact periods of 2, 5, 10, 15, and 20 min, and longer times. A portion of each sample collected was blended for 1 min in a Waring blender after dechlorination for comparison of the coliform count before and after blending, the purpose of which was to determine the effectiveness of penetration of suspended particles by chlorine. In all but one of the experiments coliform kills in excess of 99.99 percent were achieved in less than 10 min. The coliform counts in the blended samples were 5 to 10 times as great as in the unblended samples. Residual counts in the unblended samples ranged between 100 and 10,000 per 100 ml, representing kills of greater than 99.99 percent. Since these coliform kills were very much in excess of what can be expected normally from complete treatment of sewage without chlorination, the authors believe that all sewage, whether or not mixed with storm water, should be chlorinated with doses of 20 to 30 ppm and contact periods of not less than 30 min, and that all sewage subjected to complete treatment should be chlorinated with sufficient doses and contact periods to produce coliform counts in the effluent that conform to the limiting counts specified for the receiving waters.

The physical characteristics of the Public Health Service standards have no physiological significance. The purpose of these specifications is to produce an attractive and palatable water so as to discourage the use by the public of other water supplies that are not safe. The same may be said of most of the chemical substances specified in Section 5.2.1 of the standards. Sections 5.2.2 and 5.2.3 deal with toxic substances. The physiological aspects of mineral salts in water are discussed at length by Negus (*17*), by Fairhall (*18*), in the manual of *Water Quality and Treatment* (*19*), and in the appendix of the 1962 Public Health Service Drinking Water Standards (*11*).

Copper is not a significant constituent in natural waters; however, it is introduced by the solution of copper from brass and copper pipe and by the use of copper sulfate as an algicide in reservoirs. Copper is of physiological importance as a supplement to iron for hemoglobin regeneration and is an essential constituent of tissue cells, but the body requirements must be met by the food intake, since the amount of copper usually available in the drinking water is inadequate. Chronic copper poisoning is said to cause gastrointestinal catarrh and to be related to hemochromatosis, but the amount of copper required for poisoning is far in excess of the concentrations possible in drinking water. Copper in drinking water has practically no health significance. Concentrations in excess of about 1 ppm may impart a disagreeable taste to water and may result in precipitation of $Cu(OH)_2$, which will increase the turbidity of water or stain plumbing fixtures. The characteristic blue-green stain produced by copper is probably a mixture of $Cu(OH)_2$ and $CuOH$.

Copper sulfate when used as an algicide is required in concentrations ranging from 0.10 to 0.5 ppm for most organisms, with doses up to 10 ppm required for a few. Copper in low concentrations is toxic to fish (*20*). In soft waters, 0.12 ppm of copper sulfate is fatal to trout, 0.40 ppm to pickerel and catfish, 0.50 ppm to goldfish, 0.75 ppm to perch, 1.20 ppm to sunfish, and 2.10 ppm to black bass. In dosing the New York City reservoirs, fish have seldom been killed with copper sulfate doses ranging from 0.05 to 0.5 ppm.

Zinc ions are rarely present in natural waters in considerable concentration, but zinc does enter water supplies by solution of the metal from zinc galvanizing on pipes and tanks. Zinc is a normal constituent of the human body. Its presence in drinking water in concentrations up to about 40 ppm appears to have no health significance, but it does impart an astringent taste to water, and it will precipitate as $Zn(OH)_2$ or $ZnCO_3$ in alkaline waters to produce a milky turbidity. The recommended limit of 5 ppm in the standards is to avoid the taste produced by zinc. Cadmium and lead, both toxic, are common impurities in zinc galvanizing, ranging from 0.014 to 0.4 percent for Cd and from 0.24 to 0.45 percent for Pb.

Iron and manganese are both essential to the human body, but their intake through drinking water is an insignificant part of the body requirement. The low limit placed upon these metals in the standards has no health significance and is due to the fact that the ferric and manganic oxides are very insoluble. Iron oxide or rust causes *red water*, and the manganese oxides are brown or black. Both iron and manganese precipitates deposit in distribution mains, stain plumbing fixtures, and render water unsuitable for laundering, dyeing, paper making, and many other manufacturing processes. The presence of manganic manganese in excess of 0.01 ppm interferes with the orthotolidine test for residual chlorine in waters that have been disinfected with chlorine. Both iron and manganese may be present in excess in either groundwaters or surface waters, but excess iron in tap water is frequently due to corrosion of iron pipes in the distribution system or the plumbing systems.

The limits suggested by the standards on chlorides, sulfates, and total solids are for the purposes of limiting hardness, corrosivity to metals, and taste due to salinity or saltiness. Most people cannot distinguish a salty taste until the chloride or sulfate content exceeds 250 to 400 ppm. Magnesium salts (milk of magnesia and Epsom salt) and Glauber salt, Na_2SO_4, are effective cathartics, but only in large concentrations. There is no evidence of any such specific effect in dilute concentration. The *change of water* experienced by persons who travel from soft-water regions to hard-water regions, or vice versa, produces minor physiological disturbances of the digestive system in some persons. These disturbances are due to changes in osmotic pressure at the intestinal wall and to changes in ionic equilibria in the blood and body fluids. After a short period of adjustment to a new water, the digestive disturbances are usually overcome.

In the average American diet, the inorganic elements per man per day, according to Sherman (*21*), are Ca, 0.73 g; Mg, 0.34 g; K, 3.39 g; Na, 1.94 g; Cl, 2.83 g (not including table salt); S, 1.28 g; and Fe, 0.0173 g. Only a small fraction of these requirements can be met through the drinking water, even if the water is very hard. Hence, it is better to depend upon the food intake, supplemented if necessary with mineral and vitamin preparations, to supply the minerals necessary for bodily health. There is, nevertheless, a relation between the mineral content of most foods and the mineral content of the water in the regions in which the food is grown. Spinach, for example, is not high in iron content if grown in a soil that is relatively free of iron.

Goiter is prevalent in the midwest where the water supplies contain very little iodine, I^- (0.01 to 0.1 parts per billion), and is scarce along the seaboard where the iodine content of the drinking water is much higher (1.4 to 10 parts per billion). Dosing the drinking water with NaI, which has been tried in some localities, is not an effective or satisfactory means of supplying the iodide deficiency, because in regions where the iodide content of the water supplies is low it is also low in the locally grown foods. Moreover, there are some persons who are sensitive to iodine and exhibit adverse effects if they use drinking water that has been iodized. Seafoods and iodized table salt are satisfactory media for supplying iodine to those who require it.

The addition of iodides to a drinking-water supply for the reduction of goiter, a noncontagious disease, constitutes mass medication through the water supply. This is an unwarranted use of the water supply, which violates the rights of the consumers. The water purveyor is required by law to furnish safe water to all consumers. Iodides in excess concentrations in drinking water cause hyperactivity of the thyroid in some individuals. Iodine, which is not as effective as chlorine in bacterial kill, reverts to iodide in water, just as chlorine reverts to chloride. Iodine should not be used for the disinfection of drinking water if the doses used will result in excess concentrations of iodides. The 1962 Public Health Service standards make no mention of iodine or iodide.

The limit suggested by the standards on the concentration of phenolic compounds (including cresols and xylenols) has no health significance and was selected because very small concentrations of phenol produce disagreeable odors, particularly when waters containing phenols are chlorinated for disinfection. Phenolic compounds are components of a number of industrial wastes, but are not present in natural waters. Carbolic acid is poisonous, of course, but not in the dilute concentrations at which odors are the limiting factor. The toxic range for freshwater fish and lower aquatic life has been found to be 2 to 75 mg/liter for cresols and 0.1 to 50 mg/liter for phenols (9). Concentrations much higher than 1 mg/liter would be required for injury to human health.

The 1962 standards set no limits on the pH of water. The 1946 standards recommended a limit on pH of about 10.6, which should prevent any caustic taste. The *caustic alkalinity*, due to OH^- ions, is 20 ppm as $CaCO_3$ at pH 10.6 and 25°C, and according to Hale (22), 50 to 100 ppm (pH 11 to pH 11.3) are required to produce a lime-like taste. Few natural waters have pH values as high as 10.6, but softening by the lime–soda process requires a pH between 10 and 11 for effective precipitation of $Mg(OH)_2$. High pH values up to about 10.6 are usually desirable to reduce corrosion of metallic pipes, but they accelerate corrosion of aluminum, tin, and lead. A 1952 survey (23) of the chemical quality of the finished water from the public supplies of 1,328 cities in the United States indicates that 62 places (4.7 percent) receive water with a pH of 9 or higher, 29 (2.2 percent) with a pH of 9.6 or higher, 9 (0.7 percent) with a pH of 10 or higher, and 3 with a pH of 10.5 or higher. Natural waters are not likely to be too acid for human consumption, but the pH of acid waters will usually have to be raised to reduce corrosion and to precipitate iron and manganese. Waters with a pH less than 4 have been used continuously for municipal supplies, without reported harm to drinkers.

Since the 1946 revision of the standards, there has been a tremendous increase in the use of synthetic detergents in place of soap for household and industrial purposes. These detergents differ from soaps in that they do not form precipitates or curds with bivalent and trivalent metals in water. They are therefore very soluble and are difficult to remove from water by the ordinary treatment processes. Most of the surfactants in household detergents are of the anionic type. At the time of the 1962 revision of the standards, ABS (alkyl benzene sulfonate) accounted for almost three quarters of these. ABS has subsequently been found to be nonbiodegradable in treatment plants and receiving waters and has been largely replaced in the detergents with LAS (linear alkyl sulfonate), which is biodegradable. The concentration of both of these substances is determined by the methylene blue method. The recommended limit remains at 0.5 ppm in the standards, but the name of the substance has been changed to methylene blue–active substances (MBS). The concentration of MBS in municipal sewage is of the order of 10 ppm. An off-taste, described as oily, fishy, or perfume-like, will be noticeable in drinking water containing 1.0 to 1.5 ppm of MBS, and such a water will also foam or froth. Concentrations up to 50 ppm in drinking water have produced no toxic effects on humans. The recommended limit in the standards is to avoid tastes and foaming.

Numerous new organic chemicals are now being produced by the chemical and petrochemical industries; many are toxic, and some of them are washed into ground and surface waters. These substances and their waste by-products are so numerous and varied as to defy detection and identification by practical analytical methods. Many of these substances are extensively used in agriculture as pesticides and weed killers. The *chloroform-soluble carbon-filter extract* method (CCE) has been developed as a means for determining the gross concentration of some of these organics. The recommended limiting concentration of CCE in drinking water has been set at 0.2 ppm, which is about the highest concentration observed, thus far, in any of our major watercourses.

The carbon-filter technique consists of concentrating the adsorbable organics from the water sample in a carbon filter and subsequently eluting the chloroform-soluble materials from the dried carbon by repeated extraction with chloroform. A second extract may be obtained with ethyl alcohol. Not all the organics are adsorbed on the carbon. The chloroform extract contains only organics that have been adsorbed by the carbon and are extractable by chloroform. The alcohol extract contains only organics that have been adsorbed by the carbon, are not extractable by chloroform, and are extractable by ethyl alcohol. Simple hydrocarbons and chlorinated insecticides can be recovered almost quantitatively in the chloroform. LAS will be almost completely collected by the carbon filter, but it must be extracted by methylheptylamine–chloroform solution. Simple phenols are completely collected by the carbon, but are extractable by chloroform to only 60 to 70 percent. Materials such as sugars, polysaccharides, proteins, lignins, and tannins have not been observed in chloroform extracts. The toxic organics that have been recovered in the chloroform extract include chlorinated insecticides, nitriles, a substituted nitrobenzene, and aromatic ethers. Some substances of interest recovered from water by the carbon-filter technique include oily materials, toluene, xylene, DDT, pyridine, picolines, alcohols, aldehydes, ketones, esters, and organic phosphorus-containing insecticides.

Organic phosphorus-containing insecticides have been employed as agricultural pesticides in this country for more than 20 years, and there has been a continuing development of new agents of similar chemical structure. Of major importance are the phosphorothionates. These insecticides are highly toxic to fish as well as to humans. The tolerence levels in existence for food contamination by various organic phosphorus insecticides range from 0.75 to 8 ppm. Safe levels were established on experimental animals on the basis of inhibition of the activity of cholinesterase. The safe tolerance level in drinking water for man is considered to be about 0.1 ppm, and the safe tolerance level for all aquatic life (*24*) is considered to be about 0.02 ppm. The 1962 drinking water standards contain no limit on organic phosphorus, because concentrations found, thus far, in public waters are well below tolerance limits, and no practical chemical procedures are available for identifying the insecticide and determining its concentration in water. Bioassays by means of fish or houseflies to determine excessive concentrations of biologically active organic phosphorus have been considered for use, but they are impractical without supplementary analysis to identify the cause of death to the test organism.

The 1946 standards contained no limit for cyanides. Numerous agencies have set limits ranging from 0.01 to 0.2 ppm, based primarily on the toxicity of cyanide to fish rather than to man. Cyanide in reasonable doses (10 mg or less) is rapidly detoxified to thiocyanate in the liver of man. A concentration of 0.05 ppm for 5 days has been fatal to trout. The 1962 standards recommend a cyanide concentration less than 0.01 ppm, and require a concentration less than 0.2 ppm. Recent work on fish (1966) by Doudoroff (9, p. 63) has demonstrated that HCN rather than CN is the toxic component. This makes the effect of pH on cyanide toxicity of great importance, a thousandfold increase in toxicity being associated with a drop in pH from 8.0 to 6.5.

Serious and occasionally fatal poisonings in infants have occurred following ingestion of well waters containing nitrates. Wastes from fertilizer plants and field fertilization are the usual sources of nitrates, but they may be dissolved from soils containing large amounts of animal refuse. From 1947 to 1950, 139 cases of methemoglobinemia (blue babies) due to nitrates in well waters were reported in Minnesota (25) alone; this included 14 deaths. Experience indicates that the concentration of nitrate (NO_3^-) must exceed about 70 ppm to result in infantile poisoning. The 1958 International Standards for Drinking Water (World Health Organization) states that the ingestion by infants under 1 year of age of water containing nitrates in excess of 50 to 100 ppm may give rise to methemoglobinemia, but no limiting concentration was set. The recommended limit of 45 ppm for nitrates in the 1962 Public Health Service standards appears to be safe. Among the more acceptable hypotheses for the specificity of nitrate poisoning of infants is that the lower acidity in the infant's gastrointestinal tract permits growth of nitrate-reducing flora, which reduce nitrate to nitrite. Nitrite is absorbed by the blood, converting large quantities of hemoglobin to methemoglobin.

Since it is nitrite (NO_2^-) that reacts with hemoglobin, it, too, is dangerous in water supplies. Naturally occurring concentrations of nitrites are generally of no health significance, but intentional additions of either nitrates or nitrites in high concentrations (500 ppm or more) to private systems for corrosion control are dangerous to both children and adults, if such systems are interconnected with, or discharge into, drinking water supplies. A limit of 200 ppm of nitrite or nitrate in "corned" products has been set by federal regulation on the basis that 100 g of corned beef could convert, maximally, from 10 to 40 g of hemoglobin to methemoglobin (1.4 to 5.7 percent of total hemoglobin). An instance has been reported of nitrite poisoning (26) in children who ate wieners and bologna containing nitrite in excess of 200 ppm. Nitrates are known to increase in cooked spinach kept under refrigeration.

Small concentrations of lead continuously present in drinking water are known to cause lead poisoning or plumbism, which may result in serious illness or death. Only traces of lead may be found in natural waters and, then, only in waters from heavily mineralized water sheds or in waters contaminated with industrial wastes. Surface and ground waters in regions where lead arsenate spray is extensively used for control of codling moths in orchards or Japanese beetles in grassy lands may become contaminated with both lead and arsenic. The most prevalent source of lead in drinking water, particularly in soft-water regions, is lead pipe in services and plumbing systems or pipe jointing material in distribution systems.

Lead taken into the body in quantities in excess of certain low limits is a cumulative poison. Poisoning may result from one, or all, of three common sources: food, air, and water, and also from inhaled tobacco smoke. The safe limit set in the 1962 standards was 0.05 ppm. Bacterial decomposition of organic matter is inhibited by lead concentrations exceeding 0.1 ppm. Lead is highly toxic to fish: 0.1 ppm can kill small sticklebacks.

Arsenic is sometimes present in natural waters due to their contact with arsenic-bearing minerals such as arsenical pyrites. Arsenic is also a component of smelter wastes and other industrial wastes. Arsenic poisoning (27) is known to have resulted from a water containing about 0.2 ppm of arsenic. Analyses (18) of waters in areas sprayed with lead arsenate have shown lead concentrations less than 0.07 ppm and arsenic concentrations less than 0.01 ppm. It is probable that concentrations of about 0.15 ppm of arsenic in drinking water can be tolerated without injurious effect, according to Stoof and Haase (27), provided there is no other major source of arsenic intake.

Arsenic in heavy doses acts as a cumulative poison, probably because of its slow elimination from the human body. There is recent evidence that arsenic may be carcinogenic. The widespread use of inorganic arsenic in insecticides and its presence in animal foods, tobacco, and certain industrial smokes and vapors require low limits of concentration in drinking water. The 1962 standards recommended a limit of 0.01 ppm, and required a limit of 0.05 ppm in drinking water. The new standards will have a recommended limit and a rejection limit of 0.10 mg/liter.

Concentrations of 2 to 4 ppm of arsenic in streams are reported not to interfere with bacterial self-purification. Bass have tolerated 6 ppm, but 15 ppm has proved toxic to crappies and bluegills.

Cereal grain of high selenium content, which has been grown in seleniferous soil, is related to "alkali disease" in cattle. The selenium content of water from the Colorado River Basin ranges from 0 to about 0.4 ppm but, according to Miller and Byers (28), 0.4 to 0.5 ppm is too low to cause selenium poisoning in cattle. There is some information to indicate that selenium concentrations considered safe for man may be toxic to fish.

Dean and Elvove (29) in 1936 stated that the presence of fluoride, F^-, in drinking water in concentrations exceeding 0.5 ppm may result in mild endemic dental fluorosis (mottled enamel) in the teeth of growing children, and that about 0.9 to 1.0 ppm of fluoride would result in mottled enamel on the teeth of 10 to 30 percent of the children examined. Trelles (30) and McClure (31) reported, in 1938 and 1939, that 2 to 3 ppm of fluoride is required for severe mottling. Case studies, reported by Dean et al. (32) in 1941, indicated that *dental caries* (tooth decay) is reduced by fluorides in drinking water. Subsequent studies by Dean et al. (33) reported in 1942, indicated that in communities with optimum fluoride concentration in the drinking water caries rates will be reduced 60 to 65 percent below the rates prevailing in communities using water with little or no fluorides. Fluoride is a normal constituent of groundwater in heavily mineralized regions, particularly in the southwestern United States. As a result of the 1936 findings, research was directed to find effective and simple water-treatment methods for removal of fluorides. Following the 1941–1942 findings, however, the Public Health Service initiated a promotional campaign for fluoridation of water supplies to reduce dental caries. This campaign has become more intense with the passing years and is now supported by nearly all state and local health agencies throughout the country.

Fluoride is a poison. According to Hodge and Smith (34) bone changes may be expected when water containing 8 to 20 ppm of fluoride is consumed over a long period of time, and death may be expected with a single dose of 2.25 to 4.5 g. According to Roholm (35), crippling fluorosis may be expected when 20 mg or more of fluoride is consumed per day from all sources for 20 years or more. Water is not the only source of fluoride intake. Some foods, notably fish, meat, peas, and tea are high in fluoride content, and the fluoride content of foods depends upon the fluoride content of the soil and water where the food is grown. Air may be a contributing source near aluminum plants or other industrial plants in which fluorine-containing minerals and compounds are used.

In establishing the limits for toxic substances in the 1962 drinking water standards, the advisory committee used a safety factor of roughly 100, except for fluorides. In the case of fluorides, the thresholds for mottled enamel overlap the recommended concentrations for artificial fluoridation. No account has been taken of other sources of fluoride ingestion, differences in the amount of water taken by different individuals, differences in the susceptibility to fluoride poisoning, or of the fact that the entire population must drink water fluoridated for the benefit of relatively few persons (children 6 to 12 years old). Like the addition of iodides, artificial fluoridation of a drinking water supply constitutes mass medication for control of a noncontagious disease. This use of the water supply is unnecessary and violates the rights of the consumers.

Those who desire to protect their children's teeth from decay with fluorides may do so with topical applications of fluoride compounds to the teeth. Recent studies indicate the possibility of many other means of controlling caries without the dangers inherent in the use of fluorides. Studies over a 10-year period by Harris and Nizel (*36*) indicate that a number of minerals besides fluorides have caries-inhibiting properties. In studies of the diet of experimental animals, these investigators found that a twofold increase in the phosphorus content of a caries-producing diet reduced caries 95 percent and a fourfold increase eliminated caries. It has been found that one or more strains of *Streptococcus* are responsible for dental caries in humans. Dental caries (*37*) has been prevented in rats by a germ-free environment and in hamsters by a diet containing penicillin. A particular strain of *Streptococcus* has been found to produce caries in hamsters. These studies suggest the possibility of a vaccine or periodic disinfection of the mouth.

In the report of the Delaney Committee (*38*) on "Fluoridation of Public Drinking Water" in 1952, it was recommended that studies be continued to learn more about the effects of fluoridation on the total population, and that the advisability of fluoridating a public water supply be determined for itself by each community. Four cities for which fluoridation was started in 1944 to 1946 were used to study the effect of fluoridation on the prevalence of caries in children ranging in age from 6 to 16. Decayed, missing, and filled counts (DMF) were made on the children of each age group before and after fluoridation. The percentage of reduction in caries for each age group was estimated from the ratio of the difference in counts before and after to the count before. During the 1967 hearings (*39*) of the Special Commission on the Condition of Dental Health of the Commonwealth of Massachusetts, Howard M. Thomson presented graphs of the DMF counts in the four test cities, which showed that the chief effect of fluoridation appears to be a delay of 2 to 4 years in the development of caries in younger children. Much of the data shown on the graphs was presented in the minority report (*40*) House No. 3902, December 1967. The pertinent data on the Thomson graphs are shown in Figure 6-2.

It will be noted in Figure 6-2 that straight lines can be plotted through the points representing the DMF counts, and that there is a break in the slope of these lines, in most cases where data are adequate, at ages 10 to 12, the slopes being generally steeper at the older ages. This should have been anticipated by the experimenters, since children are expected to have all 28 of their permanent teeth (except wisdom teeth) at age 10 to 12. This means that before the age at which the break in slope occurs, the slopes of the lines are determined by the time rate of eruption of the permanent teeth as well as by the rate of caries occurrence. It is obvious that the DMF count procedure is faulty. Only the teeth that are present are subject to decay. The missing teeth were not present and should not have been included. Each count should be of decayed and filled teeth for a selected number of teeth (say 100) exposed to decay (and not per child).

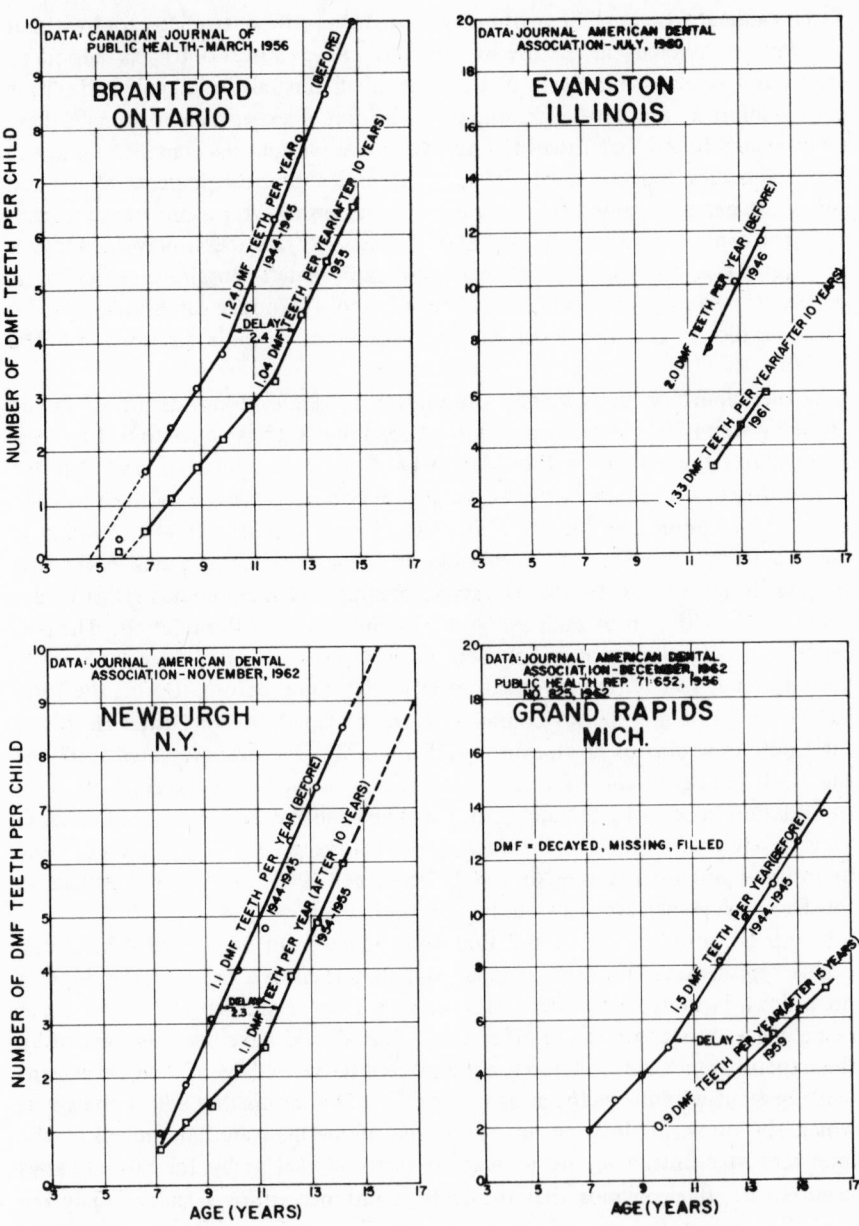

Figure 6-2
Counts of DMF permanent teeth in four test cities before and after fluoridation. (Courtesy of H. M. Thomson, who presented data in 1967 hearings, ref. *39*.)

It is obvious from the figure that the Newburgh count is erroneous because there is no break in the slope of the line. The authors have projected the lines for Brantford down to 0 DMF, which indicates that the average age of eruption of the permanent teeth was about 4.5 years before fluoridation and 6 years after fluoridation. This indicates that most of the delay in caries development pointed out by Thomson was the result of a delay in eruption of the permanent teeth. Such a delay should have been expected, because it was shown by Smith (*40*) as early as 1931 that as little as 6.3 ppm of fluoride in the diet or water fed to rats interfered with the normal development of the teeth, and 100 ppm fluoride exhibited a toxic effect on the growth of the skeletal structure. Also, Feltman (*40*), in a study to determine effects of feeding 1 mg of fluoride per day in the form of tablets to expectant mothers and their children, found that many of the children showed a marked delay in the eruption of the deciduous teeth, in many cases by as much as a year from the accepted average eruption dates.

The claim by the Public Health Service of 50 to 70 percent reduction in caries by fluoridation is erroneous because they did not take account of the delay in the eruption of the permanent teeth and did not take account of the number of teeth exposed to caries. After age 10 to 12, when the slopes of the lines on Figure 6-2 become greater, the children may be assumed to have all of their permanent teeth. The percentage of reduction in the slope before and after fluoridation that might be claimed as the maximum reduction in caries ranged from 0 for Newburgh to 40 percent DMF per year $[(1.5 - 0.9)/1.5]$ 100 for Grand Rapids. It should be noted that in the worst case at Evanston only 2 of the 28 teeth exposed, on the average, were decayed per year before fluoridation. This indicates strongly that the environment for the decayed teeth must have been quite different from the environment of the remaining teeth.

According to Mathews (*41*), the normal pH of saliva is about 7.7, and it contains about 79 ppm of calcium and 180 ppm of phosphorus. From the results of tests by Bertz (in *41*, p. 662) on tooth enamel it may be shown that 35.9 percent of the enamel is calcium and 17.8 percent is phosphorus. If it is assumed that the enamel crystals are $Ca_3(PO_4)_2$, these figures indicate that only about 4 percent of the calcium is present in some other form, probably as $CaCO_3$. The outer surfaces of the enamel of the teeth, if clean, are constantly in contact with the saliva of the mouth. Figure 6-3 shows the solubility of tricalcium phosphate as parts per million of phosphorus for various concentrations of calcium in the saliva as a function of the pH. The data for these curves were computed from the solubility product of $Ca_3(PO_4)_2$, Eq. (5-42) at 25°C. The temperature of the saliva, of course, is nearer 37°C, but no information is available on the solubility of $Ca_3(PO_4)_2$ at this higher temperature. It may be assumed, however, that the relative effect of pH will not change because of the higher temperature. It may be seen from Figure 6-3 that a change of pH from a normal 7.7 to 4.7 will

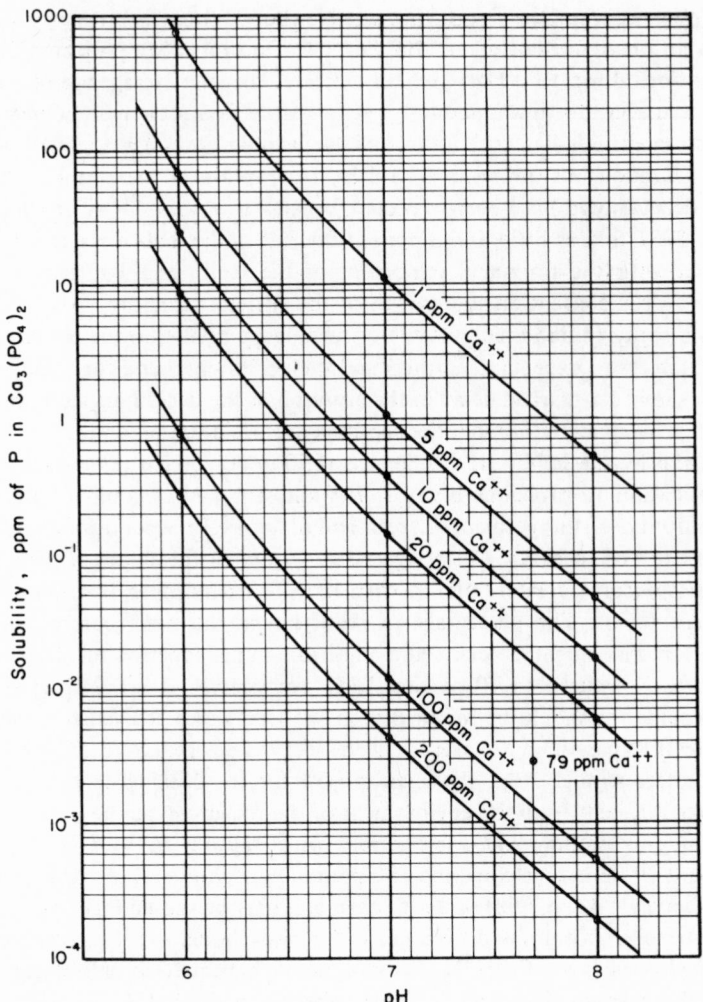

Figure 6-3
Solubility of $Ca_3(PO_4)_2$ at 25°C.

produce about a 10,000-fold increase in the solubility of $Ca_3(PO_4)_2$. Such a decrease in pH may be expected as a result of bacterial decomposition of organic matter in and under the plaque, or tartar, on teeth. Not only is the pH decreased, but the enamel under the tartar is no longer in contact with the saliva that protects clean teeth. Thus it may be expected that nearly all the caries occur as a result of plaque on the teeth.

When fluoride is ingested, most of the fluoride is attracted by the calcium in the skeletal structure and the teeth and forms calcium fluoride. This solid is very soluble, as may be seen by the solubility product [Eq. (5-39)], but its solubility is unaffected by changes in pH. In continuous fluoridation the fluoride passing through the body will approach equilibrium with the calcium fluoride in the skeletal structure and teeth. According to Shaw and Brudevold (*39*), the increase in fluoride concentration in tooth enamel increases greatly from the dentine to the outer surface of the enamel, and is about 1,000 ppm at the surface if the child has had an optimum dose of 1 ppm in his drinking water during calcification of the teeth. Shaw stated that the mottled enamel resulting from the ingestion of fluorides is caused by distortion of the crystals in the enamel, which produces different light-reflecting properties. When topical applications of fluoride solutions are employed periodically, fluoride reacts with the calcium on the surface of the teeth to produce calcium fluoride, but no mottled enamel. According to Brudevold a 50 percent reduction in caries has been observed in Boston from systematic treatments with topical applications of fluoride. A 40 percent reduction in cavities and gum diseases in older school children in Stockholm (*40*) has resulted since 1960 from supervised brushing at intervals of 2 months with a 0.5 percent solution of NaF.

Barium is rare in natural surface water and groundwaters because of the insolubility of its sulfate and other common salts. Barium carbonate occurs naturally in some mineral springs. The soluble salts of barium are injurious to man because of the deposits of insoluble barium compounds in the liver, lungs, and spleen and because of the toxic effects on the heart, blood vessels, and nerves. A rational basis (*42*) for the water standard of 1.0 ppm was derived from the threshold limit of 0.5 mg/m^3 of air set by the American Conference of Government Industrial Hygienists. Barium salts should not be employed for softening water to be used for drinking. Barium is reported to be harmful to fish in concentrations of 400 ppm and harmful to *Daphnia magna* in concentrations of 30 ppm.

Cadmium has a high toxic potential when taken by mouth. It accumulates in the soft tissues at all concentration levels down to 0.1 ppm in drinking water, resulting in anemia, poor metabolism, possible adverse arterial changes in the liver of man, and, at higher concentrations, death. Cadmium may be a water contaminant from electroplating plants. Cadmium is a common impurity in zinc used for zinc-galvanized iron and may enter solution in water distribution systems from this source. Cadmium is reported to be lethal to sticklebacks at a concentration of 0.2 ppm. The required limit of 0.01 ppm in the Public Health Service Drinking Water Standards probably has a large safety factor below the "no effect" level.

The salts of hexavalent chromium in industrial use (chromates and dichromates) are skin irritants that can produce ulcers. Chromium is known to be a carcinogenic agent for man when it is inhaled. According to Fairhall (*43*), trivalent chromium salts show none of the toxicity of hexavalent chromium, and their presence in drinking water supplies should not cause concern. Studies by MacKenzie, et al. (*44*) on rats showed that hexavalent chromium in the drinking water at concentrations of 0.45 to 25 ppm had no toxic effects, but did accumulate noticeably in the tissues at levels above 5 ppm. The toxicity concentration of chromate for fish ranges from 5 to 200 ppm and for *Daphnia magna* is reported at 0.05 ppm. The required limit of 0.05 ppm in the drinking water standards appears to have an adequate safety factor.

Because of the increased use of soluble chromates as corrosion inhibitors to form protective films on metals, caution should be exercised to prevent contamination of drinking waters with industrial and cooling waters containing high concentrations of chromates. The same may be said of glucosides and other corrosion inhibitors.

The need to set a water standard for silver arises from the possible use of ionized silver for disinfection. This germicide has gained some proponents, particularly for swimming pools, despite its high cost compared to chlorine. The chief effect of silver in the body is *argyria*, which consists of a permanent blue-gray discoloration of the skin, eyes, and mucous membranes of the victim. Most common salts of silver produce argyria when taken by mouth or injection. Silver, once absorbed, is held indefinitely in the tissues, particularly the skin, without evident loss through the usual channels of elimination. The required limit of 0.05 ppm in the drinking water standards was derived from the fact that silver in excess of 1 g for the entire adult skin, injected in the form of Ag-arsphenamine, will produce argyria. Assuming all silver from the drinking water is deposited in the skin, 0.05 ppm could be ingested for 27 years before exceeding 1 g.

According to Morris (*45*), a concentration of silver of 1 ppm or less requires several hours for effective bactericidal action; concentrations as high as 20 ppm appear to have no effect on waterborne viruses, and concentrations as high as 1,000 ppm will not kill cysts of *Endamoeba histolytica*. Morris also states that high concentrations of chlorides or proteinaceous impurities in water seriously decrease the germicidal activity of silver preparations. Silver may not be used safely, therefore, for the disinfection of drinking water, and its use for the disinfection of swimming pools is both ineffectual and hazardous.

No limitation has been placed by the standards on the chlorine concentration of drinking water, despite the fact that the use of chlorine for disinfection is one of the most prevalent causes of consumer complaint. Chlorine and chloramines are the most effective and generally satisfactory disinfectants that have been discovered, and their use should be encouraged. Complaints usually arise because of chlorinous odors, which are prohibited by the standards. There is no definite information available as to limiting concentrations of chlorine to avoid adverse physiological effects in human beings, but concentrations much in excess of 1 ppm have been used for disinfection with no ill effects. Goldfish are not injured by chlorine concentrations up to 1 ppm; cut flowers, by concentrations less than 10 ppm; and land plants, by concentrations up to 50 ppm. Another objectionable feature of high chlorine residuals is the corrosive effect of chlorine on plumbing and the piping of distribution systems.

The effects of human radiation exposure are viewed as harmful and any unnecessary exposure should be avoided. With the development of the nuclear industry, additional radiation exposure is unavoidable, but it should be minimized. The limits included in the 1962 drinking water standards are based on the guidance of the Federal Radiation Council (*46,47*) and on the recommendations of the International Commission on Radiological Protection (*48*) and the National Committee on Radiation Protection (*49*).

The limits for radium 226 and strontium 90 were derived from the recommendations of the International Commission on Radiological Protection (*48*) and the National Committee on Radiation Protection (*49*) of maximum daily intake limits of 220 and 2,200 $\mu\mu$C for continuous occupational exposure from radium 226 and strontium 90, respectively. The occupational exposure levels were reduced by a factor of 30 (*46,48*) for the general population, which leads to daily intake limits from air, food, and water of 7.3 $\mu\mu$C for radium 226 and 73 $\mu\mu$C for strontium 90.

The Federal Radiation Council guides are based upon three ranges of daily intake requiring three different control procedures. The preceding limits for the general population fall within Range II and necessitate quantitative surveillance and routine control. The daily intakes prescribed require that the dose rates be averaged over a period of 1 year. The Federal Radiation Council guides apply to normal peacetime operations.

It appeared likely to the advisory committee that intakes of radium 226 which are much above the average for the population as a whole usually result from the use of drinking water containing naturally occurring radium 226 in greater than average concentrations or from pollution of the water supply by wastewaters containing radium. With this in mind, a limit of 3 $\mu\mu$C/liter was set for radium 226 in drinking water. At a daily intake of 2 liters of water per capita, this concentration would result in an intake from water of 6 $\mu\mu$C/day. If there is evidence in a particular area that radium 226 from other sources is greater than average, the limiting concentration in water may have to be reduced below 3 $\mu\mu$C/liter.

The principal source of strontium 90 in the environment to date has been fallout from weapons tests, and human intake has been primarily from food. To allow for the difficulty of reducing the strontium 90 concentration in food, the advisory committee set the limit for water at 10 $\mu\mu$C/liter. At 2 liters/person/day, this concentration amounts to a daily intake of 20 $\mu\mu$C per capita. The limiting concentration of 10 $\mu\mu$C/liter for strontium 90 in water was substantially higher than the highest level found in public water supplies examined prior to 1961, but was exceeded by the highest values recorded for surface waters of the United States in 1962 and 1963 as reported by the National Water Quality Network (*19*). The ban on aboveground nuclear weapons testing resulted in lower readings thereafter.

Although a great variety of radionuclides may be present in drinking water, the advisory committee considered that it was unnecessary to establish limits in the 1962 standards for specific radionuclides other than radium 226 and strontium 90. Iodine 131 was not found in significant quantities in public water supplies frequently enough to require routine monitoring, and strontium 89 concentrations were not likely to be significant unless the strontium 90 concentration was also high. If significant concentrations of radioactivity are found in drinking water, it is prudent to determine all the radionuclides present and, where necessary, to reduce their concentrations to acceptable limits.

An upper limit of 1,000 $\mu\mu$C/liter of gross beta activity was set by the advisory committee (in the absence of alpha emitters and strontium 90). If this limit is exceeded, the specific radionuclides present must be identified by complete analysis to determine if the concentrations of nuclides will produce exposures above the recommended limits established in the Radiation Protection Guides (*47*).

In assessing the hazard of radionuclides whose limits were not set in the 1962 standards or for which guidance has not been provided by the Federal Radiation Council, the advisory committee suggests use of the values (MPCw for the 168-hr week) in Table I of the reports of the International Commission on Radiological Protection (*48*) or the National Committee on Radiation Protection (*49*); the values are to be adjusted by a factor appropriate for exposure of the general population. When mixtures of radionuclides are present, the permissible concentration of any single nuclide must be reduced by an amount determined through calculations described in these reports.

The work of the advisory committee was interfered with and delayed by the Atomic Energy Commission, which, at the time, appeared to have full control of all matters dealing with radioactivity, including the protection of the public from radiation hazards. The protection of the public from radiation hazards is, in the authors' opinion, in conflict with the primary objectives of the Atomic Energy Commission, that is, weapons development and the promotion of peacetime uses of atomic energy and of radioisotopes. The Public Health Service, on the other hand, is a regulatory agency charged specifically with the protection of the public health. To dissolve the impasse that was developing, the Federal Radiation Council was formed with representation from the Public Health Service and the Atomic Energy Commission.

It is to be noted that the standards finally adopted contain the following statement with regard to limiting concentrations of radium 226 and strontium 90: "When these concentrations are exceeded, a water supply shall be approved by the Certifying Authority if surveillance of total intakes of radioactivity from all sources indicates that such intakes are within the limits recommended by the Federal Radiation Council for control action." Inasmuch as the upper limits for Range II are 20 and 200 $\mu\mu C$/day for radium 226 and strontium 90, respectively, the above statement, in the opinion of the authors, constitutes a substantial relaxation of the limits of 7.3 and 73 $\mu\mu C$ originally proposed by the advisory committee.

In the future, barring unforeseen use or testing of nuclear weapons, the major concern over radioactive pollution of the environment will be related to the use of nuclear fuels in reactors and the discharge of contaminated cooling waters from these power plants. With the growing proliferation of such facilities on major waterways, the potential for radioactive insult and damage will increase. It has been estimated (50) that at present 99.9 percent of the radionuclides released to the environment are discharged from nuclear-fuel reprocessing plants. Many dangerous radionuclides in addition to strontium 90, such as cesium 137, iodine 151, and zenon 153, will require careful monitoring and control in the vicinity of such plants (51). Plutonium will be of concern if the breeder type of reactor becomes common (51).

The nuclear industry is growing rapidly, but the control of radiation hazards to the general public is lagging. The states, through their health agencies, should assume active control of the regulation of these hazards with advice and assistance from the Environmental Protection Agency and the Food and Drug Administration. It is the authors' opinion that the Atomic Energy Commission should give advice and assistance, but should be relieved of its regulatory powers on matters affecting the public health.

6-4. RAW WATER QUALITY

The 1942 Public Health Service Drinking Water Standards contained a Manual of Recommended Water-Sanitation Practice. Although it was clearly stated in the introduction to the manual that "this latter portion of the text is not to be considered as part of the standards which must be met in order to obtain certification of the water supply," it was found that at times the manual was being used as a part of the standards. To avoid this error, the Manual of Recommended Water-Sanitation Practice was not included in the 1946 drinking water standards. The manual was published separately in 1946 by the Public Health Service and was reprinted in 1958 as Public Health Service Publication No. 525 with the statement that "the Manual does not have the effect of law, regulation, or ruling." The manual was written primarily to serve as a guide to Public Health Service engineers in evaluating the sanitary features of water supplies with which they are concerned.

Included in the Manual of Recommended Water-Sanitation Practice is a classification of raw waters with respect to treatment requirements. This classification is as follows:

Group I—Water requiring no treatment: This group is limited to underground waters not subject to any possibility of contamination and meeting, in all respects, the requirements of the Public Health Service Drinking Water Standards, as shown by satisfactory, regular, and frequent sanitary inspections and laboratory tests.

Group II—Waters requiring simple chlorination or its equivalent: This group includes both underground and surface waters subject to a low degree of contamination and meeting the requirements of the Public Health Service Drinking Water Standards in all respects except as to coliform bacterial content, which should average not more than 50 per 100 ml in any month.

Group III—Waters requiring complete rapid-sand filtration treatment or its equivalent together with continuous postchlorination: This group includes all waters requiring filtration treatment for removal of turbidity and color; waters of high or variable chlorine demand; and waters polluted by sewage to such an extent as to be inadmissable to groups I and II, but containing numbers of coliform bacteria averaging not more than 5,000 per 100 ml in any one month and exceeding this number in not more than 20 percent of the samples examined in any one month.

Group IV—Waters requiring auxiliary treatment in addition to complete filtration treatment and postchlorination: This group includes waters meeting the requirements of group III with respect to the limiting monthly average coliform numbers, but showing numbers of coliform bacteria exceeding 5,000 per 100 ml in more than 20 percent of the samples examined during any 1 month and not exceeding 20,000 per 100 ml in more than 5 percent of the samples examined in any one month.

The requirements for group II waters tend to indicate that chlorination cannot be relied upon to effect a coliform kill of greater than 98 percent. The requirements for group III waters tend to indicate that rapid sand filtration with continuous postchlorination cannot be relied upon to effect a coliform kill greater than 99.98 percent. The requirements for group IV waters tend to indicate that rapid sand filtration with both pre- and postchlorination cannot be relied upon to effect a coliform kill of greater than 99.995 percent. There are a number of modern rapid sand filtration plants that regularly meet the drinking water standards despite the fact that the raw water occasionally contains much higher coliform counts than are shown by these requirements.

The recommendations in the manual were based mainly on the work of Streeter and his associates, whose studies involved the collection and analyses of data from 34 rapid filter plants in the Midwest and Middle Atlantic states (1923–1927) and from the 5-year operation of an experimental plant at Cincinnati (1924–1929). All data, except those from the experimental plant, were from plants using postchlorination only. It was found at the experimental plant that, with prechlorination, a total residual chlorine of about 0.05 mg/liter in the water applied to the filter would not destroy the biota in the filter, whereas a residual of 0.36 mg/liter or more would result in considerable sloughing of the biological slime, *schmutzdecke*, which was thought to be an essential element in the filtration process.

Since 1930, prechlorination has become common practice; coarser filter media, higher filter rates, and more effective backwashing have come into use; the schmutzdecke has been substantially eliminated, and better plant control has been secured. A study by Walton (*52*) (1954–1956) of data from about 60 plants indicates that simple chlorination may be used effectively with somewhat higher loadings than 50 coliforms/100 ml in the raw water; that coagulation, sedimentation, and filtration without chlorination, while removing less than 80 percent of the coliforms under low raw water bacterial loadings, will remove more than 99 percent under high loadings; that prechlorination, coagulation, and sedimentation are more effective in coliform removal than conventional rapid-sand filtration without chlorination; and that conventional rapid filter plants with adequate pre- and postchlorination can meet the drinking water standards with raw waters having coliform counts greatly in excess of 20,000 per 100 ml (up to 1,000,000 or more). Walton finds that chlorination, together with improvements in other processes, has made it possible to treat raw waters containing coliform bacterial loadings far in excess of the permissible loadings recommended in the manual.

The advisory committee for the 1962 drinking water standards gave long consideration to the advisability of including raw water quality standards in the drinking water standards. It decided against the inclusion of raw water quality standards because they would prohibit the use of many water supplies that might otherwise be used after satisfactory treatment, and because they would tend to discourage research and development of more effective means of water treatment. The 1962 standards state that "if the source is not adequately protected by natural means, the supply shall be adequately protected by treatment." This makes it possible to use any means of treatment provided it is adequate.

The Manual of Recommended Water-Sanitation Practice includes recommendations that 0.2 ppm, or more, of free chlorine should be maintained in the treated water after a contact period, in the distribution system, of at least 20 min beyond the point of chlorine application. It further recommends that, when chloramine treatment is used for disinfection, the residual chlorine concentration should be a minimum of 1 to 2 ppm at all points in the distribution system, and where *breakpoint chlorination* is practiced, the residual free chlorine concentration should be not less than 0.05 to 0.10 ppm at all points in the distribution system. These recommendations are intended for the purpose of killing bacteria that may be introduced into the distribution system through cross connections. It is doubtful whether the concentrations are high enough to be really effective for this purpose. Experience indicates that in many distribution systems it is not feasible to maintain the chlorine residuals at all points without frequent complaints from consumers about tastes and odors.

In the Public Health Service Drinking Water Standards and in the Manual of Recommended Water-Sanitation Practice, no mention is made of the presence or absence of dissolved oxygen as a criterion of drinking-water quality. Dissolved oxygen per se is of negligible importance to drinking water, although its complete absence may affect the taste of the water. Most waters used for drinking contain dissolved oxygen, but water pumped directly into the system from deep wells may be devoid of it. Dissolved oxygen is the most important factor influencing the rate of corrosion of metals, and its complete absence is desirable from this standpoint, although it is usually not economically feasible to remove dissolved oxygen from public water supplies for this purpose.

Most state and interstate water-pollution control agencies list the dissolved oxygen content as one of the most important criteria for the use of surface waters for public water supplies. The use of this criterion is to ensure the absence of large concentrations of polluting organic matter upon which the bacteria in the receiving waters may feed. In the feeding process the oxygen is utilized, and the rate of utilization is proportional to the concentration of food up to the limit of the rate of resupply of dissolved oxygen from the atmosphere and green plants. If this limit is reached, anaerobic decomposition of some of the organic matter will occur, resulting in the production of blackwater and malodorous gases. The objective of pollution-control agencies in using the dissolved oxygen, coliform counts, and other criteria is to control pollution rather than to prevent the use of such waters for public supplies. Pollution-control agencies usually set a minimum level of dissolved oxygen ranging from 3 to 6.5 ppm, depending upon the degree of water treatment.

Pollution-control agencies usually classify surface waters suitable for public water supplies in two or more classes, depending upon the degree of purification needed for use of the water for drinking. The highest class, requiring only chlorination, must have an average coliform MPN (most probable number) ranging from less than 50 per 100 ml in some states to less than 100 per 100-ml in others, with a maximum limit also specified in some states. When filtration and chlorination are used, the limit may be as high as 10,000 to 50,000 per 100 ml. The standards of the pollution-control agencies are further discussed in Chapter 9.

No limits or standards are suggested for the plankton content of drinking water in the Public Health Service Drinking Water Standards and none are suggested for raw water in the Manual of Recommended Water-Sanitation Practice. Algae blooms in water supply reservoirs sporadically give difficulty in shortening filter runs and producing disagreeable odors, and microorganisms in distribution systems contribute to slime growths, odors, and turbidity. It is common practice to treat reservoirs with copper sulfate or with chlorine to control excessive blooms, and to use activated carbon in treatment plants to control odors.

Biologic examinations of water are made in conformance with standard methods (8), and counts are usually reported in areal or cubic standard units or individual microorganisms per milliliter. An *areal standard unit* is 20 by 20 μm or 400 μm², and a *cubic standard unit* is 8,000 μm³. Organisms are usually identified when counted as individuals. Analyses are reported as total counts of microorganisms and total counts of amorphous matter, with counts of identified plankton when needed. According to Whipple, Fair, and Whipple (53), the oils in the microorganisms are present in sufficient concentrations to account for the odors. A concentration as low as 300 areal standard units of *Synura* in unchlorinated water is reported to be sufficient to produce the associated cucumber odor, and concentrations as low as 25 units may cause trouble in a chlorinated supply. When the total count reaches 500 organisms/ml in a reservoir, chemicals should be applied to control the bloom, particularly if the water is delivered to the consumer without filtration.

According to Whipple, Fair, and Whipple (53), most algae are strained out of the water in a sand filter in close proximity to the surface of the sand. The penetration is much deeper in rapid filters than in slow sand filters because of the larger grain size and much higher filter rate, but high concentrations of algae applied to a rapid filter will nevertheless seriously reduce the length of filter runs. The pore constrictions in a filter are about one tenth the size of the grains. In a rapid filter with 0.5-mm grains, the size of the pore constrictions is about 50 μm. Thus some *Chlorella*, 2 to 10 μm in size, and *Scenedesmus*, 7 to 10 μm in size, may pass through. *Tabellaria*, 4 to 16 μm by 20 to 80 μm, will collect near the top of the filter bed, while *Anabaena*, *Asterionella*, *Fragilaria*, *Synura*, and *Synedra* will collect in, and on, the top layer of sand.

Considerable difficulty has been experienced with the growth of algae and other microorganisms in the top sand and on top of slow sand filters. These growths shorten filter runs and produce odors; and when the organisms die, their decomposition results in additional odors and increases the ammonia content of the water. The only apparent remedy is frequent scraping of the bed. Prechlorination will kill the organisms, but this will intensify the chlorinous odors. Algae will also shorten filter runs with rapid filters, but the other difficulties are not encountered because the organisms are washed out of the beds during backwashing.

Frequent counts of microorganisms at the Billerica, Massachusetts, rapid filter plant, which takes water from the polluted Concord River, indicate that only small amounts of amorphous matter and practically no plankton penetrate through the filters. The pretreatment in 1961 consisted of aeration, breakpoint chlorination, alum coagulation, and sedimentation, with activated carbon applied most of the time. Daily counts (*54*) over a period of about 9 weeks in August, September, and October 1961, with the raw water count ranging between 160 and 10,432 areal standard units per milliliter, showed a removal by pretreatment ranging from 60.4 to 99.9 percent and averaging 91.7 percent with the sludge collectors not in operation. When the sludge collectors were in operation, the removal by pretreatment averaged only 75 percent; it was less than 50 percent on 16 percent of the samples; it reached an extreme low of 13.8 percent and was never higher than 97.8 percent. The removal of algae by filtration alone was 100 percent in 89 percent of the samples and was never less than 94.4 percent. The operation of the sludge collectors did not noticeably affect the removal by the filters, although the count in the filter influent averaged 180 with the sludge collectors in operation and only 43 with the collectors idle. Finely divided amorphous matter appeared in all filter effluent samples; the count ranged from 15 to 100 and averaged 56 areal standard units per milliliter. The amorphous matter in the filter influent averaged 1,590 with the sludge collectors operating and 640 with the collectors idle.

The filter medium at the Billerica filter plant is relatively coarse and permits deep penetration of floc and long filter runs. The material consists of an 18-in. layer of anthracite coal on top of 30 in. of sand. The grain sizes range from 1.19 mm at top to 1.68 mm at the bottom of the coal, and 0.59 mm at top to 1.19 mm at the bottom of the sand.

The water from the Boston metropolitan water supply is delivered from upland reservoirs to the distribution systems of the member communities without filtration but with chlorination. The counts (*55*) during 1960 in the high service system ranged from 2 to 180 areal standard units per milliliter, mostly *Tabellaria*, and the amorphous matter was 200 to 300 areal standard units per milliliter. Even with such relatively low counts, algae may cause difficulty in some industrial uses of water and in medical laboratories.

Poisonous algae blooms are widespread in lake waters in many parts of the world, including the United States and Canada, and have, in some cases, resulted in the violent death of cattle, horses, pigs, sheep, dogs, and birds. A study made by Olson (*56*) of poisonous algae blooms in Minnesota showed that, in 3 years (1948-1950), 49 blooms, out of 60 collected at random, were found that were toxic to test animals. The smallest lethal dose for adult mice was 0.005 ml of raw algae injected intraperitoneally, and the shortest time in which death occurred was 2 min. If the algae are taken orally, the minimum lethal dose is greater. The toxicity of a sample appeared to be directly related to the concentration of cells in a suspension, and certain species of algae such as *Anabaena lemmermanni* were more poisonous than others. Only 10 species out of 92 studied were important in producing toxicity, including *Aphanizomenon flosaquae*, *Coelosphaerium naegelianum*, *Lyngbya birgei*, *Microcystis aeruginosa*, *Microcystis incerta*, *Trichodesmium lacustre*, and four species of *Anabaena*. The authors estimate from the results of the Olson studies that about 15 ml of poisonous raw algae might be lethal to man.

Olson (57) has found that the toxic elements in poison algae blooms are produced within the cell; are soluble in water, alcohol, and acetone, but not in benzene, ether, or chloroform; can exist outside the cells in water around the algae; can pass through cellophane and animal membranes by dialysis; are nonvolatile and relatively heat stable at temperatures up to 100°C. Any means of breaking down the cell walls, such as the death of the algae, will increase the release of the toxic elements. Chlorine in doses reasonable for water supplies is not effective in destroying the toxin. Thus, it is evident that poisonous algae may become a menace to drinking water supplies and that precautions should be taken to prevent excessive blooms. It is possible that some gastroenteric outbreaks may have been caused by poisonous algae.

The problem is not limited to freshwaters. Some marine algae are capable of producing toxins, which, when the alga reaches bloom population, may result in serious economic and public health impact. "Red tide" blooms of *Gymnodinium brevis* in the Gulf of Mexico are associated with a notable mortality in fishes and animals (58). Another red tide organism is *Gonyaulax*, a dinoflagellate. The neurotoxin produced by such algae is concentrated by shellfish and may lead to complications and even death when consumed by humans. The coastal waters of New England were infected with a red tide in September 1972; illnesses were reported and the shellfish industry was temporarily curtailed. The causative organism in this case was reported as *Gonyaulax tamarensis*, which had not made its presence known in these waters before (59).

6-5. STANDARDS FOR SWIMMING AND RECREATION

The quality of water for swimming and bathing also has sanitary significance. There is a widespread suspicion among physicians and health officers that many diseases are spread through bathing waters, although convincing epidemiological evidence is lacking in most cases. Diseases may be transmitted in two ways: (1) from one bather to another in water which was safe prior to the entrance of the bathers, and (2) from outdoor bathing water that is polluted by sewage or nonfecal infectious organisms. The types of diseases whose transmission is suspected are (1) eye, ear, nose, and throat infections, (2) skin diseases such as ringworm, eczema, scabies, etc., (3) venereal infections, and (4) gastrointestinal disorders.

Since many persons swallow little or no water during bathing, intestinal infection is of much less significance than with drinking water. Nevertheless, cases of typhoid and dysentery contracted from polluted bathing waters have been reported, and there is growing evidence that sewage-polluted bathing beaches may be an important means of spreading poliomyelitis and infectious hepatitis. The diseases that appear to be most closely associated with bathing waters are skin diseases and respiratory disorders. The mineral content of bathing waters has little physiological significance.

Despite the paucity of sound epidemiological evidence that much disease is transmitted directly through polluted recreational waters, it is a fact that polio epidemics in the United States occur during, and following, the summer recreational season. Cox (2) has pointed out that the mode of transmission of about 28 percent of the numerous outbreaks of gastroenteritis (diarrhea) that were reported in the United States and Canada during 1939 was not determined. He suggested that they may have been waterborne, but discounted viruses as the cause because there was no evidence at that time that water supplies were carriers of virus infections. Many outbreaks of gastroenteritis have occurred at summer resorts. There is, however, no more reason to suspect swimming pools, or other bathing areas, than to relate the summer illnesses to much-increased individual ingestion of drinking water from such doubtful sources as bubblers, canteens, restaurants, and streams.

An outbreak of summer febrile disease (summer grippe) occurred in July and August 1958 in a circumscribed community of military families at Great Lakes Naval Training Center. A similar illness was reported in the geographic area surrounding the center. Heggie et al. (60) describe a study of about 20 percent of the population at the center which indicated that the causative agent was Coxsackie B-2 virus. Throat swabs were collected from an approximately equal number of symptomatic and asymptomatic subjects in affected and nonaffected families, and in addition, rectal swabs were obtained from children in these categories under the age of 2 years. From 361 individuals comprising 23 percent of the subjects in the study, specimens were collected for viral culture. Viruses were isolated from 12 percent (45) of the subjects cultured, and Coxsackie B-2 virus was the agent present in 78 percent of the positive cultures. The remaining 22 percent of the positive cultures contained two cases of adenovirus Type I, three of ECHO virus, four of poliomyelitis virus, and one of Coxsackie B-3 virus.

Although Coxsackie B-2 virus was isolated from only 15 percent of those who were ill during the study period, it is interesting that this virus was found in 3 percent of the subjects who were not ill. It is evident that pathogenic viruses can be harbored by persons who do not come down with the disease. Although no evidence was present that this outbreak was connected with bathing, it did occur during the recreational season, as have many other similar outbreaks.

Both rectal and throat swabs were cultured for 18 children. For five of these children, the same virus was isolated from both throat and rectal swabs, and there was no instance in which a throat swab was negative in the presence of a positive rectal swab. It is thus evident that, although the most probable seat of the virus infection is the throat, viruses are discharged into the sewers, and sewage-polluted bathing waters may be involved in the spread of the virus diseases.

There is evidence that hepatitis A virus (5) is transmitted through a fecal–oral route, but there is no evidence for a respiratory route. It is possible therefore that infectious hepatitis may be contracted at sewage-polluted bathing beaches. When person-to-person contact is the mode of transmission of hepatitis, the disease spreads leisurely through the community, usually reaching a peak in the late winter or early spring. If the outbreak occurs during, or just following, the outdoor bathing season, sewage-polluted bathing waters may be suspected. When a large number of cases occur within several weeks of each other, a common source of infection, such as water, food, or milk, must be suspected.

An unusual outbreak of waterborne viral hepatitis (61, 62) occurred in 1969 at Holy Cross College in Worcester, Massachusetts, in which 83 members of the varsity football squad and 14 others associated with the team became ill. Investigation linked the epidemic to a cross connection in close proximity to the varsity practice field; a municipal water line, terminating at this high point on the campus, and used both for drinking and irrigation purposes, developed a negative pressure during a fire in the vicinity in late August. At about the same time that fall football practice had begun five cases of hepatitis had stricken a family living in a condemned house near the field; children from this family were accustomed to play in water around the irrigation connections on the line in which negative pressure developed during the fire. The Holy Cross experience attracted a great deal of attention because the College had to cancel all but the first two of its scheduled games, the early contests having preceded the development and recognition of symptoms in the infected players.

In America the standard of quality for bathing waters established by the "Report of the Joint Committee on Bathing Places—Conference of State Sanitary Engineers and American Public Health Association" (63) has been generally accepted. For swimming pools employing recirculation, filtration, and chlorination, the committee recommends that the total bacterial count on standard nutrient agar incubated 24 h at 37°C should not exceed 200 per ml in more than 15 percent of the samples over any considerable period of time. It is further recommended that not more than 15 percent of the samples collected when the pool is in use should show a positive, confirmed test for coliform bacteria in any of the five 10-ml portions. This recommendation specifies an MPN of less than 2.2 per 100 ml. It is slightly less severe than the drinking water standard of the Public Health Service since the limit may be exceeded in

15 percent, instead of only 10 percent, of the samples. It is the authors' opinion that properly designed and operated pools should always show negative tests for coliforms when the pool is in use.

The Joint Committee recommends the following chemical and physical qualities for swimming pool water: the chlorine residual should be kept between 0.4 and 1.0 ppm when chlorine and hypochlorites are used for disinfection, and between 0.7 and 2.0 ppm when chloramines are used; when alum is used, the pH should be kept above 7; the pool water should be sufficiently clear so that a 6-in. black disk on the bottom at the deepest point is clearly visible from the side decks; in a heated pool, the water temperature should be less than 78°F and the air temperature should be about 5° warmer than the water. The use of ionized silver for disinfection is not recommended by the joint committee; ultraviolet light, ozone, or other disinfectants that do not produce a persistent residual disinfectant in the pool water are recommended only if they are supplemented by a chlorine compound.

The joint committee approves the use of high free chlorine residual for pool disinfection, with 1.0 ppm or more of free chlorine residual accompanied by a pH of 8.0 to 8.9. Advantages claimed are better bacterial kills, clearer pool water, and less irritation of the eyes of swimmers. If alum is the coagulant, it must be employed intermittently with the pH reduced to 7 or 7.5 during its use in order to produce a floc.

Iodine and bromine, which have been proposed for disinfection of swimming pools, are not discussed by the joint committee. McKee (64) has found the colicidal effectiveness of chlorine on settled Pasadena sewage to be about twice as great as iodine or bromine in terms of the required equivalent dose, and about 5 times as great in terms of the required dose by weight. Iodine and bromine are very much more expensive; their cost is many times that of chlorine for equal colicidal effectiveness. Black et al. (84) have pointed out many possible advantages in the use of iodine for swimming pool disinfection: compared to chlorine it is relatively stable and inactive, exerts very little effect on pH, maintains a free iodine residual, and provides satisfactory bacteria control. It does not control algae and results in shorter filter runs; therefore, until a satisfactory algicide is developed it cannot be recommended for general swimming pool use (84). The continued use of iodine in a swimming pool using recirculated water will result in a substantial buildup of iodides in the water, which, if swallowed in quantity, might result in hyperactivity of the thyroid in some individuals.

With the current practice of requiring bathers to take shower baths before entering a pool, there is a diminished likelihood of pollution of intestinal origin. However, this practice does not prevent contamination of the water by mouth and nasopharyngeal washings of the bathers. Since as high as 68 percent of illnesses resulting from swimming pool use are eye, ear, nose, and throat ailments (65), there is some opinion that bacteria of the nose and throat are better indices of pollutional load than coliform organisms. Mallmann (66) has studied the use of tests for streptococci in swimming pool control, and Tapley and Jennison (67) suggest the diplococcus *Neisseria catarrhalis* as an index to pollution.

Outdoor bathing places are defined by the joint committee as including small streams, rivers, lakes, and tidal waters. Fill and draw swimming pools and recirculation swimming pools readily subject to artificial purification or to constant replenishment with uncontaminated water are not included.

Flowing-through bathing pools along small streams are not recommended for a large number of bathers unless disinfection is provided. In general, disinfection is required if the flow is less than 500 to 900 gal/day per bather. Whether or not disinfection is used, every effort should be made to eliminate all sources of sewage pollution. Bathing should be limited to relatively clear waters without muddy bottoms.

The joint committee considers it neither practicable nor desirable to recommend an absolute standard of safety for the waters of outdoor bathing places. A relative scheme of classification is suggested, which may, of necessity, be varied due to differences in local conditions. Most states have standards based on limiting coliform counts, which range from 240 per 100 ml in some states to 2,400 per 100 ml in others. Some states permit the limit to be exceeded part of the time or in a certain portion of the samples collected, without restriction as to how much the limit may be exceeded. This is dangerous practice, in that it may permit slugs of excessive pollution.

The difficulties of setting standards are illustrated by the fact that, while some areas of the country are reported to be able to meet bacterial standards for natural surface waters with maximum limits of coliform MPN of not more than 240 to 500 per 100 ml, some streams which are subject to no pollution other than animal contamination may often show a coliform count of 240 to 1,000 per 100 ml. The difficulties are further compounded by the relative effects of sewage treatment on bacteria and viruses and by the difference in die-off rate of bacteria and viruses in salt water. This suggests the desirability of sufficiently effective chlorination of sewage to inactivate pathogenic viruses. It is probably not feasible to supplement coliform examination of bathing waters with virus tests, since the virus count is likely to be too small to be detected, even though the viruses are present in dangerous numbers.

Field studies (63) by the Public Health Service of Lake Michigan at Chicago showed a significant increase in illness among swimmers who used one beach on three selected days when the water had an average content of 2,300 coliforms per 100 ml. Similar studies of the Ohio River at Dayton, Kentucky, showed that, despite the relatively low incidence of gastrointestinal disturbances, swimming in river water having a median density of 2,700 coliforms/100 ml appeared to have caused a significant increase in such illnesses among the swimmers. On the other hand, bathing beaches where the coliform counts run as high as 2,400 per 100 ml have been used without reported evidence of illness.

Some water-pollution control agencies differentiate between waters suitable for bathing and recreation (boating and fishing) and waters suitable for recreation only. One state permits bathing and recreation in waters with a median coliform MPN less than 1,000 per 100 ml and a maximum MPN of less than 2,500, but permits recreation only with a median up to 20,000 and a maximum up to 50,000. Another state requires a coliform MPN of not greater than 240 per 100 ml for bathing, but will permit recreational boating and fishing in waters with no limit on the coliform count. Another pollution control agency limits the coliform MPN to 1,000 per 100 ml for waters to be used for bathing and recreation, with no higher limit for recreation alone.

It is the authors' opinion that all domestic sewage to be discharged into inland or coastal waters should be chlorinated for effective inactivation of pathogenic viruses. Since all such waters may be used for recreational purposes and even for swimming, standards should be set that will make effective chlorination of sewage compulsory. For sparsely polluted waters, a bacterial standard may suffice. For waters heavily polluted with municipal sewage, all sewage, including that mixed with storm water, should be effectively chlorinated, even though the resulting coliform count in the receiving waters might be higher than considered safe for recreational purposes.

Conventional primary settling of sewage will remove no more than about 50 percent of the coliforms in the dry weather flow, and conventional complete treatment will remove no more than an average of 90 percent. In both cases, the coliform removal is without control and varies greatly. Moreover, with treatment plants and interceptors serving combined systems and having capacities slightly greater than the peak dry weather flow, the average bacterial removal for the year may be only about 30 percent for primary treatment and only about 60 percent for complete treatment, because up to about one third the sewage solids are lost to the receiving waters in overflows during rainstorms (see p. 235). On the other hand, tests on the chlorination (68) of raw sewage indicate that coliform kills of 99.99 percent may be obtained with a 10-min contact period. The equivalent dilutions are 10 to 1 down to 6 to 1 for complete treatment and 10,000 to 1 for chlorination. In other words, chlorination alone is 1,000 to 1,700 times as effective as conventional complete treatment without chlorination. It is easier, cheaper, and more effective to reduce the spread of waterborne disease through recreational waters by effective chlorination of sewage (69) discharged to those waters than by any other method.

Schistosome dermatitis (70) (schistosomiasis) or swimmer's itch has become an important public health problem in some regions of the United States and Canada, where it renders many lakes unsuitable for bathing. This disease is caused by penetration into the bather's skin of *Schistosome cercariae*, larvae of trematode worms that are parasitic to birds and snails. The prevalence of certain types of beach snails in freshwaters is usually associated with outbreaks of the disease. It is not associated with saltwater bathing. McMullen (71) reports control measures to kill the snail hosts by the use of $CuSO_4$ crystals on the beach sands in shallow water and mixtures of $CuSO_4$ and $CuCO_3$ in deep water.

Schistosomiasis, although transitory in the United States, is endemic (72) in many parts of the world, particularly in the underdeveloped, tropical, and semitropical countries. The victims of the disease, known as snail fever, are constantly exposed to infection by wading through ponds, paddies, and ditches without boots. The disease results in fever, general debilitation, and ultimate death; it is very difficult to treat. The prevalence of the disease has increased enormously because of the flooding of lands by new reservoirs created for irrigation purposes. The human and economic toll is tremendous. In Egypt it has been estimated (73) that over 14 million people (nearly half the population) suffer at the present time from schistosomiasis, and that another 2.6 million cases can be expected as a result of the completion of the Aswan dam. The annual economic loss in Egypt, according to World Health Organization consultant M. Farooq, is $560 million (73).

6-6. STANDARDS FOR SHELLFISH GROWING AREAS

Sewage pollution of shellfish growing areas in tidal estuaries has resulted many times in outbreaks of enteric disease, particularly typhoid fever and, more recently, infectious hepatitis. To control such outbreaks, a cooperative agreement was reached in 1925 involving the Public Health Service, the states, and the shellfish industry. The State–Industry–Public Health Service shellfish certification program was first described in a "Report of Committee on Sanitary Control of the Shellfish Industry in the United States" (1925). This guide was revised in 1937, 1946, and again in 1959, when it was issued by the Public Health Service as a two-part "Manual of Recommended Practice for the Sanitary Control of the Shellfish Industry" (Publication No. 33). Part I relates to growing areas and Part II to harvesting and processing. Part I is of immediate interest since it deals primarily with water quality.

Under the cooperative program, the state must supervise all phases of the growing, harvesting, transportation, shucking–packing, and repacking of shellfish to be shipped interstate. This is done under the authority of state laws and regulations. The Public Health Service makes an annual review of each state's program and endorses it if it is satisfactory. Shellfish are defined as all edible species of oysters, clams, or mussels.

Investigations by the states and the Public Health Service from 1914 to 1925, a period when disease outbreaks attributable to shellfish were more prevalent, indicated that typhoid fever or other enteric disease would not ordinarily be ascribed to shellfish harvested from water in which not more than 50 percent of the 1-ml portions of water examined were positive for coliforms, provided the areas were not subject to direct contamination with small amounts of fresh sewage, which would ordinarily be undetected in the bacteriological examination. The bacterial limit corresponds with a coliform MPN of about 70 per 100 ml. This is the basis of the 1959 standards, but the standards also include a limit of paralytic shellfish poison (see p. 127) of 80 μg/100 g of the edible portions of raw shellfish. Unfortunately, the coliform count is probably not a satisfactory measure of the presence of the infectious hepatitis virus, since the die-off rate of the virus may be much less than that of coliforms in seawater.

Paralytic poison is collected temporarily by bivalve shellfish in some areas from free-swimming, one-celled marine plants on which these shellfish feed. The plants flourish seasonally under favorable conditions. Cases of paralytic poisoning, including several fatalities, resulting from poisonous shellfish have been reported from both the Atlantic and Pacific coasts. Epidemiological investigations indicate that 200 to 600 μg of poison will produce symptoms in susceptible persons, and one death has been attributed to the ingestion of about 480 μg. Growing areas should be closed if the poison content reaches 80 μg/100 g of the edible portions of raw shellfish.

The 1959 manual provides for a sanitary survey of each growing area prior to its approval or classification and for the reappraisal of each approved area, biennially at least. A comprehensive sanitary survey will include an evaluation of the following: all sources of actual or potential pollution on the estuary and its tributaries and the distance of such sources from the growing areas; the effectiveness and reliability of sewage-treatment works; the presence of industrial wastes or radionuclides that would cause a public health hazard to the consumers of the shellfish; the effect of wind, stream flow, and tidal currents in distributing polluting materials over the growing area. Important parts of the sanitary survey are an analysis of the bacterial quality of the growing-area water and bottom sediments and a determination of the presence and location of small sources of pollution, including boats. Growing areas are classified as (1) approved areas, (2) conditionally approved areas, (3) restricted areas, and (4) prohibited areas. An area may also be closed temporarily because of paralytic shellfish poison.

A growing area may be designated as "approved" for the taking of shellfish for direct marketing, when the sanitary survey indicates that pathogenic microorganisms (including paralytic shellfish poison), radionuclides, and/or harmful industrial wastes do not reach the area in dangerous concentration, and when this is verified, insofar as possible, by laboratory analyses. The coliform median MPN of the water should not exceed 70 per 100 ml and not more than 10 percent of the samples should exceed an MPN of 230 per 100 ml. The median MPN of 70 per 100 ml is equivalent to a dilution with about 60 million gallons of coliform-free water per day for the fecal matter from each person, based on an average discharge of 160 billion coliforms per person per day. The sanitary survey should also show that the area is protected from chance contamination by slugs of water containing heavy concentrations of fecal matter, usually from nearby sources. The presence of radionuclides in growing-area waters must be explored, since shellfish have the ability to concentrate such materials. Limits on maximum permissible concentrations of radioactive materials in food have been established (*74, 49*). The current standard should be used in evaluating the public health significance of radioactivity in shellfish.

A growing area may be designated as "conditionally approved" for harvesting shellfish for direct marketing if sanitary surveys show that the coliform MPN and other tests meet the requirements of approved areas, but also show that the maintenance of acceptable water quality is dependent upon continued satisfactory performance of sewage-treatment works discharging effluent to the area, or may be affected by seasonal population or sporadic use of dock or harbor facilities. An operational procedure for each such area must be developed to expedite closure of the area when unfavorable conditions are imminent.

A growing area may be designated as "restricted" when a sanitary survey indicates a limited degree of pollution that makes it unsafe to harvest shellfish for direct use or marketing. Shellfish from such areas may be marketed after purification or relaying under controlled conditions. The purification or relaying is to permit the shellfish to purge themselves of the excess pollution as measured by coliform counts. The same standards for radioactivity and paralytic shellfish poison apply to restricted areas as to approved areas. The coliform median MPN of the water should not exceed 700 per 100 ml, and not more than 10 percent of the samples should exceed an MPN of 2,300 per 100 ml. The median MPN of 700 per 100 ml is equivalent to a dilution with about 6 million gallons of coliform-free water per day for the fecal matter for each person, based on an average discharge of 160 billion coliforms per person per day.

A growing area is classified as "prohibited" when a sanitary survey indicates that the bacterial contamination is so great, or that the area is so contaminated with radionuclides or industrial wastes, that consumption of the shellfish might be hazardous. The bacterial contamination is excessive when the median coliform MPN of the water exceeds 700 per 100 ml or when more than 10 percent of the samples have a coliform MPN in excess of 2,300 per 100 ml.

6-7. WATER QUALITY FOR FISH AND WILDLIFE

Excessive pollution of rivers, lakes, and tidal estuaries by sewage and industrial wastes has resulted in deterioration of commercial fishing, fish kills, and injury to waterfowl. Much of the pressure for pollution abatement comes from those interested in fisheries, wildlife, and recreation.

The damage to waterfowl results from the destruction by pollution of the feeding and breeding grounds. This may be brought about by oils, greases, foams, insecticides, and excessive depletion of dissolved oxygen by decomposing organic wastes. Acid mine wastes are also destructive. The principal hazard to fish is excessive depletion of dissolved oxygen, but increases in temperature of the water, as a result of its passage through power or industrial heat-transfer units, are also damaging.

Statistics on fish catches, both commercial and sport, have often been used to show that pollution is the cause of declining catches. Tollefson (75) has shown that fisheries statistics are valuable tools in assessing the effects of pollution, but that they must be complete and must be considered along with other factors in order to avoid unreliable conclusions. He has demonstrated that declining fish catches attributed to pollution have, in many cases, been caused by other unknown factors.

Although Lake Erie is not dead, and the total amount of fish produced has not changed appreciably, the poundage of commercially important fishes has been reduced drastically over the past 50 years (76), primarily because the more desirable fish species can no longer spawn in the grossly polluted tributaries of the lake.

Most of the substances in water that are known to be toxic to fish have been discussed in connection with the drinking-water standards, because many of them are also toxic to man. Dissolved oxygen, however, is of the utmost importance to aquatic life, whereas it is of minor importance in drinking water. Fish depend upon dissolved oxygen for respiration, and they will smother and die with an inadequate supply.

The solubility of atmospheric oxygen in freshwater at sea level ranges, approximately, from 14.5 ppm near the freezing temperature of water down to 7.2 ppm at 90°F, and it is about 20 percent less in seawater, as shown by Figure 5-1. Hot water discharged from power stations and industry will further reduce the oxygen solubility. For example, at 140°F the solubility of oxygen in freshwater is only about 3.9 ppm, and in seawater it is about 3.1 ppm. The actual content of dissolved oxygen in natural waters is usually less than these saturation values, especially when decomposing organic matter is present. In daylight, however, during algae blooms, water may become supersaturated with dissolved oxygen, since green and blue-green algae convert carbon dioxide to cell tissue by photosynthesis and, thereby, release oxygen.

In evaluating the oxygen requirements of fish, a distinction must be drawn between the ability of fish to resist low oxygen levels and the oxygen level required for their optimum growth and propagation. Tarzwell (77) states that "under actual stream conditions, a fish must maintain its position against the current, find, pursue and catch its food, avoid its enemies, and reproduce. All these activities require oxygen in such amounts that oxygen levels at which the fish can just survive are unsatisfactory. Age, size, and season are also of importance. In general, fry and younger fish have a higher metabolic rate and require more oxygen than adults. Because of increased activity and their physiological condition, fish require more oxygen at the spawning season." These are the conditions of optimum growth and propagation.

In setting water quality standards for pollution control (78), on the other hand, the minimum oxygen levels at which fish can survive and pass upstream or downstream are of paramount importance. The reasons are that (1) these levels will occur only at the bottom of the oxygen sag, downstream from points of pollution, and only in the late summer and fall when the water is warmest, (2) the spawning beds are nearly always upstream from the points of pollution, and (3) most fish will pass through the points where the oxygen level is lowest in the spring and late fall when the water is cooler and the oxygen content is greater.

According to Tarzwell (77), the environmental requirements of fishes may be roughly grouped under four main headings: (1) a favorable water supply; (2) suitable spawning facilities; (3) an adequate food supply for all age groups, and (4) good pools and shelters. In pollution control we are chiefly concerned with the first requirement, a favorable water supply. If a little of a polluting material, such as sewage, is added, the discharge to the stream has a fertilizing effect that is beneficial to fish production. If too much is added, however, the dissolved oxygen will be depleted excessively and fish damage will result.

The oxygen requirements of fish are affected by the temperature and pH of the water and by its CO_2 and dissolved-solids content. The oxygen uptake of any species of fish increases two- or threefold with each $10°C$ increase in temperature. High CO_2 concentrations interfere with the ability of fishes to utilize dissolved oxygen, as do high and low pH values.

The findings in the Lytle Creek (near Cincinnati) studies (77) indicate that for a well-rounded warm-water fish population, the dissolved oxygen must not be below 5 ppm for more than 8 hr of any 24-hr period, and at no time should it be below 3 ppm. For the maintenance of a coarse fish population, the dissolved oxygen should not be below 5 ppm for more than 8 hr of any 24-hr period, and at no time should it be below 2 ppm.

According to Tarzwell (77, 78), the salmonoid fishes are not usually found where the dissolved oxygen is less than 4 to 5 ppm, and eggs and fry require a minimum of 6 ppm. Migratory trout, salmon, shad, striped bass, herring, and several other species swim up from the ocean to small streams to spawn. The upper branches of a large stream may be trout streams; the lower portions may be bass streams or even coarse fish streams because of temperature and other factors. Tarzwell (78) states that where eggs and fry are not developing, salmonoids can survive for short periods at dissolved oxygen concentrations of 1.5 to 2 ppm, and warm-water species can live for considerable periods during cold weather at dissolved oxygen levels of 1 to 2 ppm. It is evident, therefore, that the oxygen level may be reduced to about 2 ppm at the bottom of the oxygen sag when the water is warmest without damage to migratory fish, provided the oxygen concentration is somewhat higher during fish runs and provided the fish can swim through the region with low oxygen concentration in a relatively short time.

Studies reviewed by Tarzwell (77) indicate that the turbidity of water must be very high (20,000 ppm or more) to be directly harmful to fish. Game fish, however, which feed by sight, are at a disadvantage in muddy waters when competing with catfish and with carp, buffalo, and suckers, which employ a suction type of feeding. Suspended solids and turbidity reduce light penetration and thus limit photosynthesis and algal growth. Since algae are the basic food materials for aquatic life, turbidity indirectly affects fish production. Suspended solids that settle on stream bottoms lower the supply of suitable stream-bottom insects and thus reduce the food supply of fish. Such deposits also reduce the number of nest-building and spawning areas. Most of the turbidity and suspended matter in streams is produced by silt erosion during storm-water runoff.

According to Tarzwell (77), pH values from 5 to 9.5 have not been shown to be detrimental to fish, although at values below 5 and above 9 the pH seriously affects the abilities of some fishes to extract oxygen from the water. It is not surprising, therefore, that streams carrying acid coal-mine wastes or alkaline limestone wastes are not inhabited by fish. In the more productive streams, the pH is in the range of 6.5 to 8.5. Carbon dioxide concentrations under about 30 ppm, in the absence of other adverse factors, will have no harmful effects on most fish species.

As a group, aquatic animals in the temperate zone are adaptable to fluctuations in temperature (77) from 39 to 90°F. The salmonoids and many other northern fish can survive all winter under the ice without damage. Temperature increases to 94°F and higher may result in fish kills, particularly if the increase is rapid and accompanied by dissolved oxygen deficiencies. Brook trout do best in streams where the summer temperature is between 52 and 68°F, and the upper lethal temperature for young brook trout is about 78°F. Fluctuations of water temperature above 60°F during the spring are detrimental to the spawning and production of bass.

According to Tarzwell (77), total dissolved mineral solids up to about 3,000 ppm can be tolerated by most freshwater fish, if the materials in solution are the relatively nontoxic earth metals and are physiologically balanced. Physiologically balanced mixed salt solutions, such as seawater, may be harmful to freshwater organisms because of the salt concentration and osmotic pressure, rather than because of the specific toxicity of any particular ions present. The chloride content of mixed salt solutions, such as oil-well brines and other industrial wastes, is not a reliable index of osmotic strength.

Pollution control agencies usually set minimum dissolved oxygen standards ranging from 4 to 5 ppm for waters suitable for fish life. Some agencies limit turbidities to 40 ppm, others to 250 ppm except in case of heavy rain, and others to unobjectionable amounts. Some agencies limit color to 20 ppm, others to 100 ppm, and still others to unobjectionable concentrations. Some agencies set pH limits between, approximately, 6 and 9; others set no limits. Some agencies require that grease, oil, scum, and floating debris not be present in objectionable amounts. Some require no sludge deposits, and some limit the chloride content, the phenols, and the iron. The objective of these standards is to limit or reduce man-made pollution. It must be recognized, however, that practically all these numerical limits are violated by nature in many otherwise unpolluted streams or lakes.

The first of a series of annual reports on the effects of pollution was issued by the U.S. Public Health Service under the title "Pollution-Caused Fish Kills in 1960." The report covered the period from June 1, 1960, through December, 1960. The greatest number of fish were killed by industrial wastes, followed closely by agricultural poisons. Domestic sewage and wastes from mining operations were other major causes. Of the agricultural poisons, the most frequently reported causes were rotenone, DDT, 2-4-D, and endrin. Cyanide and metallic ions were the principal causative agents in industrial wastes.

According to Webb (79), fish are particularly sensitive to low concentrations of insecticides. Chlorinated compounds, including DDT, endrin, dieldrin, BHC, and toxaphene, constitute the group most toxic to fish, and these compounds are probably the most commonly used. DDT has been detected in the Mississippi River at Quincy, Illinois, and at New Orleans; in the Missouri River at Kansas City; in the Columbia River at Bonneville, Oregon; and in Lake St. Clair and the Detroit River. Chlorinated hydrocarbons are reported to be much more toxic to fish than organic phosphorus compounds, although some of the latter may be more dangerous to warm-blooded animals. Massive fish kills in the Mississippi River in 1963 led to a considerable effort to trace lethal concentrations of chlorinated hydrocarbons to specific sources; the difficulties encountered in this task have been summarized by Graham (80).

Weiss (81) has reported experimental studies of sublethal exposures of fish to organic phosphorus insecticides for periods up to 24 hr, which show that fish brain enzyme, acetylcholine esterase (AChE), is inhibited with concentrations of 0.1 mg/liter or less, depending upon the compound and the fish species. Eight compounds were studied, including parathion, Diazinon, malathion, demeton, EPN, Chlorthion, Guthion, and Hercules 528. Only parathion produced a continued decrease in AChE activity after removal of the fish from direct exposure. After removal of the fish from exposure to the other insecticides, the time required to regenerate the brain AChE to normal levels varied greatly, up to 40 days or more.

Rachel Carson, in her book *Silent Spring*, which appeared in 1962 (*82*), reached a wide audience and in effect wakened the world to the dangers of the unintelligent use of toxic pesticides. In a well-documented study, Carson showed that the lethal action of toxic sprays cannot be limited to the target pests. The spraying of crops, forests, and lakes for the destruction of insect pests kills not only the offending insect but many other insects; birds that feed upon the insects; fish that feed upon the insects; waterfowl; and wild and domestic animals that feed upon the birds and fish.

According to Carson, DDT and the other related chlorinated hydrocarbons are fat soluble and are accumulated and stored in organs rich in fatty substances, such as the adrenals, testes, and thyroid; in the liver and kidneys, and in the fat of the large, protective mesenteries that enfold the intestines. An intake of as little as 0.1 ppm in the diet results in storage of about 10 to 15 ppm, an increase of 100-fold or more. In animal experiments, 3 ppm of DDT has been found to inhibit an essential enzyme in heart muscle; 5 ppm has brought about necrosis or disintegration of liver cells; and only 2.5 ppm of the closely related chemicals dieldrin and chlordane did the same. One of the most sinister features of DDT and the related chlorinated hydrocarbons is the way they are passed on from one organism to another, through all the links of the food chain. For example, fields of alfalfa are dusted with DDT; meal is later prepared from the alfalfa and fed to hens; the hens lay eggs that contain DDT. Or the hay, containing residues of 7 to 8 ppm, may be fed to cows. The DDT will turn up in the milk in the amount of about 3 ppm, but in butter made from this milk the concentration may run to 65 ppm. As tested on quail and pheasants, dieldrin is about 40 to 50 times as toxic as DDT; aldrin is still more toxic; endrin is the most toxic of all, being about five times as poisonous as dieldrin. Endrin is 15 times as poisonous as DDT to mammals, 30 times as poisonous to fish, and about 300 times as poisonous to some birds.

Carson described the application of DDD, a close relative of DDT, to the waters of Clear Lake, California, to destroy a small gnat that had become a nuisance to fishermen. The chemical was applied with care so as to produce a concentration of about $\frac{1}{70}$ ppm in the waters of the lake. Control of the gnats was good at first, but 5 years later the treatment was repeated at a concentration of $\frac{1}{50}$ ppm. The destruction of the gnats was thought to be complete, but 3 years later a third application was required at $\frac{1}{50}$ ppm. Following the second application, the western or swan grebes on the lake began to die. After the third application, fatty tissues of one of the grebes was examined and found to contain 1,600 ppm of DDD. Plankton organisms were found to contain about 5 ppm; plant-eating fishes, from 40 to 300 ppm; a brown bullhead, 2,500 ppm. One of the most astonishing discoveries was that no trace of DDD could be found in the water shortly after the last application of the chemical. The poison had not really left the lake; it had merely gone into the fabric of the life the lake supports.

It should thus be evident that the use of the carbon chloroform extract (CCE) test, as a safeguard against the intrusion of excessive amounts of chlorinated hydrocarbons in unfiltered drinking waters, is of questionable value. These poisons might be present in an unfiltered water supply in dangerous concentrations, with all the chemical concentrated in the aquatic microorganisms and none of it remaining dissolved in the water. Under such circumstances, the chloroform extract would not contain the toxic material.

The organic phosphorus insecticides have the ability to destroy enzymes. Their target is the nervous system, whether the victim is an insect or a warm-blooded animal. A protective enzyme called cholinesterase prevents the buildup, under normal conditions, of acetylcholine in dangerous amounts, which might result in tremors, muscular spasms, convulsions, or death. Repeated exposure of the insect or warm-blooded animal to excessive amounts of organic phosphorus insecticides will lower the cholinesterase level and result in convulsions or death.

Parathion is one of the most widely used of the organic phosphorus insecticides. According to Carson, the State of California reported an average of more than 200 cases of accidental parathion poisoning annually. In many parts of the world, the death rate from parathion is startling: 100 fatal cases in India and 67 in Syria in 1958, and an average of 336 per year in Japan.

One of the few circumstances that saves us from extinction by the use of organic phosphorus insecticides is that these chemicals decompose rapidly. Their residues on the crops to which they are applied are therefore relatively short-lived, compared with the chlorinated hydrocarbons. Nevertheless, residues have been found in the peel of oranges 6 months after treatment with standard dosages.

Carson called attention to other uses of toxic chemicals that are growing rapidly and that are dangerous to the environment and to man. Among these are systemic insecticides, which are applied to seeds or to plants, making them poisonous to insects, and herbicides, which are used to kill undesired plants and include aquatic weed killers for use in lakes and reservoirs. All these chemicals are dangerous to man, both directly and indirectly.

The first exposures to DDT date from about 1945 for civilians, and by the early 1950s a wide variety of pesticidal chemicals had come into use. According to Carson, the death rate in the United States from all types of malignancies of the blood and lymph increased from 11.1 per 100,000 in 1950 to 14.1 in 1960. In all countries, the recorded deaths from leukemia at all ages are rising at a rate of 4 to 5 percent/year. Carson cites several case histories of fatal leukemia resulting from inhalation of pesticides and of the solvents in which they were carried.

Unfortunately, Rachel Carson died in 1964. The widespread impact of her courageous book, resulting in much legislation related to the use of pesticides, has been well documented by Graham (*83*).

REFERENCES

1. G. Berg, "The Virus Hazard in Water Supplies," *J. New England Water Works Assoc.*, *78*, no. 2, 79–104 (1964).
2. C. R. Cox, "Gastroenteritis and Public Water Supplies," A Symposium on Hydrobiology, Madison, Wis., University of Wisconsin Press, 1941, p. 260.
3. J. M. Dennis, "Infectious Hepatitis Epidemic in Delhi, India," *J. Amer. Water Works Assoc.*, *51*, 1288 (1959).
4. S. Kelly and W. W. Sanderson, *Amer. J. Public Health*, *48*, 1323 (1958); *50*, 14 (1960).
5. "Viral Hepatitis, Clinical and Public Health Aspects," rev. ed., Washington, Public Health Service Publ. 435, 1959.
6. S. L. Chang and G. M. Fair, *J. Amer Water Works Assoc.*, *33*, 1705 (1941).
7. S. L. Chang, "Viruses, Amebas and Nematodes and Public Water Supplies," *J. Amer. Water Works Assoc.*, *53*, 288 (1961).
8. *Standard Methods for the Examination of Water and Wastewater*, 13th ed., New York, American Public Health Association Inc., 1971.
9. *Water Quality Criteria*, Report of the National Technical Advisory Committee to the Secretary of Interior, Washington, D.C., Federal Water Pollution Control Administration, Apr. 1, 1968.
10. *Standard Methods for the Examination of Water and Wastewater*, 11th ed., New York, American Public Health Association, Inc., 1960.
11. "Public Health Service Drinking Water Standards," *Public Health Service Publ. 956*, Washington, D.C., U.S. Dept. Health, Education, and Welfare, Public Health Service, 1962.
12. L. Metcalf and H. P. Eddy, *American Sewerage Practice*, Vol. III, New York, McGraw-Hill Book Company, 1935.
13. S. L. Neave and A. M. Buswell, *J. Amer. Water Works Assoc.*, 388 (1927); A. M. Buswell and S. L. Neave, *Illinois State Water Surv. Bull. 30*, 1930; P. H. Mitchell, *General Physiology*, New York, McGraw-Hill Book Company, 1923; A. P. Mathews, *Physiological Chemistry*, New York, Wm. Wood & Co., 1930.
14. M. Bodansky, *Introduction to Physiological Chemistry*, New York, Wm. Wood & Co., 1930.
15. F. W. Gilcreas and S. M. Kelly, "Significance of the Coliform Test in Relation to Intestinal Virus Pollution of Water," *J. New England Water Works Assoc.*, *68*, 255 (1954).
16. T. R. Camp, "Chlorination of Mixed Sewage and Storm Water," *Trans. ASCE*, *127*, 452, (1962), Pt. III.
17. S. S. Negus, *J. Amer. Water Works Assoc.*, *30*, 242 (1938).
18. L. T. Fairhall, *J. New England Water Works Assoc.*, *55*, 400 (1941).
19. *Water Quality and Treatment*, 2nd ed., New York, American Water Works Association, Inc., 1950; 3rd ed., McGraw-Hill Book Company, 1971.
20. G. C. Whipple, G. M. Fair, and M. C. Whipple, *Microscopy of Drinking Water*, New York, John Wiley & Sons, Inc., 1927.
21. H. C. Sherman, *The Chemistry of Food and Nutrition*, New York, Macmillan Publishing Co., Inc., 1932.
22. F. E. Hale, *J. Amer. Water Works Assoc.*, *27*, 1199 (1935).

23. *U.S. Geol. Surv. Water Supply Papers 1299* and *1300* (1952), and *1460A* (1954).
24. J. P. Linduska and E. W. Surber, *U.S. Fish Wildlife Serv. Circ. 15*, (1948); Z. E. Parkhurst and H. E. Johnson, *Progressive Fish Culturist, 17*, 112 (1955); C. Henderson and O. H. Pickering, *Trans. Am. Fisheries Soc., 87*, 39 (1957).
25. H. M. Bosch et al., "Methemoglobinemia and Minnesota Well Supplies," *J. Amer. Water Works Assoc., 42*, 161 (1950).
26. J. D. Orgeron et al., "Methemoglobin from Eating Meat with High Nitrite Content," *U.S. Public Health Rept. 72*, 189 (1957).
27. H. Stoof and L. W. Haase, *Vom Wasser, 12*, 111 (1937).
28. J. T. Miller and H. G. Byers, *Ind. Eng. Chem. News Ed., 13*, 456 (1935).
29. H. T. Dean and E. Elvove, *Amer. J. Public Health, 26*, 567 (1936).
30. R. A. Trelles, *Bol. obras sanit. nacion., 2*, 367 (1938).
31. F. J. McClure, *Nat. Inst. Health Bull. 172* (1939).
32. H. T. Dean, P. Jay, F. A. Arnold, and E. Elvove, *U.S. Public Health Rept. 56*, 365 (1941).
33. H. T. Dean, F. A. Arnold, and E. Elvove, *U.S. Public Health Rept. 57*, Aug. 1942.
34. H. C. Hodge and F. A. Smith, *American Association for the Advancement of Science*, Washington, D.C., 1954.
35. K. Roholm, "Fluorine Intoxication: A Clinical Hygienic Study," London, Lewis and Co., 1937.
36. Department of Food Technology, Massachusetts Institute of Technology, *Boston Sunday Herald*, May 17, 1959, p. 78C.
37. P. Keyes and R. Fitzgerald, National Institute of Dental Research, *Boston Sunday Herald*, July 16, 1961, p. 36A.
38. *House Rept. 3254*, "Investigation of the Use of Chemicals in Food Products," the Delaney Committee, Government Printing Office, Washington, D.C., 1951.
39. Hearings, Special Commission on the Condition of Dental Health of the Commonwealth of Massachusetts, House No. 3902, Dec. 1967.
40. *Op. cit.*, Minority Report.
41. A. P. Mathews, *Physiological Chemistry*, 5th ed., New York, Wm. Wood & Co., 1930, p. 333.
42. H. E. Stokinger and R. L. Woodward, "Toxicological Methods for Establishing Drinking Water Standards," *J. Amer. Water Works Assoc., 50*, 515 (1958).
43. L. T. Fairhall, *Industrial Toxicology*, Baltimore, The Williams & Wilkins Company, 1957.
44. R. D. MacKenzie et al., "Hexavalent and Trivalent Chromium Administered in Drinking Water to Rats," *Amer. Med. Assoc. Arch. Ind. Health, 18*, 232 (1958).
45. J. C. Morris, "Disinfection of Water," *The Sanitarian, 16*, 221 (Mar.–Apr. 1954).
46. Federal Radiation Council, Repts. 1 and 2, "Background Material for the Development of Radiation Protection Standards," Washington, D.C., Government Printing Office, 1960 and 1961.
47. Federal Radiation Council, "Radiation Protection Guides for Federal Agencies," 25 FR 4402-4403, May 18, 1960; *Federal Register* (Sept. 26, 1961).
48. International Commission on Radiological Protection, "Report of Committee on Permissible Dose for Internal Radiation," New York, Pergamon Press, Inc., 1959.
49. National Committee on Radiation Protection, "Maximum Permissible Body Burden and Maximum Permissible Concentrations of Radionuclides in Air and in Water for Occupational Exposure." *National Bureau of Standards Handbook 69*, Washington, D.C., Government Printing Office, June 4, 1959.
50. A. Rivera-Cordero, "The Nuclear Industry and Air Pollution," *Env. Sci. & Tech., 4* (1970).
51. *Man's Impact on the Global Environment*, Report of the Study of Critical Environ-

mental Problems, sponsored by Massachusetts Institute of Technology, Cambridge, Mass., The MIT Press, 1970.

52. G. Walton, "Effectiveness of Water Treatment Processes as Measured by Coliform Reduction," Washington, D.C., *Public Health Service Publ. 898*, 1961.

53. G. C. Whipple, G. M. Fair, and M. C. Whipple, *Microscopy of Drinking Water*, 4th ed., New York, John Wiley & Sons, Inc., 1927.

54. Private communication from Barbara Hawkes, Water Chemist, Billerica, Mass., Oct. 30, 1961.

55. Private communication from W. H. Green, Jr., Senior Bacteriologist, Metropolitan District Commission, Oct. 27, 1960.

56. T. A. Olson, "Water Poisoning—A Study of Poisonous Algae Blooms in Minnesota," abst. *Amer. J. Public Health, 50*, 883 (1960).

57. T. A. Olson, private communication, Sept. 6, 1961.

58. T. J. Smayda, in *The Encyclopedia of Oceanography*, R. W. Fairbridge, ed., New York, Van Nostrand Reinhold Company, 1966.

59. New England Marine Resources Information Program, Narragansett, R.I., Nov. 1972.

60. A. D. Heggie et al., "An Outbreak of Summer Febrile Disease Caused by Coxsackie B-2 Virus," *Amer. J. Public Health, 50*, 1342 (1960).

61. *J. Amer. Med. Assoc.* (Feb., 1972).

62. *The Boston Herald-Traveler*, Feb. 7, 1972.

63. "Report of the Joint Committee on Bathing Places—Conference of State Sanitary Engineers and American Public Health Association," 10th ed., New York, American Public Health Association, Inc., 1957.

64. J. E. McKee, "Report on the Disinfection of Settled Sewage," California Institute of Technology, April 1957.

65. H. W. Wolf, "The Coliform Count as a Measure of Water Quality," in *Water Pollution Microbiology*, R. Mitchell, ed., New York, John Wiley & Sons, Inc., 1972.

66. W. L. Mallmann, "Streptococcus as an Indicator of Swimming Pool Pollution," *Amer. J. Public Health, 18*, 771 (1928).

67. G. O. Tapley and M. W. Jennison, "Swimming Pool Sanitation: Neisseria Catarrhalis as an Index of Pollution," Madison, Wis., University of Wisconsin Press, 1941, p. 355.

68. T. R. Camp, "Chlorination of Mixed Sewage and Storm Water," *J. San. Eng. Div., Am. Soc. Civil Engrs. 87* (Jan. 1961; May 1961; Sept. 1961).

69. T. R. Camp and R. H. Culver, "Objectives and Standards for Disinfection," Second Rudolfs Research Conference, New Brunswick, N.J., Rutgers University, June 1961.

70. S. Brackett, "Schistosome Dermatitis and Its Distribution," A Symposium on Hydrobiology, Madison, Wis., University of Wisconsin Press, 1941, p. 360.

71. D. B. McMullen, "Methods Used in the Control of Schistosome Dermatitis in Michigan," A Symposium on Hydrobiology, Madison, Wis., University of Wisconsin Press, 1941, p. 379.

72. *Readers Digest* (Aug. 1961).

73. F. E. McJunkin, quoted in *ESE Notes*, Department of Environmental Sciences and Engineering, School of Public Health, University of North Carolina, Chapel Hill, *8*, 4 (July 1971).

74. Atomic Energy, *Federal Register* (Jan. 29, 1957).

75. R. Tollefson, "Fisheries Statistics in Evaluating Claims of Pollution," *J. San. Eng. Div., Amer. Soc. Civil Engrs.*, *87*, 11 (May 1961).

76. A. M. Beeton, "Eutrophication of the St. Lawrence Great Lakes," *Limnol. Oceanog.*, *10*, no. 2, 240–254 (1965).

77. C. M. Tarzwell, "Water Quality Criteria for Aquatic Life," unpublished paper, 1957.

78. C. M. Tarzwell, "Dissolved Oxygen Requirements for Fishes," and Discussions, Oxygen Relationships in Streams, *W58-2, Tech. Rept.* U.S. Public Health Service 1958, pp. 15–24.

79. H. J. Webb, "Water Pollution Resulting from Agricultural Activities," *J. Amer. Water Works Assoc., 54,* 83 (1962).

80. F. Graham, Jr., *Disaster by Default—Politics and Water Pollution,* New York, M. Evans and Co., Inc., and Philadelphia, J. B. Lippincott Co., 1966, pp. 107–135.

81. C. M. Weiss, "Response of Fish to Sub-lethal Exposures of Organic Phosphorous Insecticides," *Sewage Ind. Wastes, 31,* 580 (1959).

82. R. Carson, *Silent Spring,* Boston, Houghton Mifflin Company, 1962.

83. F. Graham, Jr., *Since Silent Spring,* Boston, Houghton Mifflin Company, 1970.

84. A. P. Black et al., *Amer. J. Public Health, 60,* 3 (Mar. 1970).

CHAPTER 7

Standards of Water Quality for Agricultural and Industrial Uses

7-1. AVAILABLE FRESHWATER SUPPLY IN THE UNITED STATES

According to Hoak (*1*), the annual rainfall on the continental United States averages about 30 in., of which 21.4 in. or 71 percent returns to the atmosphere through evapotranspiration. The remaining 29 percent, averaging about 1,220 bgd (billion gallons per day), runs off to streams or is added to the groundwater supply. The use of water (not including hydropower) in the United States as of 1954 was distributed as shown in Table 7-1, which was adapted by Hoak from the "Report of the Senate Select Committee on National Water Resources" (1961). Hydropower generation (*2*) in 1950 used about 1,150 bgd, but much of this water was used several times, and substantially all the water was returned to the watercourses unchanged, except for its loss of potential energy and, in some cases, its deterioration in quality resulting from loss of dissolved oxygen by benthal decomposition on the bottom of the reservoirs.

The figures in Table 7-1 indicate that about 60 percent of the water withdrawn for irrigation is lost by evapotranspiration, and that this represents about 95 percent of the total consumptive use of water. Of the water withdrawn for municipal use, only about 12 percent is consumed; for manufacturing, only about 9 percent; for mining, about 20 percent; for steam-electric power, only about 0.5 percent.

Table 7-1
Estimated Water Use in 1954

	Total Use		Consumed		Returned	
	bgd	%	bgd	%	bgd	%
Irrigation	176.1	58.6	103.9	94.9	72.2	37.8
Municipal	16.7	5.6	2.1	1.9	14.6	7.7
Manufacturing	31.9	10.6	2.8	2.5	29.1	15.2
Mining	1.5	0.5	0.3	0.3	1.2	0.6
Steam-electric power	74.1	24.7	0.4	0.4	73.7	38.7
U.S. total	300.3	100.	109.5	100	190.8	100
Percent	100		36.4		63.6	

It is evident from Table 7-1 that irrigation is by far the largest consumptive use, accounting for about 95 percent of the total. All other uses result in large returns to the streams; about 64 percent of the total withdrawals are returned for possible reuse. Hoak (2) shows a considerable in-plant reuse of water by industry in 1954. The actual withdrawal of about 32 bgd would have been about 60 bgd without in-plant reuse.

Each use of water generally results in some impairment of its quality for reuse. Municipal and industrial uses result in pollution and temperature increases, irrigation use results in silting and increases in dissolved solids, and steam power generation results in temperature increases.

The Senate Select Committee on Natural Water Resources (3) estimates that it may ultimately be economically feasible to develop about 650 bgd in the United States on a sustained-yield basis for all water uses. Since this is more than twice the 1954 total withdrawal and about six times Hoak's estimate of the 1954 consumptive use, there should be no national water shortage in the foreseeable future. Water must be distributed and reused more efficiently, however, as the demand increases, and more adequate treatment will be required of both water and wastewater. The Senate Select Committee (3) estimates that by the year 2000 the consumptive use may increase to 156.3 bgd, of which 81 percent or 126.3 bgd will be for irrigation and 13.3 percent or 20.8 bgd will be for manufacturing.

The wasteful use of water in the United States indicates that it is regarded as a substance of low value, particularly in regions of plentiful supply. Renshaw (4) has estimated "the maximum amount people would be willing to pay for the use of water in any amount or direction rather than forego the amount or use entirely." Renshaw's figures, which are based on actual costs during 1950, are shown in Table 7-2, and show a very large gap between the value of water for domestic and industrial use and its value for all other uses studied. The value of

Table 7-2
Use Value of an Acre-Foot[a] of Water

Use	Mean ($)	Maximum ($)
Domestic	100.19	235.66
Industrial	40.73	163.35
Irrigation	1.67	27.04
Hydropower	0.71	5.90
Waste disposal	0.63	2.56
Inland navigation	0.05	1.17
Commercial fisheries	0.025	1.06

[a] 1 acre-foot = 0.326 million gallons.

water for both industrial and agricultural use depends upon the amount of water required per unit of product and the sales value of the product; hence, the value varies widely. Very costly water treatment is justified for some industrial products, whereas any cost of treatment may be prohibitive for other industrial uses and for irrigation.

Many articles have been published in recent years expressing alarm about the future adequacy of the water resources of the United States. One result of this concern has been the expenditure by the Office of Saline Water (OSW) of the U.S. Department of the Interior of considerable sums for research and development on the recovery of freshwater from seawater and brackish water. This has always been feasible, but conversion costs have been so high that it has been economically impractical except where cost is not a primary consideration. The research by OSW has achieved a considerable cost reduction, but it is unlikely that the cost of desalinization will ever be low enough to result in a significant addition to our supply of freshwater. The average U.S. cost of municipal water at the consumer's tap is 12.3 cents/1,000 gal, according to Hoak (*1*), of which the average cost of the supply from river water is only about 5 cents. The lowest cost of conversion from seawater, thus far, appears to be about 60 cents/1,000 gal, and from brackish water, with total dissolved solids about 7,000 ppm, the cost is about 30 cents/1,000 gal.

As pointed out in Article 5 of Chapter 3, research and development (see refs. *21* and *22* in Chapter 3) is in progress on the use of hexadecanol and octadecanol to reduce evaporation from water surfaces and transpiration from plants. Inasmuch as about 71 percent of the total average rainfall in the United States, or about 3,000 bgd, is lost through evapotranspiration (principally transpiration), this approach should be very much more fruitful than desalinization in increasing our available water supply. A 10 percent reduction in these losses would add about 50 percent to the sustained yield of about 650 bgd estimated by the Senate Select Committee as being ultimately economically feasible.

A reduction in evapotranspiration from land areas may also be accomplished by means of an oil-based mulch (5) recently developed. This mulch promises to increase crop yields and reduce evaporation from land areas.

7-2. AGRICULTURAL USE

The largest single agricultural use of water is for irrigation. Other uses are for watering and care of livestock and poultry, and for cleansing and other general purposes. These other uses are relatively very small. Water suitable for drinking is usually satisfactory, but for some uses special precautions must be taken. For example, Garibaldi and Bayne (6) have found that iron in excess of about 1.0 mg/liter in water used in egg-washing machines results in increased spoilage of eggs. The iron effect was probably due to the reversal of the protective action of the iron-building protein conalbumin.

The processing and packing of farm products for market, as in dairies, canneries, and packing houses, are usually considered as industrial operations. The quality of water required for such processes will be discussed with industrial uses.

The quality of irrigation water appears to be governed by four characteristics (7): (1) total concentration of soluble salts, (2) relative proportion of sodium to other cations, (3) concentrations of boron or other elements that may be toxic, and (4) under some conditions, the bicarbonate concentration as related to the concentration of calcium plus magnesium.

The Department of Agriculture customarily reports soluble salts in milliequivalents per liter (meq/liter) and total soluble salts in terms of the conductivity of the water in microsiemens (i.e., micromhos) per centimeter (μS/cm). From Chapter 4, the molal concentration of any soluble constituent, A, is

$$[A] = \frac{c'_A}{M} \times 10^{-3} \tag{7-1}$$

where c'_A is the concentration in mg/liter and M is the molecular weight. The concentration in meq/liter is

$$c''_A = c'_A \frac{z_A}{M} = z_A [A] \times 10^3 \quad \text{or} \quad [A]'' \tag{7-2}$$

where z_A is the valence of the substance or ion, A.

The ratio of the electrical conductivity of a water in μS/cm at 25°C to the cation concentration in meq/liter is about 100, as shown by Figure 7-1, prepared by the Department of Agriculture (7). The ratio may be as low as 80 for bicarbonate or sulfate waters, high in calcium and magnesium, or as high as 110 for chloride waters, high in sodium. The ratio of the total dissolved solids in parts

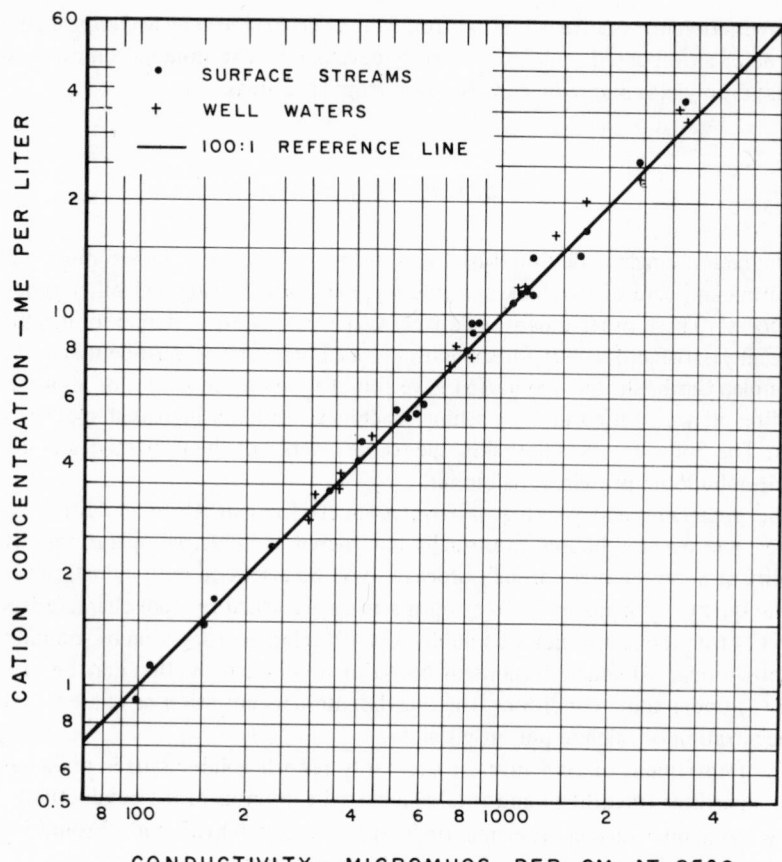

Figure 7-1
Relation between cation concentration and electrical conductivity.

per million to the conductivity in μS/cm at 25°C is about 0.64, as shown by Figure 7-2, prepared by the Department of Agriculture (7).

According to Handbook No. 60 (7), nearly all irrigation waters that have been used successfully for a considerable time have conductivities of less than 2,250 μS/cm. Saline soils are those in which the saturation extract is greater than 4,000 μS/cm. In the absence of salt accumulation from groundwater, the conductivity of the saturation extract of a soil is usually from 2 to 10 times the conductivity of the applied irrigation water. This increase in salt concentration results from continual moisture extraction by plant roots and from evaporation. Hence, the use of waters of moderate to high salinity may result in saline conditions even where drainage is good.

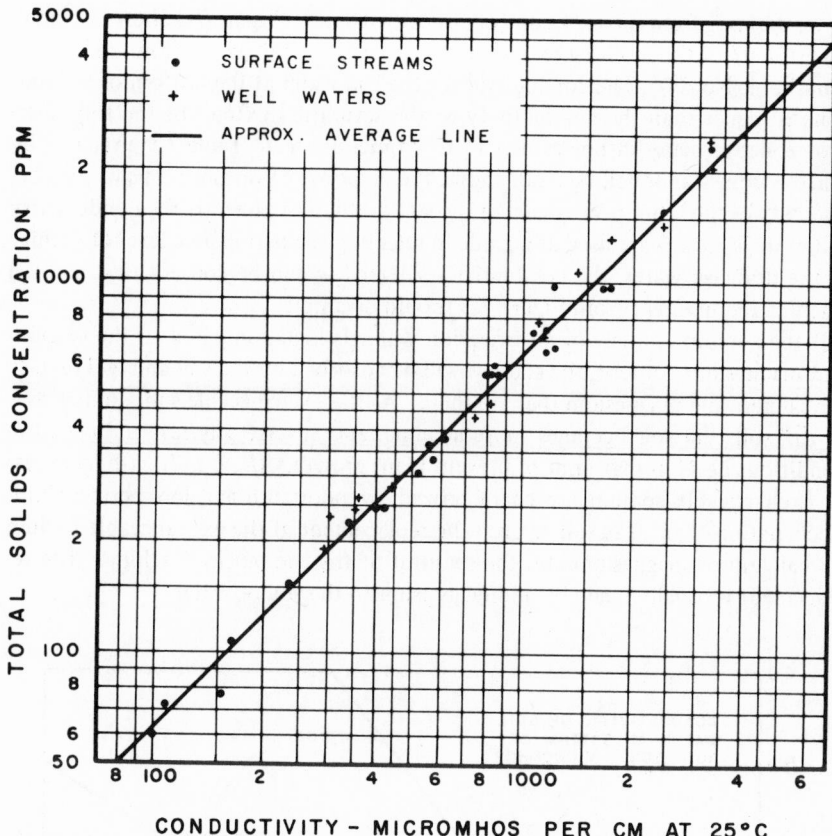

Figure 7-2
Relation between total dissolved solids and electrical conductivity.

In general, waters with conductivities below 750 μS/cm are satisfactory for irrigation insofar as salt content is concerned, although salt-sensitive crops may be adversely affected by the use of irrigation waters with conductivities from 250 to 750 μS/cm, depending upon leaching and drainage conditions. Waters in the range of 750 to 2,250 μS/cm are widely used with satisfactory crop growth under good management and drainage, but saline conditions will develop if leaching and drainage are inadequate.

The steady-state leaching requirement for soils where no precipitation of salts occurs is directly related to the conductivity of the irrigation water and the permissible conductivity of the water draining from the root zone. The *leaching requirement* is defined as the percentage of the applied irrigation water that must pass through and beyond the root zone to maintain the salt content of the water draining from the root zone at a specified value. Figure 7-3 shows a graphical estimate of the leaching requirement prepared by Wilcox (*8*) from the data of Handbook No. 60. The conductivity of the soil water at the bottom of the root zone is higher than the conductivity of the saturation extract of the soil in the root zone; in many instances it is $1\frac{1}{2}$ to 2 times as great. Table 7-3, prepared by Wilcox from Handbook No. 60, shows the response of various crops to the conductivity of the saturation extract of the soil in the root zone. If the conductivity values in Table 7-3 are used as a guide in selecting the permissible level of salinity in the drainage water, the conductivity selected would be conservative, and the leaching requirement from Figure 7-3 should be ample.

If the sodium content of an irrigation water is high compared to the calcium and magnesium content, the sodium will be adsorbed by the soil and will replace the calcium and magnesium that are there. As the exchangeable sodium increases in the soil, the soil becomes more alkaline, and adverse physical and chemical conditions develop that limit or prevent plant growth. *Alkali soils* with an excess of exchangeable sodium are characterized by poor tilth and low permeability. Reclamation of an alkali soil involves the replacement of the exchangeable sodium by calcium or magnesium and the removal of the sodium by leaching. The replacement is usually made by adding gypsum to the soil or water.

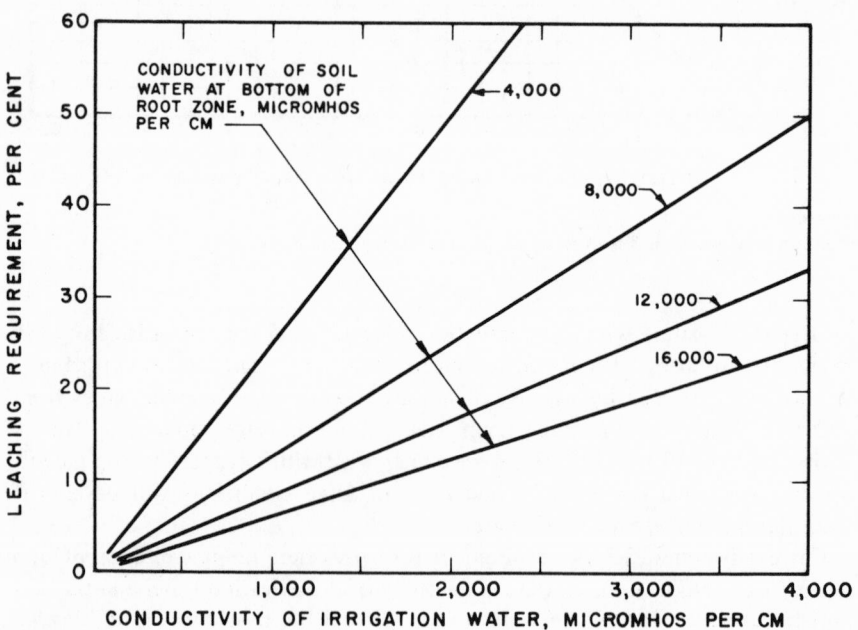

Figure 7-3
Estimate of leaching requirement for irrigation water. [After Wilcox (*8*).]

Table 7-3
Crop Response to Salinity

Conductivity at 25°C of the Saturation Extract (µS/cm)	Related Crop Response
0–2,000	Salinity effects mostly negligible.
2,000–4,000	Restricted yields of the more sensitive crops, such as
	avocado, citrus, strawberry, peach, apricot, almond, plum, prune, apple, pear;
	beans, celery, radish;
	most clover species, meadow foxtail.
4,000–8,000	Yields of many crops restricted.
	The more sensitive crops in this group include
	grape, cantaloupe;
	cucumber, squash, peas, onion, carrot, bell pepper, potato, sweet corn, lettuce.
	The more tolerant crops in this group include
	olive, fig, pomegranate;
	cauliflower, cabbage, broccoli, tomato;
	oats, wheat, rye, alfalfa, Sudan grass, Dallis grass, strawberry clover, perennial rye grass, sweet clovers;
	flax, corn, rice.
8,000–16,000	Only salt-tolerant crops yield satisfactorily. These include
	date palm;
	asparagus, kale, garden beets;
	birdsfoot trefoil, barley, many species of wheat grasses and wild ryes, Rhodes grass, Bermuda grass, saltgrass;
	some varieties of cotton;
	sugar beet.
More than 16,000	Satisfactory yields from only a few very salt-tolerant species: certain native range plants.

The *sodium* or *alkali hazard* of an irrigation water is conveniently measured by the *SAR* or *sodium adsorption ratio*, where

$$SAR = \frac{[Na^+]''}{\sqrt{\dfrac{[Ca^{2+}]'' + [Mg^{2+}]''}{2}}} \tag{7-3}$$

in which the ionic concentrations are in meq/liter. Irrigation waters of low mineral content may be used with SAR values up to about 26; if the mineral content is high, the SAR values must be lower than about 10.

Figure 7-4 is a diagram from Handbook No. 60 for the classification of irriga-

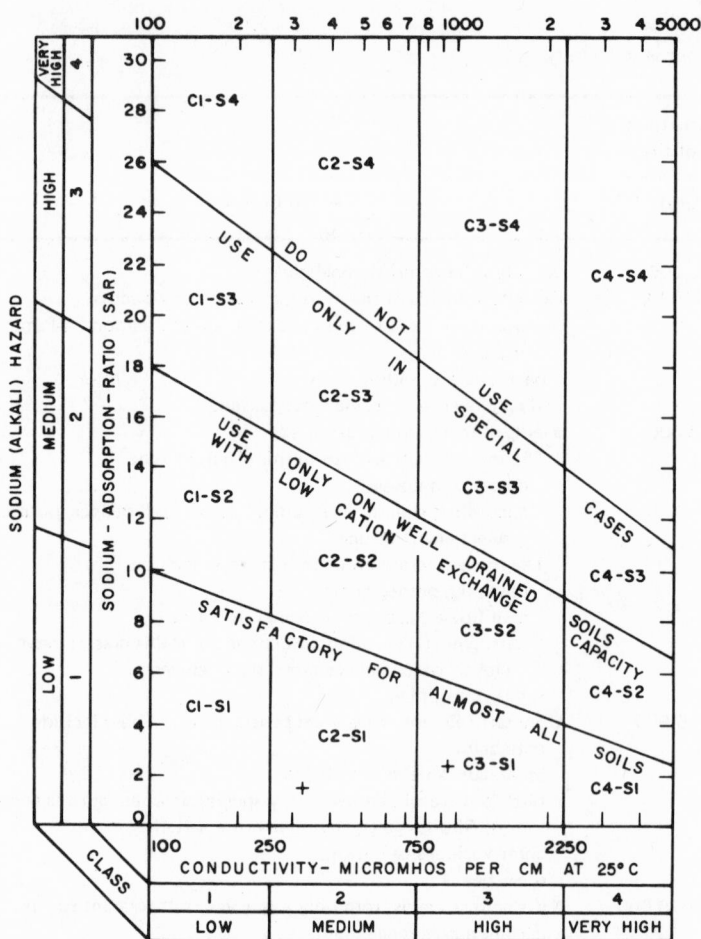

Figure 7-4
Classification of irrigation waters.

tion waters based on the salinity and SAR. Low-sodium water, S1, can be used on almost all soils; however, sodium-sensitive crops, such as stone-fruit trees and avocados, may accumulate injurious concentrations of sodium unless some gypsum is added. Medium-sodium water, S2, will present an appreciable sodium hazard in fine-textured soils having high cation-exchange capacity, especially under low-leaching conditions, unless gypsum is present in the soil. This water may be used on coarse-textured or organic soils with good permeability. High-sodium water, S3, may produce harmful levels of exchangeable sodium in most soils and will require special soil management: good drainage, high leaching, and additions of organic matter. Gypsum-bearing soils may not develop harmful levels of exchangeable sodium from this water. Chemical treatment may be required to replace the exchangeable sodium; but if the salinity is too high, such treatment may not be feasible. Very high sodium water, S4, is generally unsatisfactory for irrigation purposes except at low salinity, where the use of gypsum may be effective.

The Public Health Service Drinking Water Standards gives no consideration to the suitability of municipal drinking water for lawn sprinkling and other agricultural uses, although such uses are increasing rapidly throughout the United States. No limit has been set on sodium concentration, and the limit of 250 ppm for chlorides was based upon taste. The limit of 500 ppm for total dissolved solids was recommended with full realization that this figure will be exceeded in regions where the cost of treatment whose sole purpose is to reduce the total solids is prohibitive.

Most municipal drinking waters contain only low concentrations of both sodium and chloride unless salt is added as an industrial waste or unless seawater encroaches on the freshwater supply. A study of Figure 7-4 reveals that the limit of 250 ppm of chlorides recommended in the drinking water standards is too high for some waters, if they are to be satisfactory for sprinkling and irrigation. For example, if sodium chloride to the extent of 250 ppm of chloride is added to water having a hardness of only 10 ppm, all as calcium bicarbonate, the resulting conductivity will be about 570 μS/cm and the SAR will be about 13.4. Figure 7-4 indicates that this water will have a high sodium hazard. If the chlorides are increased to 250 ppm by seawater encroachment, the resulting conductivity will be about 735 μS/cm and the SAR will be about 12.5. Figure 7-4 indicates a similar high sodium hazard. To keep this soft water within the low-sodium-hazard zone with an SAR of 9, the chlorides must be limited to about 100 ppm.

Some general conclusions may be drawn regarding the suitability of waters of higher mineral content for sprinkling and irrigation without adding gypsum, if it be assumed that nearly all the hardness is due to calcium and nearly all the chlorides result from sodium chloride. For waters with a total solids content of 480 ppm (conductivity 750), the SAR should be less than 6, the hardness should be at least 90 ppm, and the chloride content should not exceed 200 ppm. For waters with a total solids content of 1,440 ppm (conductivity 2,250), the SAR should be less than 4, the hardness should be at least 550 ppm, and the chlorides should not exceed 330 ppm. If the chlorides are to be limited to 250 ppm, as recommended for drinking waters, the SAR should be less than 5, the hardness should be at least 200 ppm, and the total solids will then be less than 870 ppm (conductivity 1,360). Since many waters of high mineral content have high concentrations of other ions, such as sulfates and magnesium, it is unsafe for a particular water to have a sodium standard based on its chloride concentration.

Boron is a constituent of practically all natural waters, but its concentration is usually very low. Although it is essential to plant growth, boron is exceedingly toxic at concentrations only slightly above optimum. Handbook No. 60 (7) presents analyses of some river waters used for irrigation in the western United States that show boron concentrations ranging from 0.03 to 0.62 ppm. Boron toxicity occurs in limited scattered areas in arid or semiarid regions. The boron content of normal, mature leaves of plants, such as citrus, avocados, walnuts, figs, grapes, and cotton, and of alfalfa tops is about 50 ppm. Boron contents of 20 ppm or less indicate deficiency; values above 250 ppm are usually associated with boron toxicity. Stone-fruit trees, apples, and pears do not accumulate high concentrations of boron in their leaves, although these plants are sensitive to excess boron.

The effect of a given concentration of boron in the irrigation water on the boron content of the soil solution will be conditioned by soil characteristics and by management practices that influence the degree of boron accumulation in the soil. According to Handbook No. 60 (7), the approximate safe limit for sensitive crops is 0.7 ppm boron in the soil saturation extract, 0.7 to 1.5 ppm are marginal, and more than 1.5 ppm appears to be unsafe. Crops with greater tolerance can withstand higher concentrations. Scofield (9) has proposed permissible limits of boron in irrigation waters ranging from 0.33 to 1.25 ppm for sensitive crops, 0.67 to 2.50 ppm for semitolerant crops, and 1.00 to 3.75 ppm for tolerant crops. The relative tolerance to boron of certain plants, taken from Handbook No. 60, is shown in Table 7-4.

In irrigation waters containing high concentrations of bicarbonate, there is a tendency for calcium and magnesium to precipitate as carbonates as the solution becomes more concentrated in the root zones. The concentrations of calcium and magnesium are thus reduced and the relative amount of sodium is increased. The excess bicarbonate is measured by the *residual sodium carbonate*, which is defined as the concentration of carbonate and bicarbonate minus the concentration of calcium and magnesium, all expressed in meq/liter. According to Handbook No. 60 (7), waters with more than 2.5 meq/liter of residual sodium carbonate are not suitable for irrigation purposes. Waters containing 1.25 to 2.5 meq/liter are marginal, and those containing less than 1.25 meq/liter of residual sodium carbonate are probably safe.

Unchlorinated sewage and the effluents of sewage-treatment plants have long been used for irrigation of crops. Food from such crops may be ingested with comparative safety if it is well cooked, but there is danger of enteric waterborne bacterial and virus infections to the handlers of the crops and to persons who ingest uncooked food. Diseases caused by round worms, tapeworms, hookworms, and schistosomes may also be transmitted in this manner. Most of the former diseases are contracted through the ingestion of water or food containing the causative organisms, but the worm diseases are contracted primarily by wading in polluted waters or by walking on land subjected to irrigation with unchlorinated

Table 7-4
Relative Boron Tolerance of Certain Plants
(In each vertical grouping, the plants first named are considered to be more tolerant;
the last named, more sensitive.)

Tolerant	Semitolerant	Sensitive
Athel (*Tamarix aphylla*)	Sunflower (native)	Pecan
Asparagus	Potato	Walnut (black, Persian, English)
Palm (*Phoenix canariensis*)	Cotton (acala and pima)	Jerusalem artichoke
Date palm (*P. dactylifera*)	Tomato	Navy bean
Sugar beet	Sweetpea	American elm
Mangel	Radish	Plum
Garden beet	Field pea	Pear
Alfalfa	Ragged Robin rose	Apple
Gladiolus	Olive	Grape (sultanina and malaga)
Broad bean	Barley	Kadota fig
Onion	Wheat	Persimmon
Turnip	Corn	Cherry
Cabbage	Milo	Peach
Lettuce	Oat	Apricot
Carrot	Zinnia	Thornless blackberry
	Pumpkin	Orange
	Bell pepper	Avocado
	Sweet potato	Grapefruit
	Lima bean	Lemon

sewage. It is, therefore, desirable that sewage or sewage-treatment-plant effluent to be used for irrigation be subjected to heavy chlorination.

Since sewage is the spent water supply of a community, it contains, in addition to the dissolved constituents in the tap water, all substances added during the use of the water. Stone and Merrell (*10*) have studied the increase in mineral content of the sewage, comparing it with the water, of about 25 municipalities. The normal increase in total dissolved mineral solids was found to vary from about 121 to 450 ppm; mineralization resulting from oil-brine disposal from oil fields and refineries ranged up to 1,700 ppm. Seawater infiltration into low-lying sewers increased the total solids up to more than 6,000 ppm, with chlorides up to 5,000 ppm. The boron content of the municipal water ranged from 0.04 to 0.4 ppm; the sewage contained 0.04 to 1.0 ppm. The increase in sodium varied from less than 10 percent to more than 45 percent with larger increases from oil brines and seawater infiltration. The studies of Stone and Merrell indicate that sewage or sewage-treatment-plant effluent is normally quite satisfactory for irrigation from the standpoint of mineral content, provided excessive oil brines or seawater infiltration are excluded.

7-3. INDUSTRIAL USE

Many industries have set up standards of water quality for specific uses within the industry, but the first effort to formulate comprehensive standards of quality for industrial use appears to have been made by the Committee on Quality Tolerances of Water for Industrial Uses of the New England Water Works Association. The tolerances, specified in the 1940 progress report (11) of this committee, for boiler feed water are given in Table 7-5 and for cooling and processing in Table 7-6. The committee stated that since "the allowable limits of impurities in process water are affected by variations in the process and in the quality of product desired, the committee has endeavored to select values that correspond to normal practice. It is recognized that not a few industrial plants will be found in which water of higher content of impurities than that specified in this report is utilized successfully." For any specific case, the tolerances for process water may be determined by laboratory or plant scale studies in which economic factors are duly considered. The tolerances in Tables 7-5 and 7-6 have been reproduced, with some modifications, in Table II of the "Manual

Table 7-5
Suggested Limits of Tolerances for Boiler Feed Water

	Pressure (psi)			
	0–150	150–250	250–400	Over 400
Turbidity, ppm	20	10	5	1
Color, ppm	80	40	5	2
Oxygen consumed, ppm	15	10	4	3
Dissolved oxygen[a], ppm	1.4	0.14	0	0
Hydrogen sulfide[b], H_2S, ppm	5	3	0	0
Total hardness, ppm as $CaCO_3$	80	40	10	2
Sulfate–carbonate ratio, $\dfrac{\text{ppm } Na_2SO_4}{\text{ppm } Na_2CO_3}$	1:1	2:1	3:1	3:1
Aluminum oxide, Al_2O_3, ppm	5	0.5	0.05	0.01
Silica, SiO_2, ppm	40	20	5	1
Bicarbonate, HCO_3^-, ppm	50	30	5	0
Carbonate, CO_3^{2-}, ppm	200	100	40	20
Hydroxide, OH^-, ppm	50	40	30	15
Total solids[c], ppm	3,000–500	2,500–500	1,500–100	50
pH value, minimum	8.0	8.4	9.0	9.6

[a]Limits applicable only to feed water entering boiler, not to original supply.
[b]Except when odor in live steam would be objectionable.
[c]Depends on design of boiler.

on Industrial Water and Industrial Wastewater" (*12*) of the American Society for Testing Materials, which also includes standard test procedures.

It is usually neither economical nor feasible for a municipality to furnish water that is satisfactory for all industrial users, as well as for domestic consumption; however, two of the most important defects in the quality of water for industry, corrosiveness and hardness, are also important, and should be limited in the public water supply. Corrosiveness results in depreciation and loss in capacity of pipes in the distribution system and in house plumbing and heating systems. Hardness causes scale in domestic plumbing, heating, and hot-water systems, and it results in soap waste in household processes such as laundering, dish washing, cleaning, and bathing.

Ordinary sodium soaps react with hardness to produce *curd*, as illustrated by the following reaction of sodium oleate:

$$2NaCO_2C_{17}H_{33} + Ca^{2+} \longrightarrow \underline{Ca(CO_2C_{17}H_{33})_2} + 2Na^+ \qquad (7\text{-}4)$$

All the hardness-producing metals must be precipitated as curd before a lather can be obtained. The softening of water with soap is both objectionable and expensive. The curd adheres to clothes and turns them gray when ironed, it produces "rings" on bathtubs, and it combines with grease to adhere to dishes during dish washing. Fabric life may be shortened by as much as 25 percent in continued hard-water washing in which curds are produced.

It has been well stated by Hoover (*13*) that "the extravagance of removing hardness with soap can be appreciated when it is realized that a pound of lime costing approximately $\frac{1}{2}$ cent will neutralize as much hardness as 20 pounds of soap costing approximately \$3.00." Prior to 1941 it had been demonstrated (*13*) that the saving in soap for home use alone accomplished by the softening of water with a hardness exceeding about 150 ppm was sufficient to pay for the cost of softening the municipal supply. After World War II the picture changed radically as a result of the development and rapid acceptance of synthetic detergents (*syndets*) in place of soap. Syndets are soluble and do not produce curds in hard water.

Hardness is the most objectionable quality in boiler feed waters because of the formation of scale on the surfaces of tubes and metal plates as the water is evaporated. Boiler scale reduces the heat-transfer efficiency and, thus, is an economic loss. It has been estimated (*14*) that a heat loss of 7 to 16 percent results from scale 0.02 to 0.11 in. thick. Hard water necessitates frequent blowing off of boilers to eliminate sludge and periodic cleaning of tubes to eliminate scale. Corrosiveness of boiler feed water is also a serious defect because it impairs the efficiency and safety of feed-water heaters and boilers. SO_4^{2-}, Cl^-, NO_3^-, and silicates are called *incrustants* because of the tenacity of the

Table 7-6
Suggested Water Quality Tolerances for Industrial Use

Industry or Use	Turbidity (ppm)	Color (ppm)	Hardness (as CaCO₃) (ppm)	Fe (ppm)	Mn (ppm)	Total Solids (ppm)	Alkalinity (as CaCO₃) (ppm)	Odor, Taste	H₂S (ppm)	Other Requirements
Air conditioning	—	—	—	0.5[a]	0.5	—	—	Low	1	No corrosiveness or slime formation.
Baking	10	10	—	0.2[a]	0.2	—	—	Low	0.2	Potable.[b]
Brewing										
Light beer	10	—	—	0.1[a]	0.1	500	75	Low	0.2	Potable. NaCl less than 275 ppm. pH 6.5–7.0.
Dark beer	10	—	—	0.1[a]	0.1	1,000	150	Low	0.2	Potable. NaCl less than 275 ppm. pH 7.0 or more.
Canning										
Legumes	10	—	25–75	0.2[a]	0.2	—	—	Low	1	Potable.
General	10	—	—	0.2[a]	0.2	—	—	Low	1	Potable.
Carbonated beverages	2	10	250	0.2 (0.3[a])	0.2	850	50–100	Low	0.2	Potable. Org. color plus O₂ consumed less than 10 ppm.
Confectionery	—	—	—	0.2[a]	0.2	100	—	Low	0.2	Potable. pH above 7 for hard candy, low for fondant.
Cooling	50	—	50	0.5[a]	0.5	—	—	—	5	No corrosiveness or slime formation.
Distilling										
Gin and spirits	10	—	—	0.1[a]	0.1	500	75	Low	0.2	Potable. NaCl less than 275 ppm. pH 6.5–7.0.

Use										Remarks
Whiskey	10	—	—	0.1^a	0.1	1,000	150	Low	0.2	Potable. NaCl less than 275 ppm. pH 7.0 or more.
Blending	—	—	—	—	—	—	—	—	—	Distilled or special treatment.
Food, general	10	—	—	0.2^a	0.2	—	—	Low	—	Potable.
Ice	5	5	—	0.2^a	0.2	170	—	Low	—	Potable. SiO_2 less than 10 ppm.
Laundering	—	—	50	0.2^a	0.2	—	—	—	—	
Plastics, clear, uncolored	2	2	—	0.02^a	0.02	200	—	—	—	
Paper and pulp										
Groundwood	50	200	180	1.0^a	0.5	—	—	—	—	No grit or corrosiveness.
Kraft pulp	25	15	100	0.2^a	0.1	300	—	—	—	
Soda and sulfite	15	10	100	0.1^a	0.05	200	—	—	—	
High-grade light papers	5	5	50	0.1^a	0.05	200	—	—	—	No slime formation.
Rayon (viscose)										
Pulp production	5	5	8	0.05^a	0.03	100	Total 50, OH 8	—	—	Al_2O_3, SiO_2, Cu less than 8, 25 and 5 ppm, respectively.
Manufacture	0.3	—	55	0.0	0.0	—	—	—	—	pH 7.8 to 8.3.
Tanning	20	10–100	50–135	0.2^a	0.2	—	Total 135, OH 8	—	—	
Textiles										
General	5	20	—	0.25	0.25	—	—	—	—	
Dyeing	5	5–20	—	0.25^a	0.25	200	—	—	—	Constant composition, Al_2O_3 less than 0.5 ppm.
Wool scouring	—	70	—	1.0^a	1.0	—	—	—	—	
Cotton bandage	5	5	—	0.2^a	0.2	—	—	Low	—	

[a] Limit applies both to iron alone and to the sum of iron and manganese.

[b] Potable means complying with drinking water standards of the U. S. Public Health Service.

scales produced by precipitation of the corresponding calcium and magnesium salts.

The tolerances of impurities for steam generation are relatively high for low-pressure boilers, since the introduction of boiler compounds for internal water treatment is usually practicable. With operating pressures above 400 psi, however, water that more nearly approaches deaerated distilled water in quality is desirable. Demineralization plants are now widely used for boiler feed water in modern steam plants. In high-pressure boilers, silica should be very low, because conditions are favorable for the formation of dense silicate scales. The specifications in Table 7-5 for the sulfate–carbonate ratio conform to suggestions of the American Society of Mechanical Engineers, resulting from studies which show that embrittlement of boiler plate may be inhibited by maintaining satisfactory ratios of sulfate to carbonate. Foaming (or priming) in boilers is caused by high concentrations of finely divided solids and is probably accentuated by high concentrations of certain ions.

The supercritical power plants in use today, with temperatures of 1,200°F and pressures of 5,000 psi, require very high water quality standards. Table 7-7 shows typical feed-water specifications (*15*) drawn up for the Philo plant of the Ohio Power Company:

Table 7-7

Total dissolved solids	0.5 ppm max
Silica	0.02 ppm max
Iron	0.01 ppm max
Copper	0.01 ppm max
Dissolved oxygen	0.005 ppm max
pH value	9.5 to 9.6

Water of this high purity is regularly achieved by ion-exchange demineralization and deaeration by well-known methods, followed by careful handling and storage in resistant equipment.

It has been pointed out by Powell (*16*) that the temperature of water to be used for cooling, condensing, and air-conditioning is often of more economic significance than the composition and chemical quality of the water. Ground-waters are usually more satisfactory than surface waters because they are cooler and more uniform in temperature throughout the year.

In water for textile and laundry use and for making high-grade paper, iron is the most objectionable of all common impurities, and manganese is equally objectionable wherever it is found. If, however, iron is present in combination with vegetable organic matter, much higher tolerances are permissible for some uses, since in such combination it does not react with soaps or form oxides so readily.

Discolorations and stains from iron and manganese necessitate limits on these metals in nearly all industrial water, as illustrated in Table 7-6. In many cases the presence of iron and copper is due to corrosion of piping and tanks; hence, the corrosiveness of water is equally significant.

Nordell (*17*) classifies waters used in industrial plants as follows: (1) boiler feed water, (2) cooling water, (3) process water, and (4) general-purpose water. General-purpose water must usually be suitable for drinking. Nordell presents a partial list of 331 different types of industrial plants. Water-treating equipment for boiler feed has been used in plants of nearly all the types listed, and for processes and general purpose in most of the types listed. The types of treatment listed by Nordell are sodium-cation exchange, hydrogen-cation exchange, cold lime or lime–soda softening, hot lime–soda process, demineralization, filtration, and iron and manganese removal. All these treatments may be used in a single industrial plant for different process waters. In addition, there are many other special types of treatment such as deaeration, odor removal, corrosion inhibition by addition of chemicals, and chlorination for control of slimes and barnacles. Industrial water problems are so varied and complex that specific solutions must be sought for each plant. A major concern in water supply for industry is that the quality remain relatively constant. Short-time fluctuations in the concentration of impurities in process water require constant attention and expense (*18*).

REFERENCES

1. R. D. Hoak, "Are We Really Running Out of Water?", *Trusts and Estates Magazine* (Apr. 1962).

2. R. D. Hoak, "Industrial Water Conservation and Reuse," *Tappi, 44*, no. 2 (Feb. 1961).

3. Select Committee on Natural Water Resources, U. S. Senate, 86th Congress, 1961.

4. E. F. Renshaw, *J. Amer. Water Works Assoc., 50*, 303 (1958).

5. *The Lamp*, Standard Oil of New Jersey, Summer 1962.

6. J. A. Garibaldi and H. O. Bayne, "The Effect of Iron on the Pseudomonas Spoilage of Experimentally Infected Shell Eggs," *Poultry Sci., 39*, no. 6 (Nov. 1960).

7. *U.S. Dept. Agric. Handbook 60*, Feb. 1954.

8. L. V. Wilcox, "Classification and Use of Irrigation Waters," *U.S. Dept. Agric. Circ. 969*, Nov. 1955.

9. C. S. Scofield, "The Salinity of Irrigation Water," *Smithsonian Inst. Ann. Rept.*, 1936, pp. 275–287.

10. R. Stone and J. C. Merrell, Jr., "Significance of Minerals in Wastewater," *Sewage Ind. Wastes, 30*, 928 (1958).

11. *J. New England Water Works Assoc., 54*, 261 (1940).

12. *ASTM Spec. Tech. Publ. 148-D*, 1959.

13. C. P. Hoover, "Water Supply and Treatment," Washington, D.C., National Lime Association, 1941.

14. *U.S. Bur. Mines Tech. Paper 218*, Government Printing Office, Washington, D.C., 1920.

15. *Corrosion Reporter*, International Nickel Co., Inc., April 1960.
16. S. T. Powell, "The Treatment of Water for Industrial Boiler Feed Uses," *J. New England Water Works Assoc.*, *52*, 418 (1938).
17. E. Nordell, *Water Treatment for Industrial and Other Uses*, New York, Van Nostrand Reinhold Company, 1961.
18. J. E. McKee and H. W. Wolf, eds., *Water Quality Criteria*, 2nd ed., Resources Agency of California, State Water Quality Control Board, Publ. 3-A, 1963.

CHAPTER 8

Corrosiveness of Water

8-1 ELEMENTS OF THE CORROSION CELL

Water is corrosive to a solid when it tends to dissolve the solid. The solution of a solid mineral such as $CaCO_3$ is accompanied by its dispersion as positive and negative ions in solution. The solution of a solid nonpolar compound is accompanied by its dispersion as nonionized molecules in solution. Neither of these cases involves a transfer of electrons and there is no flow of electricity. The *corrosiveness* of a water to solid mineral compounds and nonpolar compounds depends entirely upon its degree of saturation with the ions or molecules of the compound.

Water is corrosive to a solid metal when it dissolves the metal as positive ions or furnishes constituents that react with the metal at the interface. Either process is *oxidative* and results in the release of negative electrons, which flow backward through the metal, as shown in Figure 8-1 (*1*). No metal can enter solution as positive ions or withdraw constituents from solution unless an equivalent reaction in the reverse direction takes place at some other point in the metal-water interface, for the solution must remain electrically neutral with equivalent concentrations of positive and negative ions. This accompanying reaction is always *reductive* and results in the capture of the electrons that flow through the metal.

For most corrosion processes to take place, there must be water or moisture in contact with the metal, and the water or moisture must contain ions to form the electrolyte. There must be anodic areas at which the oxidations take place and cathodic areas at which the reductions take place. The anode and cathode

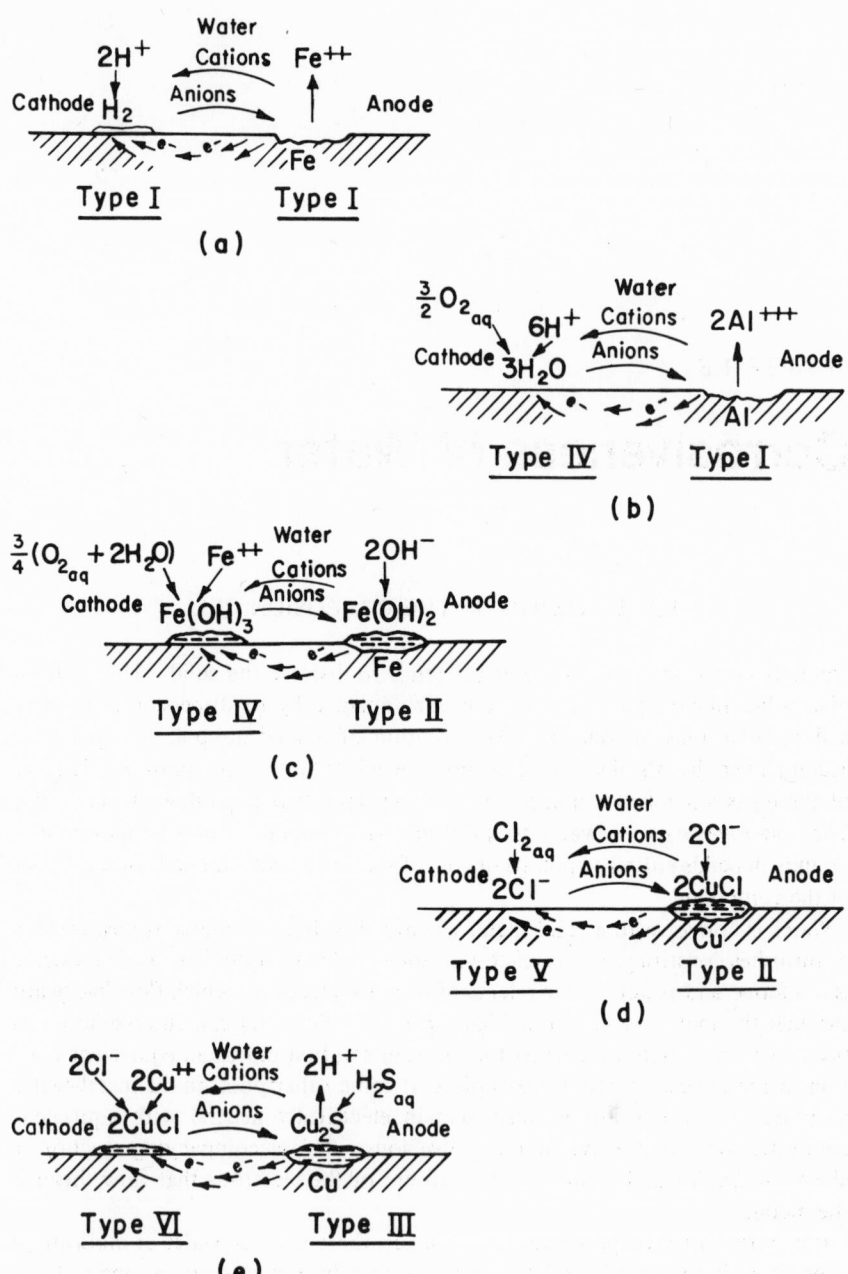

Figure 8-1
Types of reactions occurring in corrosion of metals in contact with water.

of an elementary corrosion cell must be connected by an electrical conductor, usually the metal undergoing corrosion, but also, in many cases, including the films or coatings over the metal surface. In the initial stages of the more common forms of corrosion, the anode and cathode are very close together, for the *principle of least action* requires that the path of current flow be smaller and the resistance be the least. In normal corrosion, therefore, the anode may be only a few molecules removed from the cathode; they may be so close, in fact, that the casual observer may overlook the fact that two separate electrochemical reactions are taking place.

The corrosion of a metal is therefore an electrochemical process, such as that which occurs in a galvanic cell, and it is always accompanied by a flow of electric current. The *corrosiveness* of water to a particular metal depends not only on the water's ion content of that metal, but also upon the content of other ions and molecules capable of entering into the *oxidation–reduction* reactions and upon the nature of the films covering the metal surface.

The *oxidation* reaction by which a metal corrodes is an *anodic* reaction. The equivalent *reduction* is a *cathodic* reaction. The electric current is carried by the negative electrons through the metal and is carried jointly by *cations* and *anions*, which migrate in opposite directions through the solution, as shown in Figure 8-1. By convention, the direction of flow of positive electricity (cations) is from anode to cathode through the solution and in the reverse direction, in the electric circuit, to the direction of negative electron flow through the metal. The voltage or potential drop through the solution is taken as positive in the direction of movement of positive ions.

The reaction at each electrode is called a *half-cell* or *single-electrode* reaction. The following types of anodic reactions are common in the corrosion of metals:

Type I: Metallic element $\underset{\text{Red}}{\overset{\text{Ox}}{\rightleftharpoons}}$ cations + electrons

         Examples: $\underline{Fe} \underset{\text{Red}}{\overset{\text{Ox}}{\rightleftharpoons}} Fe^{2+} + 2e^-$

                    $H_{2\,aq} \underset{\text{Red}}{\overset{\text{Ox}}{\rightleftharpoons}} 2H^+ + 2e^-$

Type II: Anions + metal $\underset{\text{Red}}{\overset{\text{Ox}}{\rightleftharpoons}}$ compound + electrons

         Examples: $2OH^- + \underline{Fe} \underset{\text{Red}}{\overset{\text{Ox}}{\rightleftharpoons}} \underline{Fe(OH)_2} + 2e^-$

                    $Cl^- + \underline{Cu} \underset{\text{Red}}{\overset{\text{Ox}}{\rightleftharpoons}} \underline{CuCl} + e^-$

Type III: Compound 1 + metal $\underset{\text{Red}}{\overset{\text{Ox}}{\rightleftharpoons}}$ compound 2 + electrons

    Examples: $H_2S_{aq} + \underline{Zn} \underset{\text{Red}}{\overset{\text{Ox}}{\rightleftharpoons}} \underline{ZnS + 2H^+ + 2e^-}$

              $HS^- + \underline{Fe} \underset{\text{Red}}{\overset{\text{Ox}}{\rightleftharpoons}} \underline{FeS + H^+ + 2e^-}$

    $\frac{1}{2}O_{2\,aq} + 2H_2O + Fe \underset{\text{Red}}{\overset{\text{Ox}}{\rightleftharpoons}} \underline{Fe(OH)_3 + H^+ + e^-}$

These reactions proceed from left to right at the anode. Reactions of the same type may occur at the cathode, proceeding from right to left. The following additional types of cathodic reactions, which do not involve the solid metal, are common:

Type IV: Compound $\underset{\text{Red}}{\overset{\text{Ox}}{\rightleftharpoons}}$ oxidant + cations + electrons

    Examples: $H_2O \underset{\text{Red}}{\overset{\text{Ox}}{\rightleftharpoons}} \frac{1}{2}O_{2\,aq} + 2H^+ + 2e^-$

          $\underline{Fe(OH)_3} \underset{\text{Red}}{\overset{\text{Ox}}{\rightleftharpoons}} \frac{3}{4}O_{2\,aq} + \frac{3}{2}H_2O + Fe^{2+} + 2e^-$

Type V: Anions $\underset{\text{Red}}{\overset{\text{Ox}}{\rightleftharpoons}}$ oxidant + electrons

    Example: $2Cl^- \underset{\text{Red}}{\overset{\text{Ox}}{\rightleftharpoons}} Cl_{2\,aq} + 2e^-$

Type VI: Compound $\underset{\text{Red}}{\overset{\text{Ox}}{\rightleftharpoons}}$ cations + anions + electrons

    Example: $CuCl \underset{\text{Red}}{\overset{\text{Ox}}{\rightleftharpoons}} Cu^{2+} + Cl^- + e^-$

Type VII: Compound + anions $\underset{\text{Red}}{\overset{\text{Ox}}{\rightleftharpoons}}$ oxidant + cations + anions + electrons

    Example: $\underline{MgCO_3} + OH^- \underset{\text{Red}}{\overset{\text{Ox}}{\rightleftharpoons}} \frac{1}{2}O_{2\,aq} + Mg^{2+} + HCO_3^- + 2e^-$

These reactions proceed from right to left at the cathode. Some anions such as nitrate and nitrite may react with the metal at the anode and also be oxidants at the cathode.

In accordance with Kirchhoff's* second law, the electromotive force of the corrosion cell must equal the voltage drop caused by the resistance around the corrosion circuit. The potential or voltage profile around the corrosion circuit is illustrated in Figure 8-2, in which the circuit includes an impressed

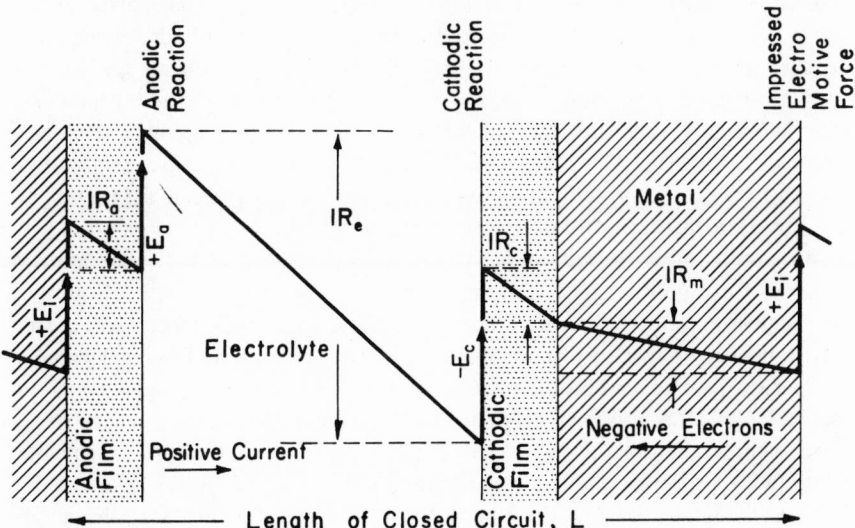

Figure 8-2
Potential profile around corrosion circuit.

electromotive force such as might be used in a test cell for studying corrosion. The direction shown for the impressed electromotive force is such as to increase the current I, but it should be noted that the current might be reduced to zero or reversed by an electromotive force of sufficient magnitude in the opposite direction. The circuit also includes, in the interest of completeness, an anodic film or coating and a cathodic film or coating. In accordance with Kirchhoff's second law, the following relations hold:

$$E_i + E_a - E_c = I(R_a + R_e + R_c + R_m) \qquad (8\text{-}1)$$

where E_i = impressed electromotive force
E_a = single electrode potential at the anode
E_c = single electrode potential at the cathode
I = corrosion current
R_a = resistance of the anodic film or coating

*Gustav R. Kirchhoff (1824–1887), German physicist.

R_e = resistance of the electrolyte

R_c = resistance of the cathodic film or coating

R_m = resistance of the metal

For a clear understanding of the corrosion process, one must keep in mind that there are two separate regions in a corrosion cell where electrochemical half-cell reactions take place and that the reactions which take place at the anode are completely independent of the reactions at the cathode, except that no current can flow without two half-cell reactions. Moreover, the electromotive force at the anode and the electromotive force at the cathode are completely independent of the rate of current flow, despite opinions expressed to the contrary (2). The potentials of the half-cell reactions depend only upon the temperature, the participating substances, and the activities of the participating substances as shown in Eqs. (8-4), (8-5), and (8-5a).

The circuit shown in Figure 8-2 is, of course, idealized to indicate complete separation of anode and cathode and a single length for the path of flow of the electric current. The actual corrosion cell is very much more complicated with numerous stream paths of different length for current flow and variable thicknesses of films. To study the process, however, it is essential to separate the variables, and this can only be done in the laboratory by means of an experimental corrosion cell in which all factors are measurable and controllable. Figure 8-2 represents such an idealized circuit.

The rate of corrosion is the time rate at which the reactions at the anode take place. The rate is the same at the cathode and throughout the corrosion circuit, and is measured by the corrosion current. In accordance with Faraday's* law, the corrosion current in amperes is equal to 96,484 times the number of equivalents per second taking part in the anodic reactions or in the cathodic reactions. For example, for a corrosion rate of 0.01 inch per year (ipy) for iron, the corrosion current is approximately 2.16×10^{-5} A/cm^2 of iron surface, provided, however, that ferrous iron is participating in all the reactions at the anode.

An examination of Eq. (8-1) or Figure 8-2 will indicate that, for a given total electromotive force, the corrosion current, and hence the rate of corrosion, depends upon the resistance of the corrosion circuit. In fundamental corrosion research, therefore, it is most important to know the resistances of the elements of the corrosion circuit. The resistance of the electrolyte may be computed from its equivalent conductance, and the resistance of the metals and films may be computed from the specific resistance of the substances if the lengths are known. Unfortunately, there is very little information in the literature relative to the specific resistance of the films encountered in the corrosion process. This is a fertile field for future research.

*Michael Faraday (1791–1867), British chemist and physicist.

8-2. PRINCIPLES FOR THE IDENTIFICATION
OF HALF-CELL REACTIONS

The reactions at the electrodes of a corrosion cell will depend upon the metals and upon the kind and concentration of ions and other solutes adjacent to the electrodes. Of all possible anodic reactions, the reaction or reactions with the highest potential will prevail, and of all possible cathodic reactions, the ones with the lowest potential will prevail. Thus the reactions at any instant will be those which produce the greatest electromotive force for the cell. At any instant there is a single electromotive force corresponding to the half-cell reactions at each electrode. Therefore, if two or more half-cell reactions are occurring simultaneously at a point in an electrode, the concentrations of the reactants must be such as to produce the same electromotive force. The above-stated principle is the basis for the identification of the corrosion reactions. Arguments in support of the validity of the principle are presented below.

As is well known, the electromotive series of the metals is a table of standard single electrode potentials arranged in order of magnitude of the electromotive forces, with those of greatest magnitude at the top. As is also well known, if any two of the metals are immersed as electrodes in an aqueous solution containing molal concentrations of the metals, the metal highest in the series will enter solution and the other will plate out. The electromotive force of the cell will be the difference of the standard single electrode potentials. If the concentrations of the ions of the two metals are different from the molal concentrations, the single electrode potentials will differ from the standard potentials; the metal with the highest potential will be the anode, and the electromotive force of the cell will be the difference of the actual potentials.

The single electrode reactions of the metals are special cases of the more general type of half-cell reaction, which must be considered in corrosion phenomena. The same laws govern all cases, and if only two half-cell reactions are possible in a corrosion cell, the one with the highest potential will be the anodic reaction. If more than two half-cell reactions are possible, it is easy to see that the direction of flow of positive current will be from the reactions of higher potential to those of lower potential; however, it is not so obvious which of the reactions will prevail at anode and cathode.

Consider a single current streamline in a corrosion cell, as shown in Figure 8-3, with the anodic reactions occurring at point A and the cathodic reactions at point C. For the corrosion current, I, at any instant, it is obvious that there can be only one value of the IR drop through the electrolyte, metal, and films because there is only one value of the total resistance along the current streamline. Since the total IR drop is equal to the difference in electromotive force of the half-cell reactions, this difference also has only one value. Since the IR drop through the electrolyte along the current streamline has only one value,

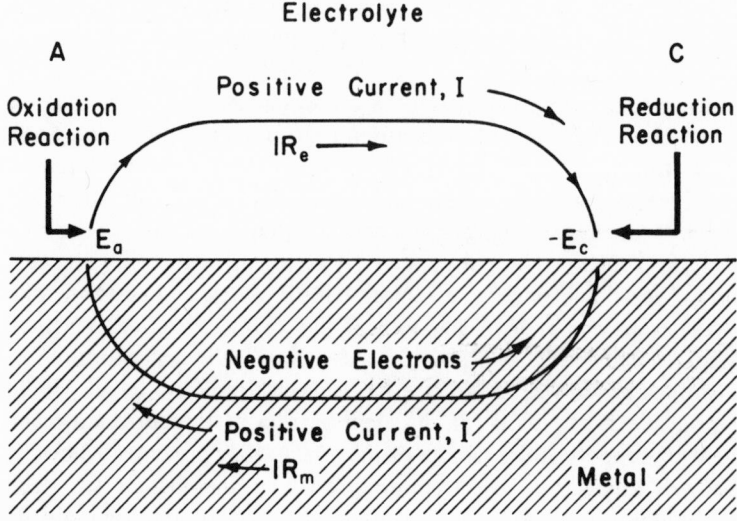

Figure 8-3
Current streamline in corrosion cell.

each half-cell has only one value for its potential. Two or more half-cell reactions may occur simultaneously at either electrode, but only if their potentials are equal.

Now which of the possible half-cell reactions do occur at the anode, those with the highest, those with the lowest, or those with intermediate potentials? Applying what we already know, the reactions with the highest potential will be anodic to those with the lower potentials and will prevail at the anode. The possible reactions at the anode with the lower potentials tend to be cathodic to the anodic reactions that prevail, but they are anodic to the prevailing reactions at the cathode. These anode reactions with lower potential cannot take place, therefore, because they would proceed in both directions at the same time and would result in a flow of current through the same point in the metal interface in both directions at the same time. Moreover, they would result in more than one value for the total IR drop. Similar reasoning will show that, of all possible cathodic reactions, the ones with the lowest potential will prevail and all others will be inactive.

Referring again to Figure 8-3, it is evident that all parallel current streamlines in the corrosion circuit must follow the same laws. Therefore, at all points in the anodic areas, the half-cell reactions with the highest potential will prevail; and at all points in the cathodic areas, the reactions with the lowest potential will prevail. All other possible reactions will be inactive.

From the preceding discussion it follows that the half-cell reactions which actually take place in a corrosion cell may be identified by comparing all possible reactions until those are found which produce the highest electromotive force (emf) for the cell.

To determine the electromotive force of a half-cell reaction, it is necessary to have available or to determine the standard electrode potential of the reaction, which is based upon an activity of unity for each of the reacting substances, and to modify this potential to conform with the actual concentrations of the reacting substances. The standard electrode potentials may be computed from the free energies of formation of the reacting substances. The free energies (*3*) of a large number of substances are available for the standard temperature of 25°C.

8-3. HALF-CELL POTENTIALS

The electrochemical reactions that occur at the electrodes of a cell involve a change in free energy. Gibbs (*4*) has shown that this change in free energy is given by the following relation:

$$-\Delta F = \frac{96,484}{4.1833} nE \tag{8-2}$$

in which

ΔF = decrease in free energy in calories per mole

E = reversible potential or emf of the half-cell in volts (E does not include the voltage drop through the electrolyte due to migration of ions or the voltage drop through the film, if present)

n = number of faradays of electricity per mole (the number of electrons transferred in the reaction)

96,484 = number of coulombs per faraday

4.1833 = number of joules per calorie

From Eqs. (4-8a) and (4-13b), the free energy change in a half-cell reaction may be represented as follows:

$$\Delta F_T = \Delta F_T^\circ + RT \ln Q \tag{8-3}$$

where ΔF_T° is the standard free energy change at temperature T, and Q is the activity function or the ratio of the product of the activities of the reaction products to the product of the activities of the reactants.

From Eq. (8-2), the *standard oxidation potential* corresponding to the standard free energy change ΔF_T° at any temperature T is

$$E_T^\circ = -\frac{4.1833}{96,484n} \Delta F_T^\circ \tag{8-4}$$

and the *oxidation potential* for any half-cell reaction is

$$E_T = E_T^\circ - \frac{4.1833RT}{96,484n} \ln Q \tag{8-5}$$

For temperature at $25°C$, the oxidation potential is given by

$$E_{25} = E_{25}^\circ - \frac{0.05916}{n} \log Q \tag{8-5a}$$

in which E_{25}° is the standard oxidation potential at $25°C$ corresponding to ΔF_{25}°.

The senior author has published a table (5) showing the standard oxidation potentials at $25°C$ of 267 different half-cell reactions that occur in corrosion. A larger list of standard oxidation potentials at $25°C$ was published by Latimer (6). Many half-cell reactions that occur in corrosion are not listed in these tables.

8-4. APPLICATION OF THEORY TO IDENTIFY HALF-CELL REACTIONS

To identify the half-cell reactions for a particular case, it is necessary to compute the single electrode potentials of the half-cell reactions that are possible with the particular metal and water. The prevailing anodic reaction will be the one with the highest potential, and the prevailing cathodic reaction will be the one with the lowest potential. To simplify the computations, the activity coefficients have been taken as unity for all reaction constituents.

To illustrate the procedure, consider anodic reactions 2 and 4 of Table 8-1 in the corrosion of iron. The change in standard free energy for reaction (2) is

$$\text{Fe} \rightleftharpoons \text{Fe}^{2+} + 2e^-$$
$$F_{25}^\circ = 0 \qquad -20,300 \qquad \Delta F_{25}^\circ = -20,300 \text{ cal}$$

From Eq. (8-4),

$$E_{25}^\circ = -\frac{4.1833}{96,484 \times 2} (-20,300) = +0.440 \text{ V}$$

For 0.1 ppm of ferrous iron in solution,

$$[\text{Fe}^{2+}] = \frac{0.1}{55.8} \times 10^{-3} = 1.79 \times 10^{-6}$$

From Eq. (8-5a),

$$E_{25} = +0.440 - \frac{0.05916}{2} \log (1.79 \times 10^{-6}) = +0.61 \text{ V}$$

The change in standard free energy for reaction 4 is

$$\begin{array}{lllll} & 2OH^- + \underline{Fe} & \rightleftharpoons \underline{Fe(OH)_2} & + 2e^- & \\ F^\circ_{25} = -75{,}190 & 0 & -115{,}570 & & \Delta F^\circ_{25} = -40{,}380 \text{ cal} \end{array}$$

From Eq. (8-4),

$$E^\circ_{25} = -\frac{4.1833}{96{,}484 \times 2} (-40{,}380) = +0.877 \text{ V}$$

For pH values of 8 and 10, for example,

$$[OH^-] = 10^{-6} \quad \text{and} \quad 10^{-4}$$

From Eq. (8-5a),

$$E_{25} = +0.877 - \frac{0.05916}{2} \log [-(10^{-6})^2] = +0.522 \text{ V} \quad \text{at pH 8}$$

$$= +0.877 - \frac{0.05916}{2} \log [-(10^{-4})^2] = +0.64 \text{ V} \quad \text{at pH 10}$$

A comparison of these results indicates that, with 0.1 ppm of ferrous iron in solution, reaction 2 with a potential of +0.61 V prevails over reaction 4 at pH 8 with a potential of +0.522 V. However, at pH 10, reaction 4 with a potential of +0.64 V prevails over reaction 2 with a potential of +0.61 V. Figure 8-4 shows graphically the anodic potentials of both reactions for various pH values and concentrations of ferrous iron.

Consider, now, cathodic reactions 6 and 11 of Table 8-1 in the corrosion of iron. The change in standard free energy for reaction 6 is

$$\begin{array}{llllll} \underline{Fe(OH)_3} & \rightleftharpoons \tfrac{3}{4}O_{2\,aq} & + Fe^{2+} & + \tfrac{3}{2}H_2O & + 2e^- & \\ F^\circ_{25} = -166{,}000 & +2{,}960 & -20{,}300 & -85{,}035 & & \Delta F^\circ_{25} = +63{,}625 \text{ cal} \end{array}$$

From Eq. (8-4),

$$E^\circ_{25} = -\frac{4.1833}{96{,}484 \times 2} (+63{,}625) = -1.386 \text{ V}$$

For 0.1 ppm of ferrous iron and 10 ppm of oxygen in solution,

$$[Fe^{2+}] = 1.79 \times 10^{-6} \quad \text{and} \quad [O_{2\,aq}] = \tfrac{10}{32} \times 10^{-3} = 3.12 \times 10^{-4}$$

From Eq. (8-5a),

$$E_{25} = -1.386 - \frac{0.05916}{2} \log \left[+(3.12 \times 10^{-4})^{3/2} (1.79 \times 10^{-6})^2 \right] = -1.138 \text{ V}$$

The change in standard free energy for reaction 11 is

$$H_2O \rightleftharpoons \tfrac{1}{2}O_{2\,aq} + 2H^+ + 2e^-$$

$$F_{25}^\circ = -56{,}690 \qquad +1{,}973 \qquad 0 \qquad \Delta F_{25}^\circ = +58{,}663 \text{ cal}$$

From Eq. (8-4),

$$E_{25}^\circ = -\frac{4.1833}{96{,}484 \times 2} (+58{,}663) = -1.272 \text{ V}$$

For 10 ppm of dissolved oxygen and pH 7,

$$[H^+] = 10^{-7} \quad \text{and} \quad [O_{2\,aq}] = 3.12 \times 10^{-4}$$

From Eq. (8-5a),

$$E_{25} = -1.272 - \frac{0.05916}{2} \log \left[+(3.12 \times 10^{-4})^{1/2} (10^{-7})^2 \right] = -0.806 \text{ V}$$

A comparison of the preceding results shows that, with 10 ppm of dissolved oxygen and 0.1 ppm of dissolved ferrous iron, at a pH of 7, reaction 6 with a potential of -1.138 V will prevail at the cathode over reaction 11 with a potential of -0.806 V. Figure 8-4 shows graphically the cathodic potentials of both reactions for various pH values and concentrations of dissolved oxygen and ferrous iron.

Computations have been made by the authors of the single electrode potentials of the principal half-cell reactions that take place in the corrosion of the commonly used metals. These computations were made, as illustrated above, from the standard free energies at 25°C. The activity coefficients have been taken at unity for all constituents of the reactions. The results of these computations are presented in the following articles, together with the interpretations that may be made relative to the nature of the corrosion process.

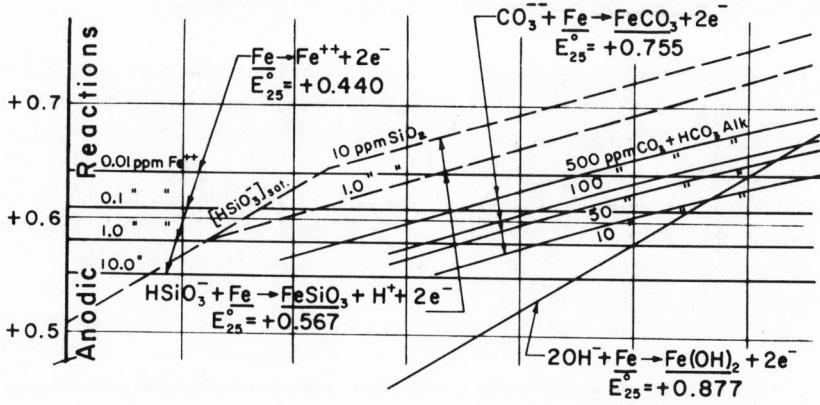

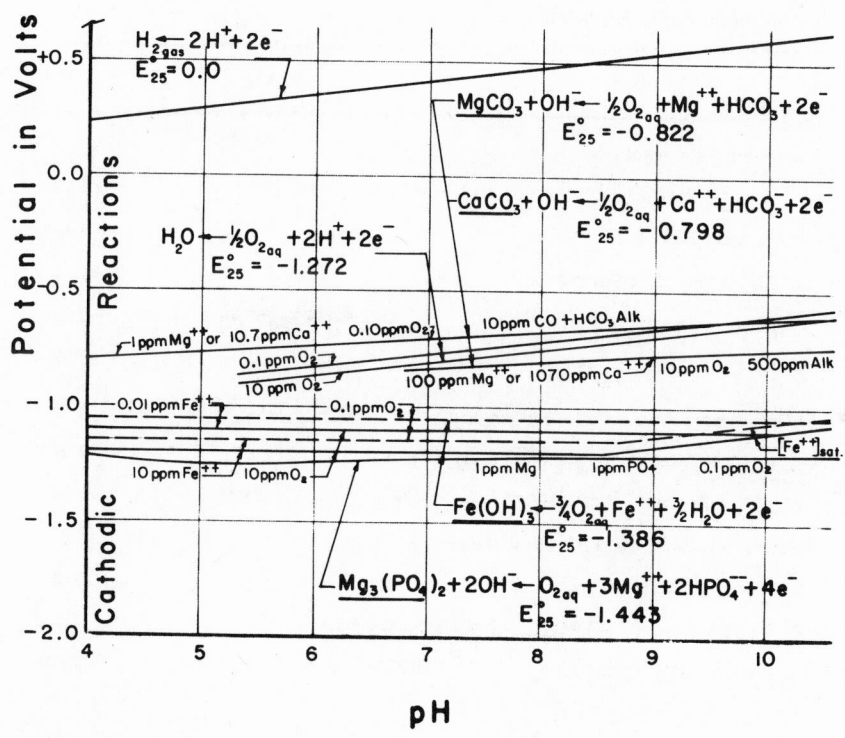

Figure 8-4
Single electrode potentials in iron corrosion at 25°C.

8-5. CORROSION OF IRON IN FRESHWATER

The principal anodic reactions and cathodic reactions in the corrosion of iron in freshwater are shown in Table 8-1, together with the standard single electrode potentials for each half-cell reaction. Figure 8-4 shows graphically the potentials for most of these reactions within the pH range from 4 to 11. Reaction 1 is not shown in Figure 8-4, because the potential is too high for the scale of the figure. The potential for reaction 1 with 10 ppm of dissolved oxygen ranges from +2.50 at pH 4 to +2.86 at pH 10 and is about 0.06 V lower with 0.1 ppm of dissolved oxygen.

The importance of dissolved oxygen in the corrosion reactions is at once apparent from Table 8-1 and Figure 8-4. In the absence of dissolved oxygen or any other oxidizing agent, such as chlorine, nitrate, nitrite, and dichromate, a number of reactions may take place at the anode in the corrosion of iron, but

Table 8-1
Corrosion of Iron in Freshwater

	E_{25}° (V)
Principal anodic reactions	
1. $\frac{1}{2}O_{2\,aq} + 2H_2O + \underline{Fe} \rightleftharpoons Fe(OH)_3 + H^+ + e^-$	+2.37
2. $\underline{Fe} \rightleftharpoons Fe^{2+} + 2e^-$	+0.440
3. $CO_3^{2-} + \underline{Fe} \rightleftharpoons FeCO_3 + 2e^-$	+0.755
4. $2OH^- + \underline{Fe} \rightleftharpoons Fe(OH)_2 + 2e^-$	+0.877
5. $HSiO_3^- + \underline{Fe} \rightleftharpoons FeSiO_3 + H^+ + 2e^-$	+0.567
Principal cathodic reactions involving iron	
6. $Fe(OH)_3 \rightleftharpoons \frac{3}{4}O_{2\,aq} + Fe^{2+} + \frac{3}{2}H_2O + 2e^-$	−1.386
7. $3Cl^- + Fe(OH)_3 \rightleftharpoons 3HOCl + Fe^{2+} + 5e^-$	−1.58
Principal cathodic reactions—any metal	
8. $H_{2\,gas} \rightleftharpoons 2H^+ + 2e^-$	0.00
9. $\underline{MgCO_3} + OH^- \rightleftharpoons \frac{1}{2}O_{2\,aq} + Mg^{2+} + HCO_3^- + 2e^-$	−0.822
10. $\underline{CaCO_3} + OH^- \rightleftharpoons \frac{1}{2}O_{2\,aq} + Ca^{2+} + HCO_3^- + 2e^-$	−0.798
11. $H_2O \rightleftharpoons \frac{1}{2}O_{2\,aq} + 2H^+ + 2e^-$	−1.272
12. $Cl^- + H_2O \rightleftharpoons HOCl + H^+ + 2e^-$	−1.49
13. $\underline{Mg_3(PO_4)_2} + 2OH^- \rightleftharpoons O_{2\,aq} + 3Mg^{2+} + 2HPO_4^{2-} + 4e^-$	−1.443

only one reaction can take place at the cathode, the plating out of hydrogen gas, reaction 8. The reaction cannot continue unless the hydrogen gas film is released, and this cannot take place without overcoming the *hydrogen over-voltage.* Since the required hydrogen overvoltage is greater than the algebraic sum of the potential of any of the anodic reactions of Figure 8-4 and the potential of the hydrogen reaction, the corrosion of iron is effectively stopped in the absence of dissolved oxygen or other oxidizing agents at pH values above about 5.0. It follows, therefore, that if the dissolved oxygen can be removed from the water, the rate of corrosion will be markedly reduced in most cases. This is usually practicable only in closed systems where the water is not exposed to the atmosphere. In a closed system where the water is not replenished with oxygen-containing water, the rate of corrosion will be drastically reduced as soon as all the dissolved oxygen originally present is utilized in the corrosion reactions.

In the presence of dissolved oxygen, reaction 1 prevails at the anode to form a hydrous ferric oxide film. If this film stays intact, as it probably does over most of the surface of the metal, the metal that took part in the formation of the film becomes *passive* and can no longer take part in an anodic half-cell reaction. The film probably does take part, however, in the cathodic reactions with further building up of hydrous ferric oxide, with the precipitation of ferrous iron from solution as indicated by reaction 6. Ferric hydroxide is very insoluble at pH values above 4. The solubility is about 0.01 ppm of ferric iron at pH 3.8 and 3 ppm at pH 3. At pH values less than 3, ferric hydroxide is increasingly soluble.

Since the density of ferric hydroxide is from 3.4 to 3.9, whereas the density of metallic iron is 7.86, it is obvious that the hydrous ferric oxide molecule, formed in reaction 1 from the reaction of an atom of metallic iron, must occupy more space at the interface than is occupied by the atom of iron. It is not possible, therefore, for all the metallic iron at the interface to react with dissolved oxygen to form hydrous ferric oxide. Some unreacted metal must remain under the hydrous ferric oxide film. If the film is porous to water or breaks down at points, the other anodic reactions in Table 8-1 and Figure 8-4 may take place beneath the film, provided, of course, the dissolved oxygen is effectively used up in the cathodic reactions that take place on top of the film. Corrosion may, therefore, continue with reaction 1 no longer taking part in the process. Most of the observed phenomena in the corrosion of iron in freshwater may be explained in terms of reactions 2 to 13 of Table 8-1.

The principal source of hydrous ferric oxide rust is the cathodic reaction 6 of Table 8-1, in which the rust is formed by the reaction of ferrous ions with dissolved oxygen. Unlike reaction 1, this reaction may take place at considerable distance from the metal surface, provided the precipitate forms on top of previously deposited rust through which the captured electrons may travel from the metal to the reaction. The types of rust growth, which consist of oriented crystals or whiskers on an underlying oxide film, appear to be dependent on temperature, environment, and the metallurgical history of the specimens.

Figure 8-4 indicates that the principal anodic reactions in the corrosion of iron, which take place under the cathodic film where the oxygen is absorbed, are the solution of the metallic iron as ferrous iron below pH values of 8 to 10, and the plating out of carbonate or hydroxide to form ferrous carbonate or ferrous hydroxide at higher pH values. Where bisilicate is present, it may plate out at the anode to form ferrous silicate at pH values above about 6 to 7.

If the reaction of the metallic iron with carbonate or hydroxide is not sufficiently violent to pull the atoms of iron out of the solid metal interface, a protective film should be formed to make the metal surface passive, as in the case of reaction 1 with dissolved oxygen. Very little is known about the effectiveness of the ferrous carbonate and the ferrous hydroxide films as protective coatings, but experience indicates that some corrosion of iron does take place at pH values above 10. However, it should be kept in mind, in interpreting Figure 8-4, that all the anodic reactions must take place beneath the cathodic film which shields the metal surface from dissolved oxygen. This same protective film forms a barrier between the main body of water where the pH is measured and the water adjacent to the metal. The reactions themselves result in a lowering of the pH value at the metal–water interface.

As the pH value increases above pH 9, it is well known that the rate of corrosion of iron decreases. This tends to indicate that the anodic films of ferrous carbonate and ferrous hydroxide, formed by reactions 3 and 4, become more effective with higher pH values. Ferrous hydroxide is soluble to the extent of 1 ppm of iron at pH 9 and 0.01 ppm of iron at pH 10, and with only 10 ppm of alkalinity, ferrous carbonate is soluble to the extent of 0.1 ppm of iron at pH 9.2 and 0.01 ppm of iron at pH 10.4. Most natural waters contain silica, but it is not known whether this is in the form of colloidal silica or in solution as a true bisilicate. The presence of bisilicate may explain why some natural waters are less corrosive to iron than others. Bisilicate may take part in the anodic reaction 5 of Table 8-1. With only 1 ppm of bisilicate silica, ferrous silicate is soluble to the extent of 0.1 ppm of iron at pH 6.2 and 0.01 ppm of iron at pH 7.2.

As previously indicated, all cathodic half-cell reactions involve dissolved oxygen or some other oxidizing agent. In Table 8-1 hypochlorous acid has been included among the oxidizing agents because most drinking waters are chlorinated. The cathodic potential with hypochlorous acid is always somewhat lower than it is for dissolved oxygen, and it is well known that chlorinated waters are more corrosive than unchlorinated waters. A reaction similar to reaction 1 may also take place with chlorine at the anode, but this reaction has no more significance in the continuation of corrosion than does reaction 1.

Nitrate that is sometimes present in natural waters is also an effective substitute for dissolved oxygen in the cathodic reactions as well as in an anodic reaction analogous to reaction 1. The same is true of nitrites and dichromates, which, though not present in drinking-water supplies, are frequently used in high concentrations as corrosion inhibitors in recirculated cooling water systems. Inhibitors and passivators will be discussed at greater length.

A study of the cathodic reactions of Table 8-1 and Figure 8-4 indicates that as long as dissolved oxygen is continuously replenished from the water supply and ferrous iron is present in solution, either from the supply or from the corrosion reactions, hydrous ferric iron rust will continue to plate out at the cathode, as indicated by reaction 6, while corrosion continues. This is the prime cause of the formation of rust scale and tubercles in the corrosion of iron. In the absence of dissolved ferrous iron at a cathodic area, magnesium carbonate will plate out at the cathode by means of a half-cell reaction involving dissolved oxygen, reaction 9 of Table 8-1. This reaction has been encountered by the senior author in a case where magnesium carbonate crystals were found at cathodic areas on the exterior of a steel pipe that had been subjected to stray current electrolysis. It may be shown that magnesium carbonate plates out in preference to calcium carbonate at 25°C, unless the concentration of calcium in parts per million is more than 10.7 times the magnesium concentration.

Most of the reactions in Table 8-1 and Figure 8-4 are not to be found in the conventional literature, and experimental proof of their occurrence is desirable. Experiments by Larson and King (7) furnish confirmatory evidence for some of the reactions. In their work, an experimental corrosion cell was devised with iron electrodes; the nine compartments between the electrodes were separated by porous Alundum plates. Tap water with a pH of 7.4, a hardness of 250 ppm, and an alkalinity of 330 ppm was used for the electrolyte and was allowed to flow continuously through the middle compartment. The water had a trace of iron and manganese, 60 ppm of calcium, 24 ppm of magnesium, and 6.0 ppm of dissolved oxygen. Electric current was allowed to flow through the cell in one direction with an impressed electromotive force. The objective was to study ion migration.

Progressive quality changes occurred in each compartment with continuation of electric current flow. In the cathode compartment the pH rose to 11, the calcium and magnesium decreased almost to the vanishing point, and the alkalinity remained almost constant. These phenomena are fully explained by reactions 9 and 10 of Table 8-1. In the anode compartment the pH decreased slightly and the alkalinity decreased to approximately one sixth of its original value. These changes may be explained by the plating out of carbonate at some anodic points, according to reaction 3, concurrently with the solution of ferrous iron at other anodic points in accordance with reaction 2 of Table 8-1. Silicates and polyphosphates (when added) were found to decrease in both anode and cathode compartments in confirmation of reactions 5 and 13 of Table 8-1. Unfortunately, the changes in dissolved oxygen and ferrous iron concentration were not reported.

In the corrective treatment of water by the use of the calcium carbonate equilibrium, alkali is added to slightly supersaturate the water with calcium carbonate. It is widely believed that the precipitation of calcium carbonate by an ordinary chemical reaction forms a protective coating on the pipe wall. It is evident from reaction 10 that calcium carbonate films are formed by an electrochemical reaction involving dissolved oxygen. It is apparent, however, that protection against corrosion is afforded primarily by the anodic reaction and not by the calcium carbonate film, which is permeable to water. The important feature is the high pH within the film which retards the solution of ferrous iron and results in the plating out of carbonate or hydroxide at the anode. As further evidence of this point, iron or steel in contact with strong solutions of sodium carbonate or sodium hydroxide (about 6 percent) do not corrode appreciably. The effectiveness of cement lining, which is also permeable to water, in protecting iron pipe from corrosion may be explained by the fact that the pH of the pore water adjacent to the metal is about 12.4 in newly lined pipe.

In the conventional literature on corrosion, the plating out of hydrogen ions, reaction 8 of Table 8-1, is considered as a principal cathodic reaction. It is assumed that, if the water contains no dissolved oxygen, hydrogen is liberated as hydrogen gas, and that if the water contains dissolved oxygen, the molecular hydrogen produced at the cathode reacts with dissolved oxygen to produce water. This assumption is contrary to the knowledge that molecular hydrogen does not react with molecular oxygen except at high temperatures. The production of water at the cathode is best explained by reaction 11 of Table 8-1. As previously indicated, reaction 8, the plating out of hydrogen ions to form hydrogen gas, does not take place in ordinary corrosion reactions except in the complete absence of oxygen or other oxidizing agents. Even with the complete absence of oxidizing agents, this reaction does not take place in the corrosion of iron except at very low pH values.

Phosphates, metaphosphates, and silicates are used in the treatment of drinking water to retard the rate of corrosion of metals. Inadequate information is available about the free energies of some of the various polyphosphates and silicate compounds; therefore, it is not possible to determine the nature of all the reactions involved with phosphate and silicate treatment. Figure 8-4 indicates that silicates plate out at the anode to form ferrous silicate at pH values above about 6. Experiments by Eliassen and Skrinde (8) on the corrosion of iron in distilled water, to which sodium hydroxide and sodium silicate were added, show that reaction 5 in Table 8-1 does take place, and that 8 ppm of bisilicate silica will reduce the rate of corrosion about 63 percent at pH 9. The formation of biferrous silicate at the anode is not likely because the standard potential, +0.35 V, is not high enough as compared to the potential of reaction 5. The formation of ferric phosphate at either the anode or cathode is not likely because the standard potentials of these reactions are +0.152 and -1.29 V, respectively. The formation of ferrous or biferrous silicate

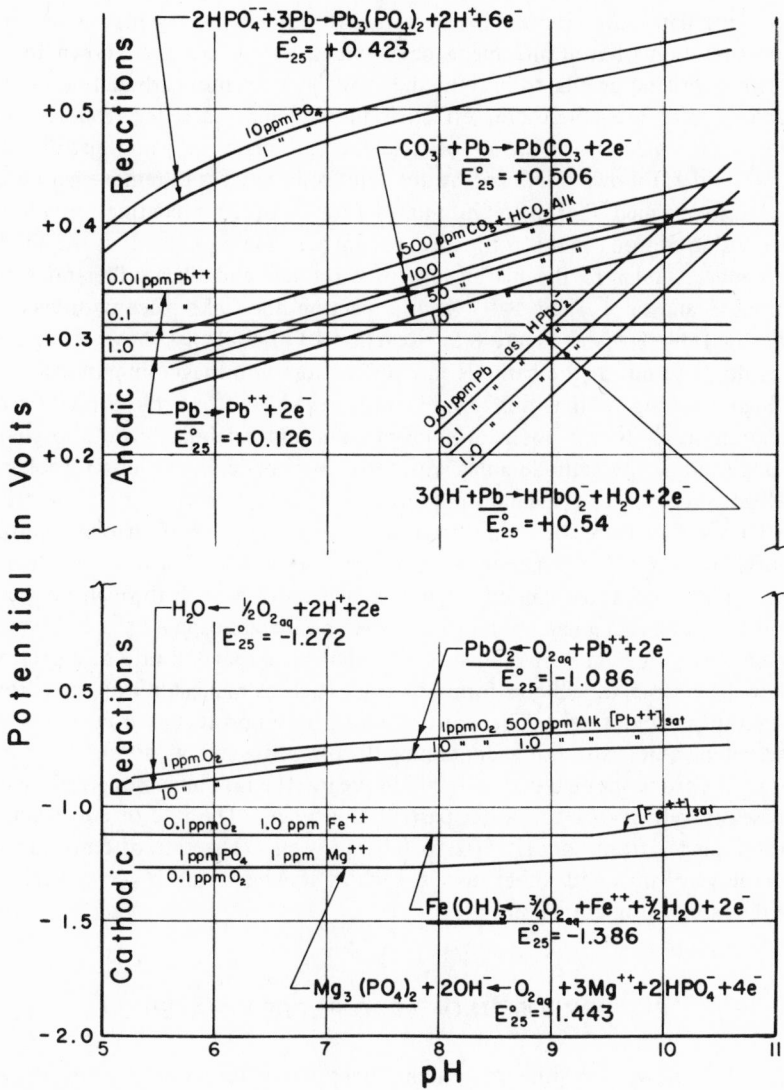

Figure 8-5
Single electrode potentials in lead corrosion at 25°C.

at the cathode is unlikely because the standard potentials are −0.985 and −0.975 V, respectively.

A study of reaction 13 of Table 8-1 shows that this reaction will prevail at the cathode in the corrosion of iron even with very low concentrations of magnesium, dissolved oxygen, and phosphates, as shown by the curve on Figure 8-4. Reaction 13 will always prevail at the cathode over a similar reaction producing calcium phosphate, if magnesium is present.

Metaphosphates are known to be effective complexing agents to prevent the precipitation of iron and manganese oxides. They are also known to be effective peptizing agents to prevent the growth of calcium carbonate crystals. Metaphosphates are, therefore, effective in reducing red-water troubles and retarding the *after-precipitation* of calcium carbonate in water-distribution systems, which follows treatment by the lime–soda process to remove hardness.

It is also claimed that metaphosphates form protective coatings to retard the corrosion of iron. If this is the case, phosphate should plate out at the anode in a manner similar to the plating out of carbonate and silicate illustrated by reactions 3 and 5 of Table 8-1. Such a reaction does take place with lead, as shown by Table 8-1 and Figure 8-5. Reaction 13 of Table 8-1 shows that phosphates do plate out at the cathode in combination with magnesium or calcium, and Figure 8-4 shows that this reaction will prevail even over the formation of ferric iron rust. In the absence of magnesium and calcium, ferric phosphate will plate out at the cathode along with ferric hydroxide, in a reaction involving dissolved oxygen. The standard potential of this reaction is −1.29 V; at pH 7 with 1 ppm phosphate, 0.1 ppm dissolved oxygen, and 1 ppm ferrous iron, the potential is −1.03 V. Experience with metaphosphates indicates that in unlined cast-iron pipes, the amount of iron in the water passing through the pipelines is increased to a greater extent than when no metaphosphate is used.

Extensive testing on corrosion specimens indicates that the rate of corrosion is about the same for all the forms of iron, such as malleable iron, cast iron, wrought iron, and steel. The corrosion rate of unprotected iron and steel immersed in quiet soft water containing dissolved oxygen, at normal temperature, ranges from about 0.002 to 0.006 in./year. The rate increases greatly with increases in velocity of the water past the specimen. The rate of corrosion in seawater ranges from about 0.002 to 0.01 in./year. The rate of corrosion of many alloys of iron with other metals, such as stainless steel, is very much less than the rate of iron corrosion.

8-6. CORROSION OF LEAD IN FRESHWATER

Table 8-2 shows the principal anodic reactions in the corrosion of lead in freshwater and the principal cathodic reactions in which lead is a reactant.

Table 8-2
Corrosion of Lead in Freshwater

	E°_{25} (V)
Principal anodic reactions	
1. $\frac{3}{2}O_{2\,aq} + H_2O + 3Pb \rightleftharpoons Pb_3O_4 + 2H^+ + 2e^-$	+2.10
2. $Pb \rightleftharpoons Pb^{2+} + 2e^-$	+0.126
3. $3OH^- + Pb \rightleftharpoons HPbO_2^- + H_2O + 2e^-$	+0.54
4. $CO_3^{2-} + Pb \rightleftharpoons PbCO_3 + 2e^-$	+0.506
5. $2HPO_4^{2-} + 3Pb \rightleftharpoons Pb_3(PO_4)_2 + 2H^+ + 6e^-$	+0.423
Principal cathodic reactions involving lead	
6. $PbO_2 \rightleftharpoons O_{2\,aq} + Pb^{2+} + 2e^-$	−1.086
7. $2Cl^- + 2H^+ + PbO_2 \rightleftharpoons 2HOCl + Pb^{2+} + 2e^-$	−1.54

Cathodic reactions 9 to 13 inclusive of Table 8-1 may also take place in the corrosion of lead. The plating out of molecular hydrogen at the cathode, reaction 8 of Table 8-1, cannot take place because the potential is higher than the potentials of the other cathodic reactions.

Figure 8-5 shows a comparative study of the anodic and cathodic reactions in the corrosion of lead over a pH range from 5 to 10.6 and over a wide range of concentration of the dissolved constituents of the reactions. Reaction 1 is not shown because the potential is too high for the scale of the figure. The potential for reaction 1 with 10 ppm of dissolved oxygen ranges from +2.24 at pH 5 to +2.56 at pH 10, and is about 0.09 V lower with 0.1 ppm of dissolved oxygen.

In the presence of dissolved oxygen, reaction 1 prevails at the anode at the start of corrosion to form a lead oxide film. If this film stays intact, as it probably does over most of the surface of the metal, the metal that took part in the formation of the film becomes passive and can no longer take part in an anodic half-cell reaction. This reaction is analogous to reaction 1 of Table 8-1 in the corrosion of iron; as in the case of iron, the other anodic reactions must take place below this oxide film or in the complete absence of dissolved oxygen. The density of metallic lead is 11.34, whereas the density of the lead oxide is 9.1. The volume occupied by the oxide is, therefore, somewhat greater than the volume occupied by the atom of lead entering into the reaction, although the difference is not as great as with iron.

The lead oxide formed by reaction 1 probably dissolves as plumbous hydroxide and plumbic hydroxide, with two thirds of the lead in the form of the plumbous ion. The plumbic portion is very insoluble over a wide pH range; the plumbous portion is soluble to the extent of about 1 ppm at pH 8, and it is much more soluble at lower pH values. It is quite probable, therefore, that reaction 1 at the anode does not form a very stable oxide film except at pH values above about 9, where the solubility of plumbous lead is 0.01 ppm. The rate of corrosion of lead in freshwater saturated with oxygen at normal temperatures is usually less than 0.0004 in./year. In distilled water devoid of carbonates but containing dissolved oxygen, the rate may increase to 0.01 in./year. It is thus obvious that carbonates play an important part in the formation of protective films for lead.

It will be noted from Figure 8-5 that, in the absence of phosphates, the prevailing anodic reaction in the corrosion of lead at pH values less than about 7 is the solution of lead, reaction 2 of Table 8-2, and that the prevailing reaction at the anode at pH values above 9.5 is the solution of lead as biplumbate ion, reaction 3 of Table 8-2. In the pH range between 7 and 9.5, the prevailing reaction at the anode is the plating out of carbonate to form lead carbonate, reaction 4. The higher the carbonate and bicarbonate alkalinity, the wider will be the pH range over which this protection occurs. With over 100 ppm alkalinity, lead carbonate will form at the anode over a pH range from about 6 to 10. It will also be noted from Figure 8-5 that, where phosphates are present in concentrations from 1 to 10 ppm, lead is protected by the plating out of phosphate at the anode to form triplumbous phosphate, reaction 5 of Table 8-2. This protection is quite effective over the pH range from 5 to 10.6. Triplumbous phosphate is very insoluble. A study of a similar reaction with silicates indicates that lead silicate does not form at the anode.

A comparative study of the cathodic reactions in the corrosion of lead, as shown by Figure 8-5, indicates that at pH values less than about 7 and in the absence of phosphates the prevailing reactions will be the formation of water by the reaction of hydrogen ions with dissolved oxygen, reaction 11 of Table 8-1. At pH values above about 7, the prevailing cathodic reaction will be the reaction of lead ions with dissolved oxygen to form lead oxide, reaction 6 of Table 8-2. The concentration of lead ions available for this reaction is limited by the solubility of lead as determined by the solubility product of lead carbonate. In highly alkaline waters, the prevailing reactions will be the plating out of magnesium or calcium carbonate, reactions 9 and 10 of Table 8-1. If chlorine is present, a reaction similar to reaction 6 but involving chlorine, reaction 7 of Table 8-2, will prevail over reaction 6. If phosphate is present, reaction 13 of Table 8-1 or a similar reaction with calcium will prevail at the cathode. The formation of magnesium or calcium phosphate will prevail over the formation of lead phosphate. Lead silicates will not form at the cathode because the potentials are not low enough. If iron is present, in the absence of phosphates, it will plate out at the cathode, reaction 6 of Table 8-1.

The rate of corrosion of lead in freshwater saturated with oxygen at normal temperatures ranges from about 0.00002 to 0.0004 in./year. In distilled water devoid of carbonates but containing dissolved oxygen, the rate may increase to 0.01 in./year. The rate of corrosion of lead in seawater ranges from about 0.0004 to 0.0006 in./year.

8-7. CORROSION OF COPPER IN FRESHWATER

Table 8-3 shows the principal half-cell reactions in the corrosion of copper in freshwater in which copper is a reactant. Reactions 9 to 13 inclusive of Table

Table 8-3
Corrosion of Copper in Freshwater

	E°_{25} (V)
Principal anodic reactions	
1. $\frac{1}{2}O_{2_{aq}} + H_2O + 4\underline{Cu} \rightleftharpoons 2Cu_2O + 2H^+ + 2e^-$	+0.33
2. $\underline{Cu} \rightleftharpoons Cu^{2+} + 2e^-$	-0.337
3. $\underline{Cu} \rightleftharpoons Cu^+ + e^-$	-0.521
4. $2OH^- + 2\underline{Cu} \rightleftharpoons Cu_2O + H_2O + 2e^-$	+0.358
5. $2NH_{3_{aq}} + \underline{Cu} \rightleftharpoons Cu(NH_3)_2^+ + e^-$	+0.125
6. $2Cl^- + \underline{Cu} \rightleftharpoons CuCl_2^- + e^-$	-0.208
Principal cathodic reactions involving copper	
7. $\underline{Cu(OH)_2} \rightleftharpoons \frac{1}{2}O_{2_{aq}} + Cu^{2+} + H_2O + 2e^-$	-1.01
8. $2Cl^- + \underline{Cu(OH)_2} \rightleftharpoons 2HOCl + Cu^{2+} + 4e^-$	-1.36

8-1 may also be involved as cathodic reactions. Reaction 8 of Table 8-1, the plating out of molecular hydrogen at the cathode, cannot take place with copper because all the anodic reactions, except reaction 1, have lower potentials.

Figure 8-6 shows a comparative study of the anodic and cathodic reactions in the corrosion of copper over a pH range of 5 to 10.6 and over a wide range of concentration of the dissolved constituents of the reactions. Reaction 1 is not shown because the potential is too high for the scale of the figure. The potential for reaction 1 with 10 ppm of dissolved oxygen ranges from +0.57 at pH 5 to +0.87 at pH 10, and is about 0.03 V lower with 0.1 ppm of dissolved oxygen.

In the presence of dissolved oxygen, reaction 1 prevails at the anode at the

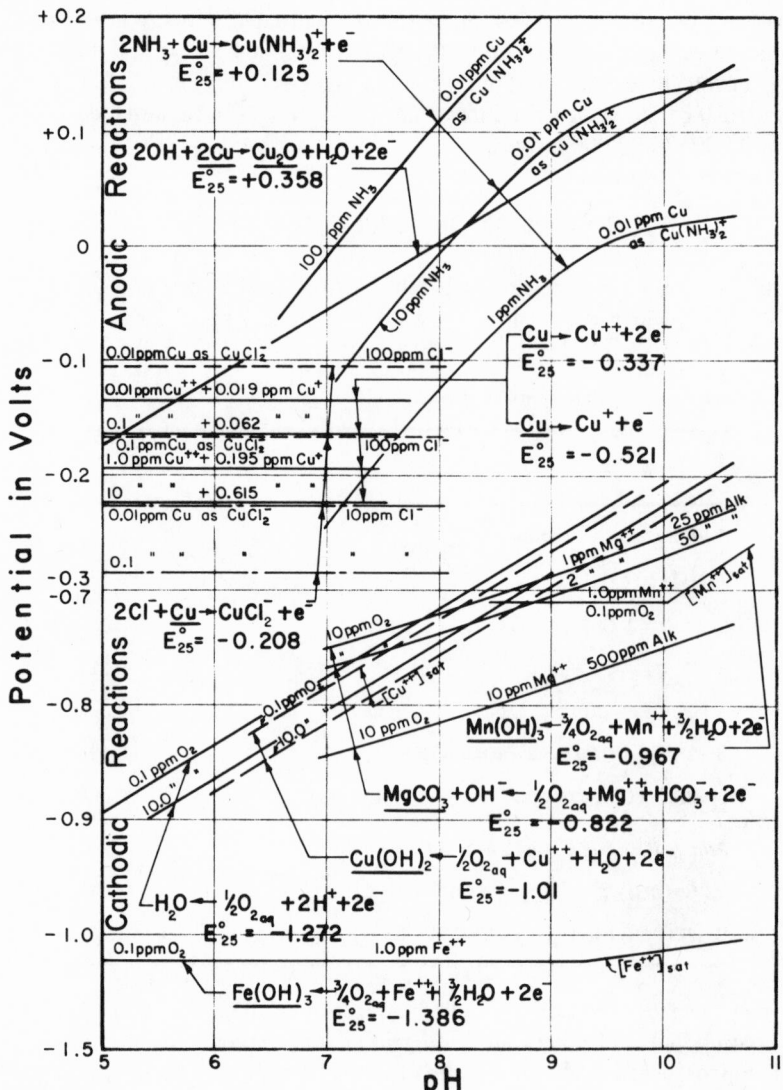

Figure 8-6
Single electrode potentials in copper corrosion at 25°C.

start of corrosion to form a cuprous oxide film. If this film stays intact, as it probably does over most of the surface of the metal, the metal that took part in the formation of the film becomes passive and can no longer take part in an anodic half-cell reaction. This reaction is analogous to reaction 1 of Table 8-1 in the corrosion of iron, and, as in the case of iron, the other anodic reactions must take place under this oxide film or in the complete absence of dissolved oxygen. The density of metallic copper is 8.92, and the density of cuprous oxide is 6.0. Hence, the volume of the oxide is greater than the volume of the atom of copper that formed the oxide, and some of the metal under the film is, therefore, left unoxidized. Cuprous oxide is soluble to the extent of less than 0.1 ppm of copper at pH 5 and is considerably less soluble at higher pH values.

Figure 8-6 indicates that the prevailing reactions at the anode at pH values less than about 6 are the solution of copper as cupric and cuprous ions or, where the chloride concentration is high, as cuprous chloride ions. A comparative study of reactions 2 and 3 indicates that both cupric and cuprous ions enter solution at the anode, and that the equilibrium concentration of cuprous ions is 7.7×10^{-4} times the square root of the concentration of cupric ions at $25°C$. The concentrations of cuprous and cupric ions are equal when the concentration of cupric ions is about 0.038 ppm. Reaction 6, the solution of copper as cuprous chloride, is not important unless the chloride concentration exceeds about 50 ppm and the pH is less than 6.

At pH values greater than about 6, reaction 4 will prevail at the anode with the plating out of hydroxide to form cuprous oxide, unless there are large amounts of ammonia in solution. With less than about 8 ppm of total ammonia, including both ammonium ions and the gas, reaction 4, the plating out of hydroxide to form cuprous oxide, will prevail at the anode. With 10 ppm of total ammonia, reaction 5 will prevail at the anode between pH values of 8.5 to 10.3, in which case copper will go into solution as a complex cuprous ammonia ion. With an increase in concentration of total ammonia, this reaction will prevail over a wider pH range, and with 100 ppm of total ammonia, reaction 5 will prevail above a pH of about 6.5.

It will be noted from Table 8-3 that no anodic reaction involving carbonate has been listed. A study of the reaction indicates that anodic protection of copper by carbonate is ineffective. The presence of carbonate ion, however, does not interfere with the plating out of hydroxide at the anode.

In the corrosion of copper below a pH of about 8, the principal cathodic reactions are the formation of water, reaction 11 of Table 8-1, and of cupric hydroxide, reaction 7 of Table 8-3. A comparative study of the cathodic reactions with dissolved oxygen indicates that cupric hydroxide will be formed, provided cupric ions are present in solution near saturation as determined from the solubility product of cupric hydroxide. At pH values above 9 in soft waters and above about 6 in very hard waters, magnesium carbonate or calcium carbonate is formed at the cathode by reactions involving dissolved oxygen, reactions 9 and 10 of Table 8-1. The cathodic reactions in chlorinated water are similar to those with iron and take precedence over the corresponding reactions involving dissolved oxygen. If phosphates are present, reaction 13 of Table 8-1 will prevail at the cathode. If iron is present in solution, it will plate out at the cathode to form hydrous ferric oxide, reaction 6 of Table 8-1. If manganese is present in solution in the absence of dissolved iron, it will plate out at the cathode as manganic hydroxide (9) above pH 8.5., as shown in Figure 8-6.

When water is corrosive to copper, bluish-green stains usually appear on white enamel plumbing fixtures. These stains result from a mixture of cupric and cuprous hydroxide. As the water residue left on the enamel evaporates after use of the plumbing fixture, the concentration of cupric and cuprous ions increases until the water is supersaturated with the hydroxides and precipitation takes place. At a pH of 7, cupric copper is soluble to the extent of about 1 ppm and cuprous copper to about 0.01 ppm. At pH 8 cupric copper is soluble to only 0.01 ppm and cuprous copper to only 0.001 ppm. At higher pH values, both hydroxides are so insoluble that the quantity precipitated out of the solution would hardly be sufficient to produce a noticeable stain on the enamel.

A series of tests on the corrosion of copper was conducted under the senior author's direction using specimens of $\frac{3}{4}$-in. copper pipe 36 in. long. The specimens were filled with water samples containing dissolved oxygen and were stoppered at both ends to exclude air. Samples of water were allowed to stand in the pipe specimens for 48 hr. Analyses were then made of the water to determine the pH, the residual dissolved oxygen, and the dissolved copper. Some of the samples were dosed with alkali, either lime or sodium hydroxide. The results were expressed in terms of the copper dissolved in 48 hr. By separate tests

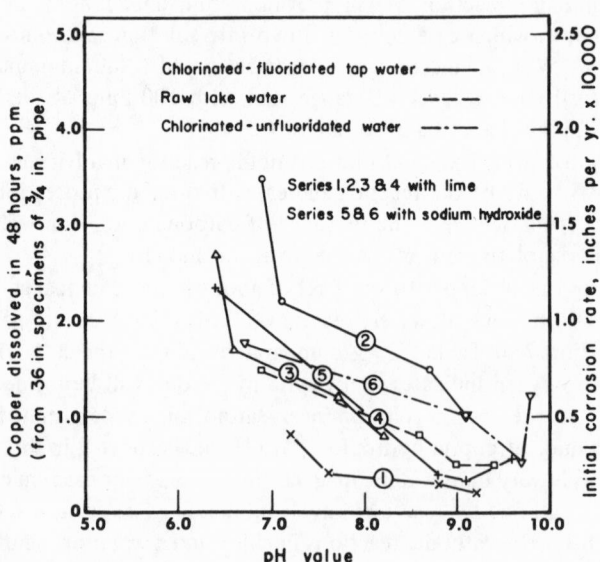

Figure 8-7
Effect of pH value on rate of corrosion of copper.

Table 8-4
Cold-Water Corrosion of
Copper Tubing

pH	Cu (ppm)	in./year
3	5	0.0027
4	2.7	0.0014
5	1.3	0.0007
6	0.8	0.00042
7	0.38	0.00020
8	0.2	0.00011
9	0.1	0.00005

the initial corrosion rate was found to be about 12 times the average rate for 48 hr. These results are shown in terms of the pH value in Figure 8-7. This figure indicates that the rate of corrosion increases markedly at pH values below about 6.5, and that the rate will decrease about 75 percent with a change in pH from 7 to 10.

The rate of corrosion of copper is normally only a small fraction of the rate of corrosion of iron or steel. Figure 8-7 indicates an average rate ranging from 0.000002 to 0.00001 in./year in still water at pH values of 7 to 9, and rates in flowing water were about 12 times greater. The rate of corrosion of copper in seawater ranges from about 0.0001 to 0.004 in./year and averages about 0.002 in./year.

Experiments were conducted by a task group (10) of the American Water Works Association on the corrosion of copper tubing in cold water, using $\frac{3}{4}$-in. tubing 60 ft long with flow at the rate of $\frac{1}{15}$ gal/min. The copper increase in the water was used as a measure of the rate of corrosion. From the results of these experiments the corrosion rates in inches per year have been computed, as shown in Table 8-4. These rates are somewhat higher than the average of the initial rates shown by Figure 8-7.

Copper–silicon alloys that contain 95 to 98.25 percent Cu, 1 to 4 percent Si, and small amounts of other metals are known as silicon bronzes. The silicon bronzes give good resistance to corrosion in both fresh and seawater, the rate of corrosion usually being less than 0.002 in./year in seawater. Other copper alloys will be discussed.

8-8. CORROSION OF ZINC IN FRESHWATER

The principal half-cell reactions in the corrosion of zinc in freshwater, which involve zinc as a constituent, are shown in Table 8-5 together with the cor-

Table 8-5
Corrosion of Zinc in Freshwater

	E_{25}° (V)
Principal anodic reactions	
1. $\frac{1}{2}O_{2\,aq} + 3H_2O + 2\underline{Zn} \rightleftharpoons 2\underline{Zn(OH)_2} + 2H^+ + 2e^-$	+2.11
2. $2OH^- + \underline{Zn} \rightleftharpoons Zn(OH)_2 + 2e^-$	+1.245
3. $\underline{Zn} \rightleftharpoons Zn^{2+} + 2e^-$	+0.763
4. $CO_3^{2-} + \underline{Zn} \rightleftharpoons ZnCO_3 + 2e^-$	+1.06
5. $HSiO_3^- + \underline{Zn} \rightleftharpoons ZnSiO_3 + H^+ + 2e^-$	+0.952
Principal cathodic reactions involving zinc	
6. $\underline{Zn(OH)_2} \rightleftharpoons \frac{1}{2}O_{2\,aq} + Zn^{2+} + H_2O + 2e^-$	−0.927
7. $2Cl + \underline{Zn(OH)_2} \rightleftharpoons 2HOCl + Zn^{2+} + 4e^-$	−1.323

responding standard electrode potentials. Reactions 8 to 13 inclusive of Table 8-1 may also be involved as cathodic reactions. Reaction 8, the plating out of hydrogen, cannot take place except in the complete absence of oxygen or other oxidant. Figure 8-8 shows a comparative study of the anodic and cathodic reactions in the corrosion of zinc over a pH range of 5 to 10.6 and over a wide range of concentration of the dissolved constituents of the reactions. Reaction 1 is not shown because the potential is too high for the scale of the figure. The potential for reaction 1 with 10 ppm of dissolved oxygen ranges from +2.35 at pH 5 to +2.65 at pH 10, and is about 0.03 V lower with 0.1 ppm of dissolved oxygen.

In the presence of dissolved oxygen, reaction 1 prevails at the anode at the start of corrosion to form a zinc hydroxide film. Since zinc hydroxide is quite soluble below pH 8.8, where the solubility is about 0.1 ppm of zinc, this film gives no protection below that pH. At pH values above 8.8, zinc hydroxide is also formed at the anode by reaction 2 beneath the film formed by reaction 1. At higher pH values, the reaction may become so violent that the zinc hydroxide precipitate is pulled out of the metal surface to form a fluffy white deposit of considerable thickness.

Figure 8-8 indicates that the prevailing reaction at the anode in the absence of silicates and dissolved oxygen, at pH values less than about 8, is the solution of zinc as zinc ions, reaction 3 of Table 8-5. In waters of low carbonate and bicarbonate alkalinity and in the absence of silicates and oxygen, the prevailing anodic reaction above a pH of about 8.8 is the plating out of hydroxide ions to

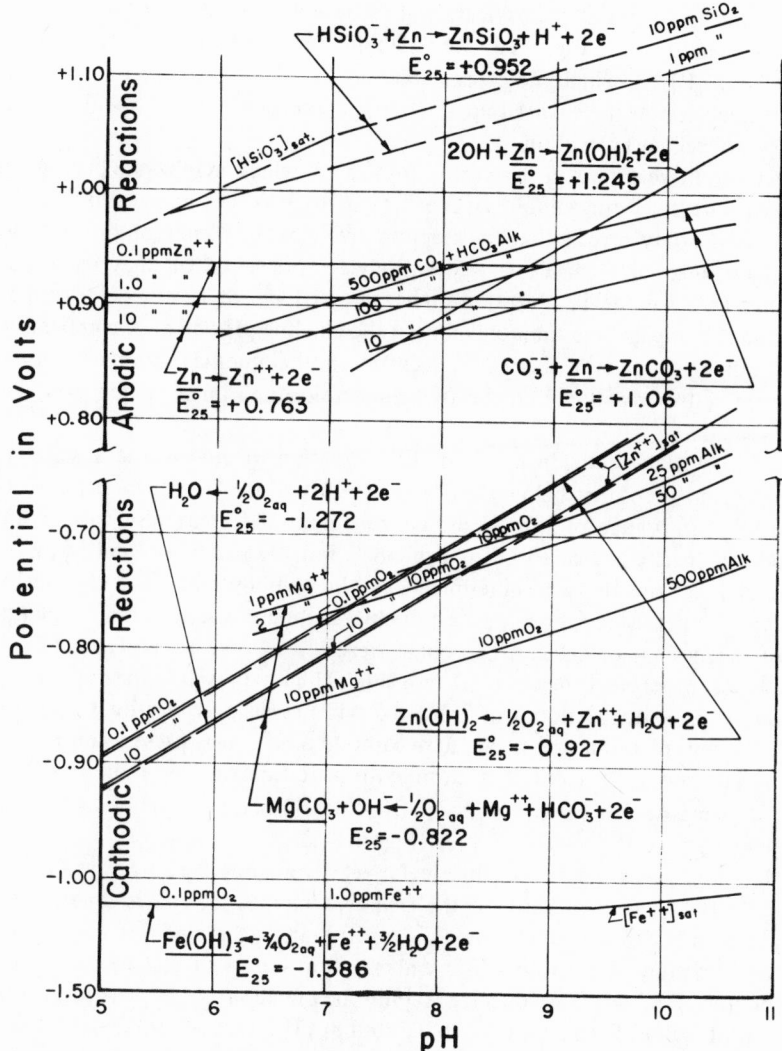

Figure 8-8
Single electrode potentials in zinc corrosion at 25°C.

form zinc hydroxide, reaction 2. With waters of high carbonate and bicarbonate alkalinity, carbonate may plate out at the anode to form zinc carbonate in a pH range above 8, reaction 4 of Table 8-5. Figure 8-8 indicates, however, that carbonate and bicarbonate alkalinity is of little importance in the anodic reactions, unless the concentration exceeds 100 ppm. Figure 8-8 indicates that when silicates are present zinc metasilicate will form at the anode over the pH range above 5, reaction 5 of Table 8-5. The solubility product of zinc metasilicate is about 3.5×10^{-21}, which indicates a solubility of only 0.02 ppm of zinc at pH 5 and much less at higher pH values. The zinc metasilicate film should, therefore, give excellent protection.

Other anodic reactions in the corrosion of zinc, which involve the plating out of hydroxide to form soluble zinc oxides or hydroxides, are possible, but a comparative study reveals that the reaction involving the formation of solid zinc hydroxide, reaction 2 of Table 8-5, will always prevail. Zinc may react with ammonia at the anode, in a reaction similar to reaction 5 of Table 8-3 for copper, to produce complex zinc–ammonia ions in the alkaline pH range. A comparative study indicates, however, that the concentration of ammonia must exceed about 300 ppm if this reaction is to be of comparable importance to reaction 2 of Table 8-5.

The principal cathodic reaction in the corrosion of zinc in the absence of phosphates at pH values below about 8 is the formation of water, reaction 11 of Table 8-1. In waters of low alkalinity, zinc hydroxide may also form at the cathode by means of reaction 6 of Table 8-5, but the reaction is limited by the availability of zinc ions as determined by the solubility product of zinc hydroxide. At pH values above 8, the principal cathodic reactions are the plating out of magnesium or calcium carbonate, reactions 9 and 10 of Table 8-1. If phosphates are present, reaction 13 of Table 8-1 may prevail at the cathode. If chlorine is present, reaction 7 of Table 8-5 will prevail over reaction 6 and over the formation of water and magnesium carbonate. If iron is present in solution, it will plate out at the cathode to form hydrous ferric oxide, reaction 6 of Table 8-1. If manganese is present without iron, it will plate out at the cathode as manganic hydroxide above pH 8.5 (see Figure 8-6).

Zinc is used to galvanize iron and steel by electroplating the zinc onto the iron or steel. The galvanizing protects the iron or steel until the zinc is gone and the metallic iron is exposed to the water. In soft waters, the zinc dissolves as zinc ions fairly rapidly at pH values below about 8; in waters of high carbonate and bicarbonate alkalinity, zinc dissolves rapidly below about pH 7. The protection is fair at pH values from about 7.5 to 9.5, but at pH values above 9.5, where zinc hydroxide is formed at the anode, the reaction may become so violent that the zinc hydroxide precipitate is pulled out of the metal to form a fluffy white deposit.

The rate of corrosion of zinc in quiet freshwater containing dissolved oxygen, at pH values from 4 to 7 and at room temperature, ranges from about 0.009 to 0.07 in./year and is greatest at lower pH values. At pH values from 7 to 9, the rate varies from 0.0007 to 0.002 in./year. The rate is much less in hard water and water in which silicates are present. The rate of corrosion of zinc in seawater ranges from about 0.001 to 0.004 in./year.

Brass is an alloy of copper and zinc, the relative amounts ranging from 60 percent copper and 40 percent zinc in Muntz metal to more than 90 percent copper in red brass. The corrosion of brass usually takes place by *dezincification* (zinc solution), which is greatest in soft waters of low pH value and in brass having a high zinc content. Other copper–zinc alloys are aluminum brass (76 percent Cu, 22 percent Zn, 2 percent Al), yellow brass (67 percent Cu, 33 percent Zn), Admiralty metal (70 percent Cu, 29 percent Zn, 1 percent Sn, 0.05 percent As or Sb), arsenical aluminum brass (76 percent Cu, 22 percent Zn, 2 percent Al, 0.05 percent As) and red brass (85 percent or more Cu, 15 percent or less Zn). Admiralty metal and arsenical aluminum brass generally resist dezincification in soft corrosive waters.

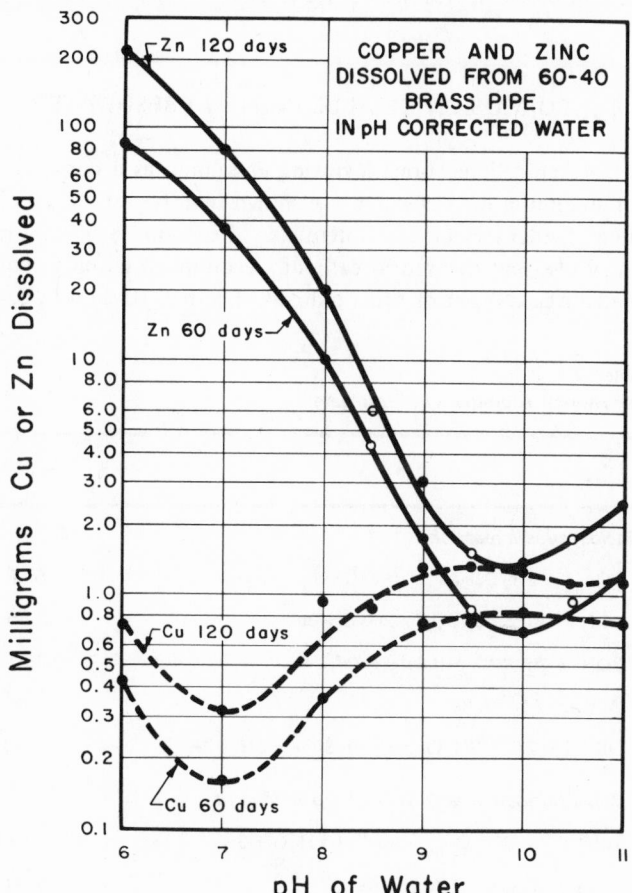

Figure 8-9
Effect of pH value on dezincification of 60–40 brass.

In experimental studies by Moore (*11*) on the solution of copper and zinc from 60–40 brass, it was found that the rate of solution of the two metals was about the same at pH 10, but that zinc dissolved about 240 times as fast as copper at pH values of 6 and 7. The rate of solution of zinc at pH 6 was about 150 times as great as the rate at pH 10, but the rate of copper solution was less at pH 6 and 7 than at pH 10. The results of these experiments are shown in Figure 8-9. Over a period of 60 days the average rate of solution of copper from the red brass was less than 0.000002 in./year at pH 10 where the rate of solution was the highest. In the same period, the average rate of solution of zinc varied from about 0.000002 in./year at pH 10 to about 0.0003 in./year at pH 6. The rate of corrosion of copper–zinc alloys, in the absence of dezincification, ranges from about 0.0001 to 0.001 in freshwater and from about 0.0003 to 0.004 in./year in seawater.

8-9. CORROSION OF ALUMINUM IN FRESHWATER

The principal half-cell reactions involving aluminum as a constituent, in the corrosion of aluminum in freshwater, are shown in Table 8-6, together with the corresponding standard electrode potentials. Reactions 8 to 13 inclusive of Table 8-1 may also be involved as cathodic reactions, reaction 8 requiring the absence of dissolved oxygen or other oxidant. Figure 8-10 shows a comparative

Table 8-6
Corrosion of Aluminum in Freshwater

	E_{25}° (V)
Principal anodic reactions	
1. $\frac{1}{2}O_{2\,aq} + 2H_2O + Al \rightleftharpoons Al(OH)_3 + H^+ + e^-$	+6.98
2. $4OH^- + Al \rightleftharpoons AlO_2^- + 2H_2O + 3e^-$	+2.35
3. $3OH^- + Al \rightleftharpoons Al(OH)_3 + 3e^-$	+2.31
4. $Al \rightleftharpoons Al^{3+} + 3e^-$	+1.66
5. $HSiO_3^- + 2Al + 2H_2O \rightleftharpoons Al_2SiO_5 + 5H^+ + 6e^-$	+1.90
Principal cathodic reactions involving aluminum	
6. $Al(OH)_3 \rightleftharpoons \frac{3}{4}O_{2\,aq} + Al^{3+} + \frac{3}{2}H_2O + 3e^-$	−1.083
7. $AlO_2^- \rightleftharpoons O_{2\,aq} + Al^{3+} + 4e^-$	−0.948
8. $Al(OH)_3 + 3Cl^- \rightleftharpoons 3HOCl + Al^{3+} + 6e^-$	−1.40

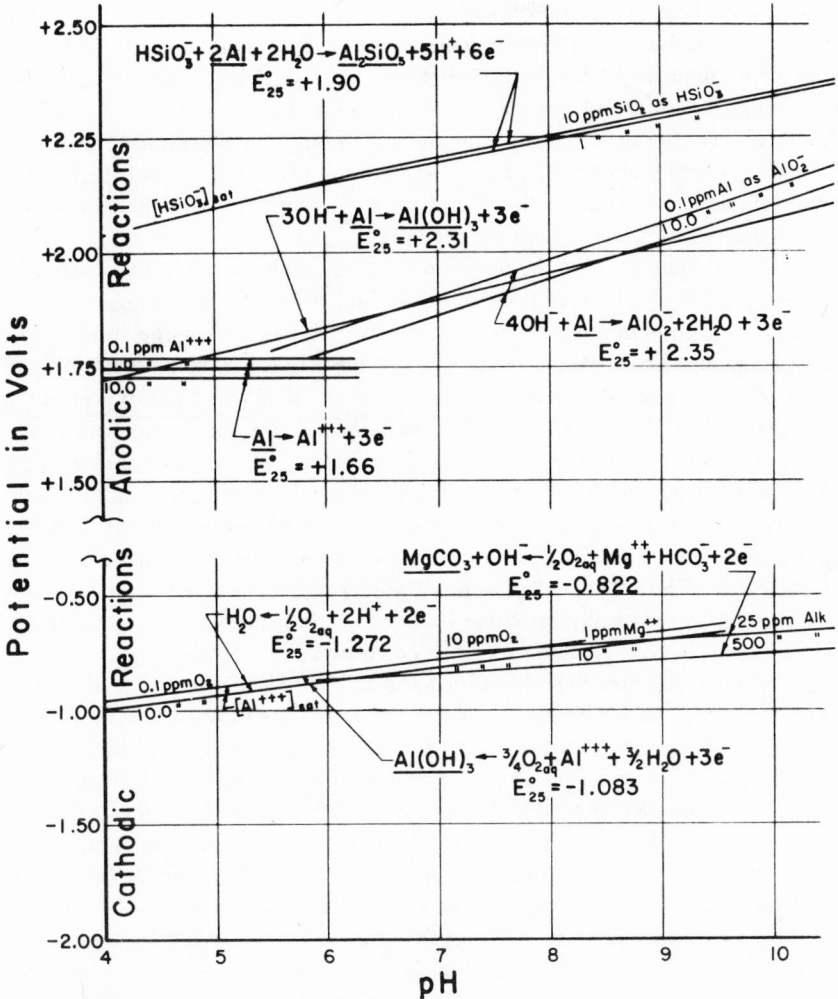

Figure 8-10
Single electrode potentials in aluminum corrosion at 25°C.

study of the anodic and cathodic reactions in the corrosion of aluminum over a pH range of 4 to 10.6 and over a wide range of concentration of the dissolved constituents of the reactions. Reaction 1 is not shown because the potential is too high for the scale of the figure. The potential for reaction 1 with 10 ppm of dissolved oxygen ranges from +7.11 at pH 4 to 7.47 at pH 10, and is about 0.06 V lower with 0.1 ppm of dissolved oxygen.

In the presence of dissolved oxygen, reaction 1 prevails at the anode at the

start of corrosion to form an aluminum hydroxide film. Below pH 5, aluminum hydroxide dissolves to yield aluminum ions and hydroxide ions; the solubility product is about 4×10^{-33} at $25°C$. Above pH 7, $Al(OH)_3$ produces hydrogen and aluminate ions, and the solubility product is about 5.5×10^{-13} at $25°C$. The solubility products indicate that the aluminum hydroxide film is soluble to the extent of only about 0.1 ppm of Al between pH 5 and 7, and is very much more soluble at pH values below 5 and above 7. Hence, the film gives little protection against corrosion except between pH 5 and 7.

Figure 8-10 indicates that the prevailing reaction at the anode in the absence of silicates and dissolved oxygen, at pH values from 5 to 7, is the plating out of hydroxide to form aluminum hydroxide. Below pH 5 the prevailing anodic reaction in the absence of silicates and oxygen is the solution of aluminum ions, and above pH 7 the prevailing reaction in the absence of silicates and oxygen is the plating out of hydroxide to form aluminate ions. Thus, aluminum corrodes at both low and high pH values, but is protected in the pH range of 5 to 7. Carbonates are of no significance in the anodic reactions in the corrosion of aluminum, since aluminum carbonates are soluble. If silicates are present, bialuminum silicate will be produced at the anode over the entire pH range, reaction 5 of Table 8-6. The calculated solubility product of bialuminum silicate indicates that the film is practically insoluble over the entire pH range and should, therefore, be very effective protection against corrosion.

Figure 8-10 indicates that the principal cathodic reaction in the corrosion of aluminum at pH values below about 8 is the formation of water, reaction 11 of Table 8-1. Aluminum may also plate out with dissolved oxygen to form aluminum hydroxide at the cathode, in accordance with reaction 6 of Table 8-6. Figure 8-10, however, indicates that this potential is about the same as that for the reaction resulting in the formation of water, even with aluminum ions present in saturation concentration as determined from the solubility product of aluminum hydroxide. A comparative study of reactions 6 and 7 of Table 8-6 indicates that both may take place at the cathode in the corrosion of aluminum at pH values below about 8. The potentials are the same for the two reactions if the water is saturated with both aluminum ions and aluminate ions. The saturation concentrations of aluminum ions and aluminate ions, as determined from the solubility products of aluminum hydroxide, are about equal at pH 5.4, but the solubility of aluminum in both forms between pH 5 and 7 is less than 0.15 ppm. The saturation concentration of aluminate ion is approximately equal to 8×10^{-10} divided by the cube root of the saturation concentration of the aluminum ions. If chlorine is present, reaction 8 will prevail at the cathode at pH values below about 8. If silicates are present, aluminum is far less soluble, since its solubility is determined from the solubility product of bialuminum silicate.

Figure 8-10 indicates that at pH values above about 8 the principal cathodic reactions in the corrosion of aluminum are the plating out of magnesium and calcium carbonates. At high alkalinities, these reactions may be the principal cathodic reactions, even down to a pH value of 6. If iron is present in solution, it will plate out at the cathode to form hydrous ferric oxide, reaction 6 of Table 8-1. If manganese is present without iron, it will plate out as manganic hydroxide at the cathode above pH 8.5 (see Figure 8-6).

The rate of corrosion of aluminum in freshwater free of silicates in the pH range of 5 to 7.5 is very low, probably on the order of 0.00004 to 0.001 in./year, although reliable test data are sparse. At low and at high pH, the rate of corrosion may increase to 0.06 to 0.6 in./year in the absence of silicates. As previously indicated, silicates give protection over a wide range of pH. The rate of corrosion of aluminum in seawater has been reported by one observer at 0.0034 in./year.

8-10. CORROSION OF TIN IN FRESHWATER

The principal half-cell reactions that occur in the corrosion of tin in freshwater and that involve tin as a constituent are shown in Table 8-7, together with the corresponding standard electrode potentials. Figure 8-11 indicates the reactions that will prevail at both the anode and the cathode, except reaction 1, whose potential is too high for the scale of the figure. It will be noted that the potentials for reactions 5 and 6 at the cathode are both so low that where dis-

Table 8-7
Corrosion of Tin in Freshwater

	E_{25}° (V)
Principal anodic reactions	
1. $\frac{3}{4}O_{2\,aq} + \frac{5}{2}H_2O + \underline{Sn} \rightleftharpoons \underline{Sn(OH)_4} + H^+ + e^-$	+3.86
2. $\underline{Sn} \rightleftharpoons Sn^{2+} + 2e^-$	+0.136
3. $2OH^- + \underline{Sn} \rightleftharpoons \underline{Sn(OH)_2} + 2e^-$	+0.921
4. $3OH^- + \underline{Sn} \rightleftharpoons HSnO_2^- + H_2O + 2e^-$	+0.909
Principal cathodic reactions involving tin	
5. $\underline{Sn(OH)_4} \rightleftharpoons O_{2\,aq} + Sn^{2+} + 2H_2O + 2e^-$	−2.43
6. $\underline{Sn(OH)_4} \rightleftharpoons O_{2\,aq} + HSnO_2^- + 3H^+ + 2e^-$	−2.79
7. $\underline{Sn(OH)_4} + 2Cl^- \rightleftharpoons 2OCl^- + Sn^{2+} + 2H_2O + 2e^-$	−3.31

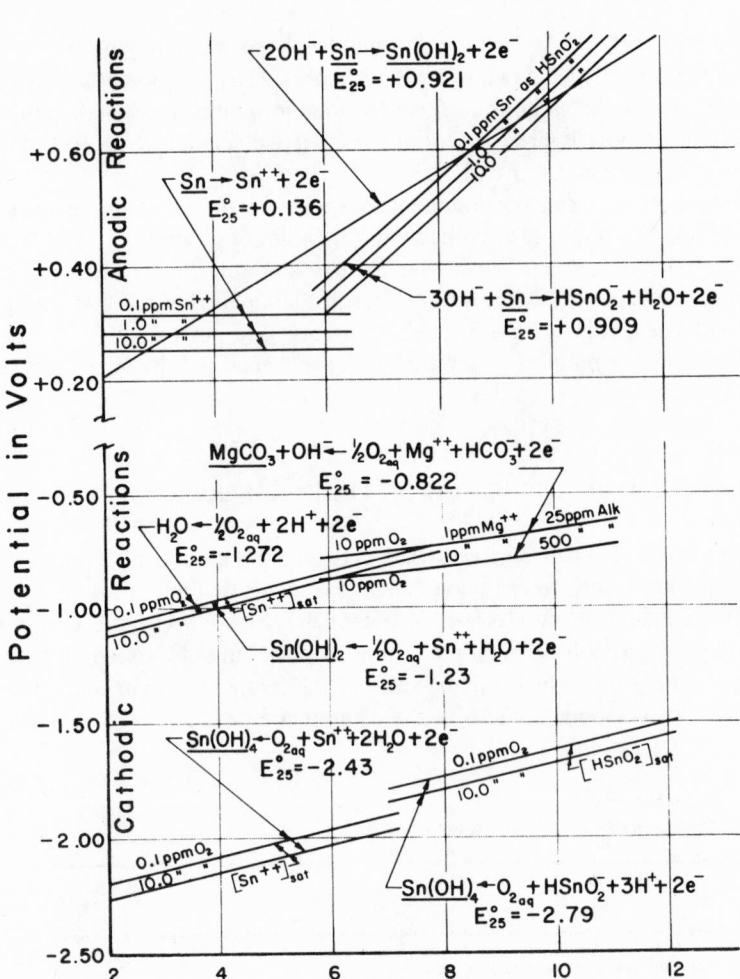

Figure 8-11
Single electrode potentials in tin corrosion at 25°C.

solved oxygen is present, these reactions should prevail over all others in the absence of chlorine, including reactions 6 to 13 of Table 8-1.

The potential of reaction 1 with 10 ppm of dissolved oxygen ranges from +3.82 at pH 2 to +4.30 at pH 10; it is about 0.09 V lower with 0.1 ppm of dissolved oxygen.

In the presence of dissolved oxygen, reaction 1 prevails at the anode at the

start of corrosion to form a stannic hydroxide film. Stannic hydroxide is amphoteric, dissolving to yield stannic ions at low pH and stannate ions at high pH. The solubility product of about 10^{-57} indicates that the saturation concentration of stannic ions is only about 0.0001 ppm of tin at pH 2 and is much less at higher pH values. The solubility product at high pH of about 5×10^{-23} indicates that the saturation concentration of stannate ion, $Sn(OH)_6^{2-}$ is about 0.06 ppm of tin at pH 8 and very much less at lower pH values. At pH values above 8, the saturation concentration of stannate ion increases greatly, being about 6.0 ppm of tin at pH 9 and 600 ppm at pH 10. The stannic hydroxide film gives excellent protection, therefore, over the pH range from 2 to 8, but is of no value below pH 1 and above pH 8.

Figure 8-11 indicates that in the absence of dissolved oxygen or beneath the stannic hydroxide film, tin goes into solution at the anode as stannous ion at pH values below about 3.5 and as stannite ion at pH values above about 8.5. Between pH values of approximately 3.5 and 8.5, hydroxide plates out at the anode to form stannous hydroxide. In this pH range, tin is very resistant to corrosion with distilled water or soft freshwaters. It is for this reason that tin has been used so successfully in the canning industry as a plating on steel.

Figure 8-11 indicates that the prevailing cathodic reactions in the corrosion of tin will be the formation of stannic hydroxide by means of reactions 5 and 6, both involving dissolved oxygen. Reaction 7 will prevail if chlorine is present. Figure 8-11 indicates that the reaction with dissolved oxygen which results in the formation of stannous hydroxide has approximately the same potential as the reaction resulting in the formation of water. The potentials for these reactions and the reaction involving the formation of magnesium carbonate are shown on Figure 8-11. It is obvious from the figure that reactions 5, 6, and 7 of Table 8-7 will prevail at the cathode.

The rate of corrosion of tin in soft freshwaters is so low that no consistent values are available for it. Some experiments indicate a gain in weight of test specimens, probably because of the addition of hydroxide as indicated by reactions 1, 3, 5, and 6. Tests indicate that the rate of corrosion of tin in seawater is approximately 0.00003 to 0.00009 in./year.

The copper–tin alloys are known as bronzes, and since they contain a small amount of phosphorus, they are also called phosphor bronzes. Some of the bronzes also contain zinc, lead, and other metals. A few of the commonly used alloys are 88 percent Cu, 8 percent Sn, 4 percent Zn; 87.0 percent Cu, 8 percent Sn, 1.0 percent Pb, 4 percent Zn; 92 percent Cu, 8 percent Sn; 95 percent Cu, 5 percent Sn; 88 percent Cu, 10 percent Sn, 2 percent Zn; 85 percent Cu, 5 percent Sn, 5 percent Zn, 5 percent Pb; 88 percent Cu, 4 percent Sn, 4 percent Zn, 4 percent Pb. These alloys show good corrosion resistance to both freshwater and seawater.

8-11. CORROSION OF NICKEL IN FRESHWATER

The principal half-cell reactions in the corrosion of nickel in freshwater for which thermodynamic data are available and that involve nickel as a constituent are shown in Table 8-8, together with the corresponding standard electrode potentials. Figure 8-12 shows a comparative study of these anodic and cathodic reactions, except reaction 1, over a pH range of 2 to 12 and over a wide range of concentration of the dissolved constituents of the reactions. Reaction 1 is not shown because the potentials are too high for the scale of the figure. The reactions shown in Table 8-8 and Figure 8-12 are not an adequate picture of the corrosiveness of nickel. Nickel is known to be highly resistant to corrosion over a wide range of pH values, and yet Figure 8-12 tends to indicate that the metal will corrode readily.

The potential of reaction 1 with 10 ppm of dissolved oxygen ranges from +1.13 V at pH 2 to +1.60 V at pH 10; it is about 0.03 V lower with 0.1 ppm of dissolved oxygen. In the presence of dissolved oxygen, reaction 1 prevails at the start of corrosion to form a nickelous hydroxide film. The solubility product of nickelous hydroxide indicates, however, that the film is quite soluble below pH 8.5 to 9. It is difficult, therefore, to account for the high resistance of nickel to corrosion at low pH values on the basis of the nickelous hydroxide film. Other more insoluble films may be formed at the anode, possibly of higher oxides. The free energy of nickelic ion is not known, and there is some evidence for the existence of tetravalent nickel, although the free energy of this ion is not available. The reported free energies of nickelic hydroxide and of NiO_2 are not

Table 8-8
Corrosion of Nickel in Freshwater

	E_{25}° (V)
Principal anodic reactions	
1. $\frac{1}{2}O_{2\,aq} + 3H_2O + \underline{2Ni} \rightleftharpoons 2Ni(OH)_2 + 2H^+ + 2e^-$	+1.06
2. $\underline{Ni} \rightleftharpoons Ni^{2+} + 2e^-$	+0.25
3. $2OH^- + \underline{Ni} \rightleftharpoons Ni(OH)_2 + 2e^-$	+0.72
4. $3OH^- + \underline{Ni} \rightleftharpoons HNiO_2^- + H_2O + 2e^-$	+0.595
Principal cathodic reactions involving nickel	
5. $Ni(OH)_2 \rightleftharpoons \frac{1}{2}O_{2\,aq} + Ni^{2+} + H_2O + 2e^-$	−0.912
6. $\underline{Ni(OH)_2} + 2Cl^- \rightleftharpoons 2HOCl + Ni^{2+} + 4e^-$	−1.315

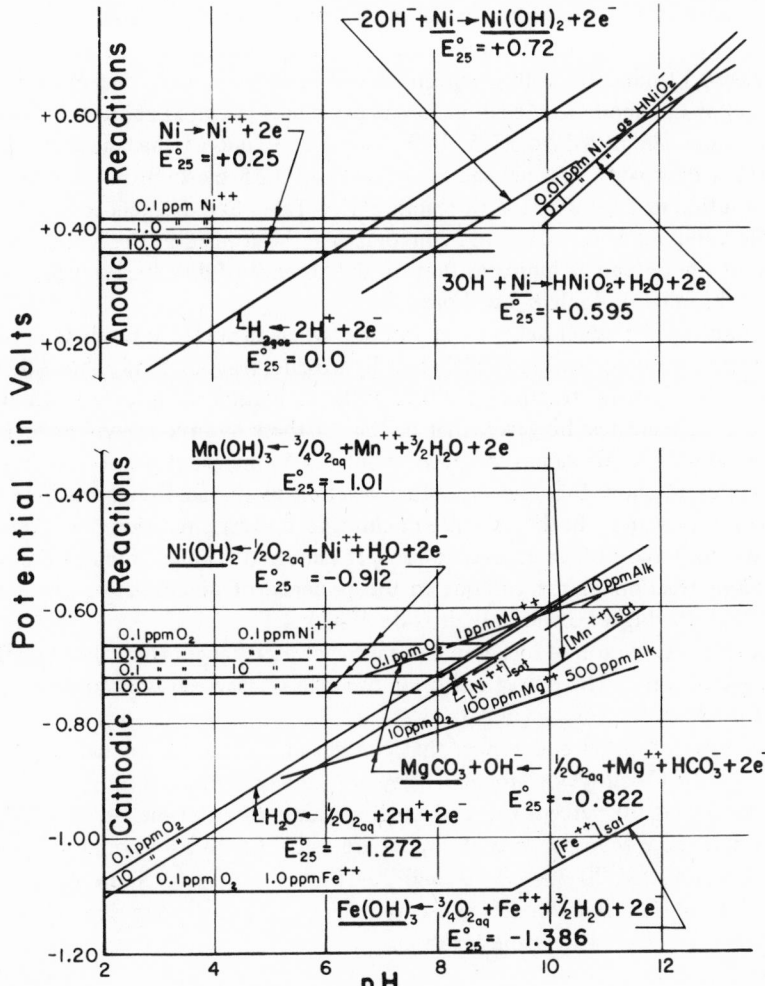

Figure 8-12
Single electrode potentials in nickel corrosion at 25°C.

large enough for reactions involving them to prevail over similar reactions, 1, 3, and 5 of Table 8-8 involving nickelous hydroxide.

Figure 8-12 indicates that in the absence of dissolved oxygen, the prevailing anodic reaction below a pH of about 8.5 is the solution of the metal as nickelous ion and that the prevailing cathodic reaction below a pH of about 8 is the formation of water by the combination of hydrogen ions with dissolved oxygen. Thus no oxide film would be formed at either anode or cathode to protect the metal.

It is possible that some higher oxide of nickel may be formed at the cathode, as well as at the anode, to give the protection the metal is known to have.

Figure 8-12 indicates that, in the absence of dissolved oxygen, nickel goes into solution at the anode as nickelous ion at pH values less than about 8.5. At pH values from about 8.5 to 12.5, hydroxide plates out at the anode to form nickelous hydroxide. At pH values above about 12.5 the hydroxide goes back into solution in accordance with reaction 4 of Table 8-8. Nickel carbonate and soluble compounds of nickel and ammonia may be formed at the anode, but a study of the potentials indicates that, in the absence of dissolved oxygen, reactions 2, 3, and 4 of Table 8-8 will prevail.

A study of the potentials at the cathode when dissolved oxygen is present indicates that the formation of nickelous hydroxide, reaction 5 of Table 8-8, will prevail over a similar reaction resulting in the formation of nickelic hydroxide. Figure 8-12 indicates, however, that neither of these hydroxides will be formed at the cathode at pH values less than about 8. As indicated previously, Figure 8-12 tends to show that the prevailing reaction in the acid region will be the formation of water. In the presence of chlorine, however, reaction 6 will prevail over the formation of water, even at low pH values. It is quite probable that the prevailing reaction at the cathode in the presence of dissolved oxygen is the formation of a higher oxide or hydroxide of nickel.

The rate of corrosion of nickel in natural freshwaters ranges from about 0.0001 to 0.001 in./year. The rate of corrosion of nickel in seawater is usually less than 0.005 in./year.

The alloys of nickel and copper that contain more than 50 percent nickel are known as Ni–Cu alloys or Monels. They are usually composed of 63 to 67 percent Ni, about 30 percent Cu, and small amounts of other metals. The alloys containing more than 50 percent copper are called Cu–Ni alloys or cupronickel. Cupronickels with 70–30, 80–20, and 90–10 ratios of Cu to Ni are used. The corrosion rates of both Monels and cupronickels are usually less than 0.001 in./year in natural freshwaters and seawater.

8-12. CORROSION OF CHROMIUM IN FRESHWATER

Table 8-9 shows the principal half-cell reactions in the corrosion of chromium in freshwater in which chromium is involved as one of the constituents. Figure 8-13 shows a comparative study of these reactions, except reaction 1, over a pH range of 2 to 12 and over a wide range of concentration of the dissolved constituents of the reactions. For comparative purposes, the single electrode potentials of the cathodic reactions in which water and magnesium carbonate are formed are also shown in Figure 8-13. Reaction 1 is not shown because the potentials are too high for the scale of the figure.

Table 8-9
Corrosion of Chromium in Freshwater

	E_{25}° (V)
Principal anodic reactions	
1. $\frac{1}{2}O_{2\,aq} + 2H_2O + \underline{Cr} \rightleftharpoons Cr(OH)_3 + H^+ + e^-$	+4.50
2. $\underline{Cr} \rightleftharpoons Cr^{2+} + 2e^-$	+0.913
3. $3OH^- + \underline{Cr} \rightleftharpoons Cr(OH)_3 + 3e^-$	+1.484
4. $Cl^- + \underline{Cr} \rightleftharpoons CrCl^{2+} + 3e^-$	about +1.28
Principal cathodic reactions involving chromium	
5. $\underline{Cr(OH)_3} \rightleftharpoons \frac{3}{4}O_{2\,aq} + Cr^{2+} + \frac{3}{2}H_2O + 2e^-$	−1.98
Principal cathodic reactions for stainless steel	
6. $\underline{Fe(CrO_2)_2} \rightleftharpoons \underline{Fe} + 2O_{2\,aq} + 2Cr^{2+} + 4e^-$	−2.62
7. $\underline{Fe(CrO_2)_2} \rightleftharpoons Fe^{2+} + 2O_{2\,aq} + \underline{2Cr} + 2e^-$	−6.42

The potential of reaction 1 with 10 ppm of dissolved oxygen ranges from +4.52 V at pH 2 to +5.05 V at pH 12; it is about 0.05 V lower with 0.1 ppm of dissolved oxygen. In the presence of dissolved oxygen, reaction 1 prevails at the start of corrosion to form a chromic hydroxide film. The solubility product of chromic hydroxide is about 3.2×10^{-38} at 25°C, which indicates that the film is very insoluble at pH values above 3.4, where it is soluble to the extent of about 0.1 ppm of chromic ion. Therefore, this film should give excellent protection at pH values above 3.4.

Figure 8-13 indicates that underneath the chromic hydroxide film formed by reaction 1 or in the absence of dissolved oxygen and other oxidants, and in the absence of halides, the principal anodic reaction in the corrosion of chromium at pH values less than about 7 is the solution of chromium as chromous ion, reaction 2. At pH values above about 7, the principal anodic reaction is the plating out of hydroxide to form chromic hydroxide, reaction 3 of Table 8-9. Chromium also goes into solution at the anode as chromic ion at pH values less than about 7, but the equilibrium concentration of chromic ion in relation to chromous ion is so low as to be negligible. The equilibrium concentration of chromic ion is approximately 3×10^{-9} times the $\frac{3}{2}$-power of the concentration of chromous ion.

Numerous complex chromic ions are formed with the halides, ammonia, cyanide, and nitrite, but little is known about their free energies. Chromium is sub-

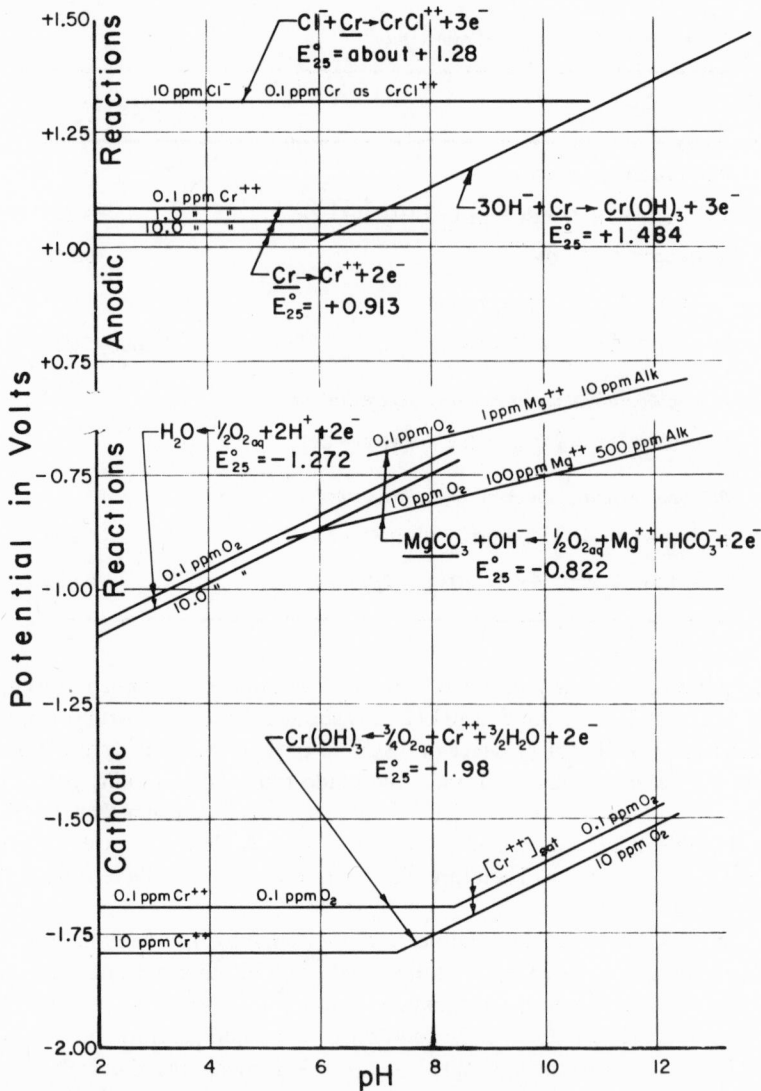

Figure 8-13
Single electrode potentials in chromium corrosion at 25°C.

ject to attack in waters containing these substances and is, therefore, not resistant to seawater. Reaction 4 of Table 8-9 represents anodic attack by chlorides. Figure 8-13 indicates that this reaction will prevail over reactions 2 and 3 of Table 8-9 at pH values up to about 11. The complex ion is very stable and will not enter, to any appreciable extent, into a cathodic reaction to form chromic hydroxide.

Figure 8-13 indicates that the principal cathodic reaction in the corrosion of chromium is the plating out of chromous ion, together with dissolved oxygen, to form chromic hydroxide, reaction 5. At pH values above about 8, the potential of this reaction is limited by the saturation concentration of chromous ion as determined from the solubility product of chromous hydroxide, about 9.1×10^{-18}. Figure 8-13 indicates that the potential of reaction 5 of Table 8-9 is so low that it may be expected to prevail over all other cathodic reactions involving chromium ions, including a similar reaction with chlorine.

Chromium, although very corrodible in air containing seawater droplets, is known to be very highly resistant to corrosion when exposed to aqueous solutions containing dissolved oxygen or other oxidants. Undoubtedly, this is due to the formation of chromic hydroxide at both the anode and the cathode in accordance with reactions 1, 3, and 5 of Table 8-9. The rate of corrosion in natural freshwaters is generally less than 0.001 in./year.

It is probable that the resistance of stainless steels to corrosion is due primarily to their chromium content. It will be noted from Figures 8-4, 8-12, and 8-13, that the potentials of the anodic reactions in the corrosion of chromium are higher than the potentials of the corresponding reactions in the corrosion of iron and nickel. Hence, the solution of chromous ions should prevail over the solution of metallic iron or nickel. The prevailing cathodic reaction should be the formation of ferrous chromite with the metallic iron, reaction 6 of Table 8-9. If ferrous ions are present, ferrous chromite may also be formed on metallic chromium, reaction 7 of Table 8-9. The solubility product of ferrous chromite, about 1.6×10^{-35}, indicates that this film is practically insoluble.

The most common austenitic stainless steel is 18–8, which contains, in addition to iron, about 18 percent chromium and 8 percent nickel. Other types are 25–12 and 25–20. The stainless steels are said to derive their corrosion resistance from their passivity. Their resistance is excellent in aerated and oxidizing media, but it is lower in reducing media and in media containing halides. Manufacturers of stainless steels *passivate* the alloys as a final treatment by using an oxidizing agent such as concentrated nitric acid. The result is doubtless the formation of a chromic hydroxide film at the anode and a ferrous chromite film at the cathode.

8-13. INFLUENCE OF ELECTROLYTES, FILMS, AND COATINGS ON RATE OF CORROSION

The previous discussion indicates the nature of the half-cell reactions that prevail at the anode and cathode in the corrosion of various metals. Many of these reactions result in the formation of hydroxide (or hydrous oxide) and other solid films at the anode and cathode. As previously shown, these films are of the utmost importance in corrosion, but very little of a specific nature is known about the effectiveness of these films in retarding corrosion. Studies by Hatch and Rice (*12*) on the corrosion of steel indicate that the anions in the electrolyte influence the effectiveness of the oxide films.

The experiments by Hatch and Rice (*12*) included some tests with distilled water containing dissolved air. It was shown that the corrosion rate with 5-day exposure in quiescent distilled water was about 0.0047 in./year, whereas the rate in agitated distilled water exposed to the atmosphere was only about one-thirteenth as high, or 0.00036 in./year. The pH of the distilled water ranged from about 4.8 to 5.9. Table 8-1 and Figure 8-4 indicate that under the conditions of these experiments a passive film of ferric hydroxide forms at the anode, and that under this film iron goes into solution as ferrous ions. The ferrous ions plate out at the cathode as ferric hydroxide. The experiments indicated that the faster dissolved oxygen was supplied to the interface, the better was the protective film.

The corrosion rate during a 5-day period of exposure in agitated distilled water containing varying amounts of calcium bicarbonate (up to about 100 ppm) was even less than in the distilled water. The initial pH of the bicarbonate-treated waters was 6.5 to 6.9. The pH rose quite rapidly, reaching 7.8 to 8 at 25 ppm of calcium bicarbonate and 8.2 to 8.4 at higher bicarbonate concentrations. Figure 8-4 indicates that, in this set of experiments, the anodic reaction beneath the passive film was the solution of iron as ferrous ions and the cathodic reaction was the same as that with distilled water, the formation of ferric hydroxide. The increase in pH value indicates a liberation of hydroxide like that which occurs if calcium carbonate is plated out at the cathode, reaction 10 of Table 8-1. An analysis of the amount of hydroxide liberated, however, indicates that no more than about 4 percent of the current carried in the corrosion reactions can be accounted for by the plating out of calcium carbonate. The remainder is associated with the plating out of ferric hydroxide. The conductivity of the distilled water used was reported by Hatch and Rice as 1.5 to 2.5 μS at 25°C. The addition of the calcium bicarbonate would greatly increase the conductivity and should, therefore, increase the rate of corrosion, other things being equal. Since the rate of corrosion was decreased, it is evident that the presence of the bicarbonate substantially improved the protective action of the film.

Tests were also made by Hatch and Rice in which calcium sulfate solutions and calcium chloride solutions were added to the distilled water and agitated throughout the test period. With the sulfate solution, the corrosion rate during the 5-day period was increased to about 0.04 in./year, more than 100 times the rate in agitated distilled water. The rate of corrosion was about 0.033 in./year for a sulfate concentration as low as 2 ppm, and increased to about 0.04 in./year with concentrations ranging from 25 to 100 ppm. The rate of corrosion during the 5-day period with calcium chloride added to the distilled water increased to about 0.017 in./year at 5 ppm of chloride and increased further to about 0.033 in./year at a chloride concentration of 100 ppm.

The increased rate of corrosion accompanying the presence of sulfate and chloride ions cannot be accounted for wholly in terms of the increase in conductivity. It is estimated that the conductivity with 100 ppm of anion was about 34 times the conductivity of the distilled water for the sulfate solution, and about 90 times the conductivity of the distilled water for the chloride solution. The increase in the rate of corrosion was about equal to the increase in conductivity for the chloride solution, but it was about three times the increase in conductivity for the sulfate solution. It is evident, therefore, that the sulfate ions interfered with the effectiveness of the ferric hydroxide protective coating.

Further tests were made by Hatch and Rice with sulfate and chloride solutions, to which varying concentrations of calcium bicarbonate were added. It was found that, with increases in concentration of bicarbonate, the rate of corrosion in both the chloride and the sulfate solutions was materially decreased. One hundred ppm of calcium bicarbonate reduced the rate of corrosion in the chloride solution about 75 percent and in the sulfate solution about two-thirds. Examination of the test specimens indicated that a substantial portion of the carbonate added was deposited as calcium carbonate.

It has long been known that certain metals may be made *passive* or more resistant to corrosion by various means. Most authorities now believe that passivity is the result of the formation of an oxide film on the metal surface. Well-known chemical *passivators* for iron are concentrated nitric acid, sodium nitrite, and potassium dichromate. Table 8-10 shows the probable half-cell reactions that take place at the anode with each of these passivators. It will be noted that the standard potentials of these anodic reactions are very high. The actual potentials will be somewhat lower, depending upon the concentration of the reactants; but they will nevertheless be high. Since most waters will be saturated with nitrogen gas from the atmosphere, the nitrogen gas formed in reactions 1, 2, 4, and 5 may have to be released from solution. The half-cell potentials will thus be reduced by the amount of the nitrogen overvoltage. The effect of these passivators is to produce a film at the anode similar to the film produced by dissolved oxygen.

Table 8-10 also shows that films will be produced at the cathode which are similar to the films produced by dissolved oxygen. A comparison of reaction 7 of Table 8-10 with reaction 6 of Table 8-1 indicates that nitrates are about as effective as dissolved oxygen in producing ferric hydroxide at the cathode. A similar comparison with reaction 8 of Table 8-10 shows that nitrites are much more effective than dissolved oxygen. Reaction 9 with dichromate, however, will not prevail over the reaction with dissolved oxygen to form ferric hydroxide at the cathode. In the case of copper, the reaction with nitrite will prevail over a similar reaction with dissolved oxygen to produce cupric hydroxide at the cathode. Neither nitrates nor dichromates will produce cupric hydroxide at the cathode, because reaction 11 of Table 8-1, the formation of water, will prevail.

Table 8-10
Passivators in Iron and Copper Corrosion

	E_{25}° (V)
Anodic reactions	
1. $NO_3^- + 3H_2O + \underline{2Fe} \rightleftharpoons 2Fe(OH)_3 + \frac{1}{2}N_2 + e^-$	+5.88
2. $3NO_2^- + 6H_2O + \underline{4Fe} \rightleftharpoons 4Fe(OH)_3 + \frac{3}{2}N_2 + 3e^-$	+4.32
3. $Cr_2O_7^{2-} + 3H_2O + \underline{3Fe} \rightleftharpoons Fe(CrO_2)_2 + \underline{2Fe(OH)_3} + 2e^-$	+3.56
4. $NO_3^- + \underline{6Cu} \rightleftharpoons 3Cu_2O + \frac{1}{2}N_2 + e^-$	+3.41
5. $NO_2^- + \underline{4Cu} \rightleftharpoons 2Cu_2O + \frac{1}{2}N_2 + e^-$	+2.68
6. $Cr_2O_7^{2-} + 3H_2O + \underline{8Cu} \rightleftharpoons 4Cu_2O + 2Cr(OH)_3 + 2e^-$	+2.06
Cathodic reactions	
7. $\underline{2Fe(OH)_3} + \frac{1}{2}N_2 \rightleftharpoons NO_3^- + 3H_2O + 2Fe^{2+} + 3e^-$	−1.375
8. $\underline{4Fe(OH)_3} + \frac{3}{2}N_2 \rightleftharpoons 3NO_2 + 6H_2O + 4Fe^{2+} + 5e^-$	−1.89
9. $\underline{2Fe(OH)_3} + \underline{Fe(CrO_2)_2} \rightleftharpoons Cr_2O_7^{2-} + 3Fe^{2+} + 3H_2O + 4e^-$	−1.12
10. $\underline{2Cu(OH)_2} + \frac{1}{2}N_2 \rightleftharpoons NO_2^- + 2Cu^{2+} + 2H_2O + 3e^-$	−1.16

The passivators appear to be effective in reducing corrosion only when they are used in relatively high concentrations, for example, 100 to 1,000 ppm or more. It should be noted that corrosion in water may be effectively inhibited either by the complete absence of dissolved oxygen or other oxidants or by heavy concentrations of oxidizing agents. The concentration of dissolved oxygen is usually limited by its saturation concentration in water, ranging for freshwater at atmospheric pressure from about 14 ppm near the freezing point to about 8 ppm at 25°C. It is not usually feasible, therefore, to use oxygen as a passivator. In the absence of dissolved oxygen, low concentrations of a passivator may produce corrosion. In the senior author's practice, a well water, devoid of dissolved oxygen but containing from 1 to 30 ppm of nitrate, about 35 ppm of chloride, and 180 ppm of sulfate, was found by test to corrode iron at a rate of about 0.006 in./year.

It has been shown that the effectiveness of cement lining (on the interior surfaces of iron pipes) in reducing corrosion is primarily the result of the high pH, about 12.4, in the pore water of the lining. Since the soluble constituents of cement slowly leach out of the lining, it is common practice to protect cement lining with a bituminous seal coat. If the seal coat wears away, the effectiveness of the cement lining may be expected to diminish with time. The leaching process is discussed in more detail in Section 8-14.

The effectiveness of seal coats and pipe linings of types other than cement depends upon their porosity and thickness. Rubber linings and some synthetic rubber linings are impervious to water. If they adhere well and do not deteriorate with age, they will prevent corrosion. Bituminous enamel linings absorb small amounts of water and are, therefore, porous. In time, water will penetrate through such a lining to the metal wall of the pipe, which becomes the anode of the corrosion cell. Metallic ions will enter into solution and diffuse outward through the pore water to the lining–water interface, where the cathodic reaction will take place. The rate of corrosion is, therefore, controlled by the rate of molecular diffusion through the pores, and is least for thick linings of low porosity.

8-14. CORROSION OF CONCRETE, CEMENT MORTAR, AND ASBESTOS-CEMENT

Concrete, cement mortar, and asbestos-cement in contact with water are subject to corrosion. The corrosion is of two types: (1) *leaching* of the cement into the water and (2) *absorption* from the water of substances such as sulfates that react with soluble materials in the cement pores to damage the hydrated cement. The rate of corrosion depends upon the character and amount of soluble materials in the hydrated cement, the porosity of the concrete or cement product, and the composition of the water in contact with the structure.

The rate of leaching is substantially independent of the quality of the water for most natural waters, unless the water is extremely hard, with very high pH, or is very acid. The materials leaching from concrete, cement mortar, or asbestos-cement come from the interior after the solubles are dissolved from the interface. The leaching process thus increases the porosity. The face of good concrete or mortar may not be visibly affected by leaching because the pores are too fine to be seen. The rate of leaching is controlled by Fick's* law of diffusion, Eq. (3-2). The substances dissolved from the hydrated cement diffuse out through the pores to the water in contact with the structure.

All cement products exposed to water leach continuously at slowly decreasing rates over long periods of time. The substances that leach out consist primarily of the lime in the hydrated cement and some silicates and aluminates. In a pipeline carrying water, leaching is evidenced by increases in alkalinity, pH, and calcium in the water as it flows along the pipe. With accurate solids balances, from water analyses, at both ends of a pipeline, the rate of leaching in inches of solids per year may be estimated from the flow, the length and diameter of the

*Adolf Fick (1829–1901), German physiologist.

pipe, and the specific gravity of the solids leached. The specific gravity of the hydrated cement ranges from about 2.1 for concrete cast in molds to about 2.3 for centrifugally cast concrete and mortar. The specific gravity of hydrated cement from asbestos-cement pipe is probably less than 2, since the pipe material weighs from 108 to 115 lb/ft^3.

After about 3 years use with soft freshwater, a 30-in. prestressed reinforced concrete cylinder pipe, without seal coat, was found to be leaching at a rate estimated to be about 0.0057 in./year. The pH of the water increased from 6.9 to 8.6, and the total alkalinity increased from 6.5 to 9.0 ppm from the upstream end through a length of about 37,400 ft. Inasmuch as the material dissolved was principally hydrated cement, the bonding agent comprising about one fourth the volume of the leaching layer, the effect of the leaching on a thin-walled pipe was much more serious than might be assumed from a leaching rate of 0.0057 in./year. The senior author's studies indicated that about two thirds of the hydrated cement had been dissolved at the concrete–water interface and that leaching had penetrated to a depth of about 0.2 in., as shown on Figure 8-14.

Concrete and other products prepared with portland cement are porous and will absorb water when immersed. The setting process of cement consists of chemical reactions between the cement and water, known as *hydration*. Hence, the surfaces of the solid particles of cement and the interior pores of the mortar have great affinity for water. The hydration process consists of a solution in the water of some of the solid cement to form a cementing paste, which hardens with supersaturation of the soluble solid material to bond the concrete or mortar together.

If an excess of water is supplied to keep the dissolved solid materials relatively low, the solution of the soluble solids will continue until they are all dissolved. Experiments were made by Lerch and Bogue (*13*) to determine the solubility of di- and tricalcium silicates by intermittent extraction with water, which was removed every 2 or 3 days. These experiments showed that about 95 percent of the lime in the principal constituents of portland cement could be dissolved in a period of about 100 days. An analysis of the residues showed that they consisted of nearly pure hydrous silica. Further tests were conducted which showed that hydrolytic equilibrium is reached with a pH value of about 12.4. It may be assumed, therefore, that the pH value of the pore water within the body of concrete or cement product is about 12.4.

The calcining process in the manufacture of portland cement volatilizes and drives off all the water and carbon dioxide. The solids left are oxides of calcium, silicon, aluminum, iron and magnesium. In addition, there is sulfur trioxide up to 2.3 to 4.0 percent, depending upon the type of cement. The sulfur is added as gypsum, $CaSO_4 \cdot 2H_2O$, or plaster of paris, $CaSO_4 \cdot \frac{1}{2}H_2O$, in order to control the rate of setting. Good cements contain about 63 percent calcium oxide, 22 percent silica, 6 percent alumina, and smaller amounts of the other

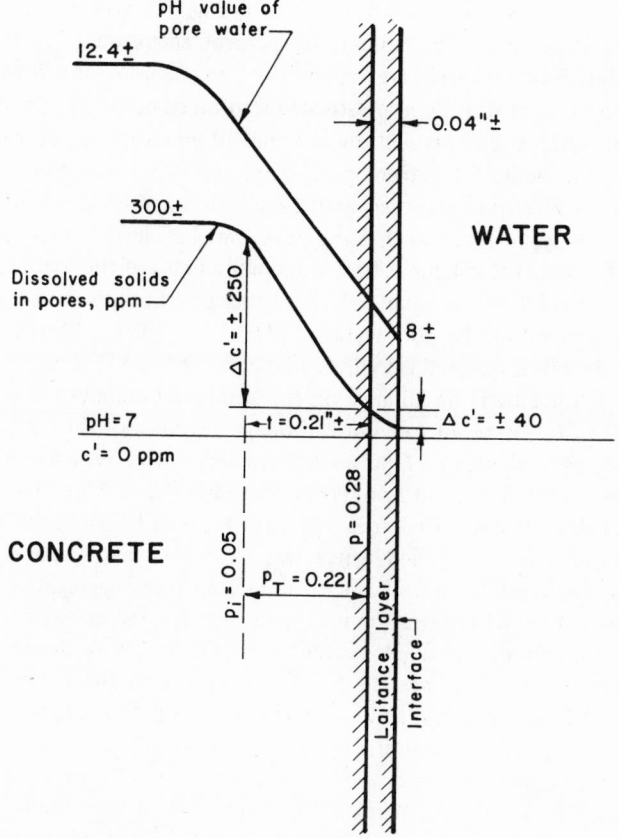

Figure 8-14

Enlarged section of concrete pipe wall showing approximate pH grade line and concentration grade line for dissolved solids in pore water.

constituents. When high sulfate resistance is required, the tricalcium aluminate, $3CaO \cdot Al_2O_3$, should be less than 5 percent (1.9 percent Al_2O_3). During hydration the calcium and aluminum oxides and silica react with water and are bonded together, the bonding material having a glassy like structure. The resulting materials are thought to be di- and tricalcium silicates, tricalcium aluminate, and some hydrated lime. The gypsum reacts with aluminates to form calcium sulfoaluminate. Doubtless much of the silica in the body of the particles of cement remains as solid silica. The surface of the solid mortar or concrete exposed to the atmosphere or to water will absorb carbon dioxide to form calcium carbonate.

Tests made under the senior author's direction on concrete used for reinforced concrete pressure pipe indicated that, after setting, the wet concrete weighed 155 lb/ft^3 and contained by volume 67 percent aggregate, 27.1 percent hydrated cement, 5 percent pore water, and 0.9 percent pore air. When this concrete was placed centrifugally in prestressed reinforced concrete cylinder pipe, it weighed 160 lb/ft^3 and contained, by volume, 70.3 percent aggregate, 25.1 percent hydrated cement, 4.1 percent pore water, and 0.5 percent entrained pore air. These tests indicate that, even with excellent concrete, the porosity is 4 to 5 percent. It may be assumed that the pore water of new concrete is saturated with lime and with the calcium silicates and aluminates of the hydrated cement. A study of the solubility products of the various constituents indicated that the total solids content of the pore water at pH 12.4 is 300 to 400 ppm and that most of the dissolved material is hydrated lime.

Analyses indicate that the laitance on the surface of concrete or mortar contains from 33 to 72 percent calcium carbonate. Studies of the rate of solution in CO_2-free distilled water of pulverized concrete, neat cement, and laitance show that the solubility in 48 hr of neat cement was 9.3 to 14.1 percent; of concrete, 6.2 to 7.5 percent; of laitance, 2.1 to 3.4 percent. The solubility of the pulverized aggregate in distilled water was 0.275 to 0.48 percent in 48 hr. Crushed trap rock, gneiss, and silica sand were used as the aggregate. About half the dissolved constituents from the neat cement and concrete was calcium, and form the laitance only about 30 percent was calcium. The silicate content of the dissolved constituents was 20 to 36 percent from the laitance, 5 to 10 percent from the neat cement, and 4 to 13 percent from the concrete.

Asbestos-cement pipe is manufactured with about 20 percent asbestos fiber and 80 percent cement, or with about 50 percent cement and 30 percent finely ground silica. According to Manson and Blair (*14*), if pipe made with finely ground silica is autoclave cured at pressures of 100 to 200 psi under saturated steam conditions for 16 or more hours, the silica reacts with the free lime to form calcium silicate hydrate, thus removing the free lime. Asbestos-cement pipe weighs about 108 to 115 lb/ft^3. No information is available on its porosity, but the weight indicates that it is highly porous.

According to Bogue (*13*), the probable composition of hydrated portland cement is close to $3CaCO \cdot 2SiO_2 \cdot aq$. When the hydrated cement dissolves in leaching, the solubility within the pores where diffusion takes place is determined by the solubility products of lime, silicic acid, and the calcium silicates. The solubility of calcium from hydrated lime increases 100-fold by a reduction of pH from 12.4 to 11.4. The solubility of calcium in monocalcium silicate is approximately doubled by a similar lowering of pH, and the solubility of di- and tricalcium silicates is also greatly increased. On the other hand, calcium carbonate, a constituent of laitance, has a solubility of only about 7 ppm at pH values from 13 down to 11, and the solubility increases to about 100 ppm at pH 8. Silica, ionizing as bisilicate, becomes increasingly soluble as the pH is increased above 9.

Figure 8-14 is an enlarged section of a concrete pipe wall in contact with water at a pH of about 8. The section shows the approximate pH grade line in the pore water from pH 12.4 at the back of the leaching layer to pH 8 at the interface with the flowing water. Also shown is an approximate concentration grade line from a dissolved solids concentration of about 300 ppm at the back of the leaching layer to about 10 ppm in the flowing water.

According to Fick's law, the rate of diffusion of the dissolved solids in the direction of decreasing concentration is directly proportional to the concentration gradient, which, from Figure 8-14, is approximately equal to $\Delta c'/t$.

To express the rate of leaching from the solid surface in terms of the rate of diffusion, it is necessary to take account of the porosity of the leaching layer and the specific gravity of the solid that has been dissolved. The values of the diffusion coefficients shown in Table 3-2 indicate that an approximate value of D_c of 10^{-5} cm^2/sec or 50 in.^{2}/year may be used for leaching studies of cement products. If the concentration of the solute is expressed in parts per million, the rate of leaching is given approximately by the following equation:

$$R_T = 50 \times 10^{-6} \frac{p_T}{\rho_1} \frac{\Delta c'}{t} \qquad (8\text{-}6)$$

in which R_T is the rate of leaching in inches of solid per year after T years; p_T is the average porosity after T years of a layer through which leaching takes place; ρ_1 is the specific gravity of the dissolving solid or the density in g/cm^3; $\Delta c'$ is the reduction in concentration in parts per million of solute through the thickness t inches; t is the thickness in inches of the layer through which leaching takes place.

It is evident from Eq. (8-6) that the rate of leaching is proportional to the ratio p_T/t. In a leaching layer both p_T and t increase with time as leaching progresses, but the ratio decreases since the rate of leaching decreases. If $\Delta c'$ is known, or can be estimated independently, both p_T and t may be estimated from Eq. (8-6), provided the initial porosity p_i and the total volume of solid leached out in time T are known. If the rate of leaching is determined at intervals throughout the period T, the total volume leached in inches is given by

$$R_{av} T = (p_T - p_i)t \qquad (8\text{-}7)$$

where R_{av} is the average rate of leaching during T years, p_i is the initial porosity, and t is the thickness of the leaching layer after T years. Equations (8-6) and (8-7) may be solved simultaneously for p_T and t. Figure 8-14 shows values of t and p, which were estimated in this manner. The porosity of the laitance layer shown in Figure 8-14 was about 40 percent prior to leaching, and the contribution of this layer to the leaching after the 3-year period was negligible.

Figure 8-15 is an enlarged section of concrete pipe wall showing the effect of a seal coat in retarding leaching. For a seal coat to be effective in retarding

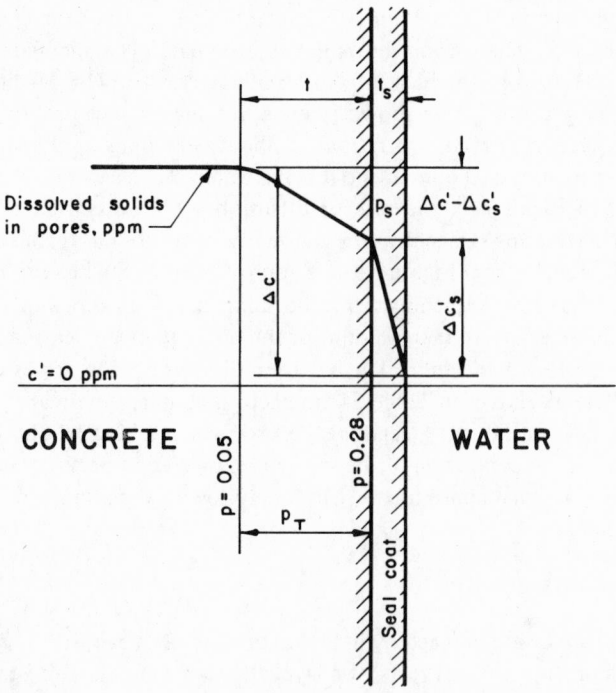

Figure 8-15
Enlarged section of concrete pipe wall showing effect of seal coat on concentration grade line for dissolved solids in pore water.

leaching, it is necessary that the laitance be fully removed so that the seal coat may be bonded directly to the concrete, mortar, or cement asbestos. The porosity of the seal coat must remain substantially less than the porosity of the concrete or cement product. This means that the seal-coat material should not be soluble in water and should contain considerably less than 1 percent water-soluble matter.

If the porosity of the seal coat is substantially less than the porosity of the concrete, the concentration grade line during leaching will be steep through the seal coat and relatively flat in the concrete, as shown by Figure 8-15. The resulting flatter grade line will retard the rate of leaching proportionately.

If it be assumed that the thickness and porosity of the seal coat remain constant and p_s and t_s are known, $\Delta c_s'$ may be computed from Eq. (8-6) for a measured rate of leaching. The rate of leaching in terms of $\Delta c'$ and t, where a seal coat is present, is given by

$$R_T = 50 \times 10^{-6} \frac{\Delta c'}{\rho_1} \frac{p_s/t_s}{1 + (p_s/t_s)(t/p_T)} \tag{8-8}$$

If p_s, t_s, R_T, R_{av}, p_i, and $\Delta c'$ are known or can be estimated, Eqs. (8-7) and (8-8) may be solved simultaneously for p_T and t.

If p_s is not known, it may be estimated from the initial rate of leaching, R_i, by rearranging the terms in Eq. (8-8) as follows:

$$\frac{p_s}{t_s} = \frac{1}{[(50 \times 10^{-6}\, \Delta c')/\rho_1 R_1] - (t/p_i)} \tag{8-9}$$

At the start of leaching through a seal coat, it may be assumed that the concentration of dissolved solids in the pores of the mortar, concrete, or cement asbestos immediately behind the seal coat is the saturation concentration, and that the entire drop in concentration is through the seal coat. The thickness t of the leaching layer is zero at the start, and the term t/p_i in Eq. (8-9) is also zero.

The senior author has analyzed the results of leaching tests conducted throughout a 5-year period, starting March 11, 1947, by the Gregg Company of Riverton, New Jersey, on two specimens of 6-in.-diameter, cement-lined, cast-iron pipe, 1-ft long. The lining of one specimen was left uncoated and the other specimen was coated with Gregg's seal coat. The first 3 years of these tests were described by Wilson (*15*). The specimens were placed upright in containers sealed at the base with wax; the containers were filled with distilled water and the upper ends were covered with glass slabs. The total dissolved solids content of the distilled water was about 4 ppm. Conductivity readings were taken on the water at the end of 3- and 4-day periods or twice weekly, and the water was changed after each reading. The conductivity readings were converted to total dissolved solids, and the results were plotted in the form of a graph for the 5-year period. A correlation was made between soap hardness and conductivity during the first 3 days of testing, which indicated that the total hardness was approximately 60 percent of the conductivity.

In the study of the results of the Gregg test, the hardness was adopted as a better measure of total solids than the conductivity, and all total solids measurements by conductivity were converted by multiplying by the factor 0.6. To eliminate the effects of initial purging, the solids dissolved during the first 14 days were neglected. The results of the analysis of the Gregg tests are shown in Figure 8-16.

The cement lining of the test specimens was placed centrifugally by the pipe manufacturer and had an average thickness of $\frac{3}{16}$-in. (0.187); it was made with 1 part type I portland cement to 1.5 parts dry sand, by volume. If each bag of cement is assumed to use 3.5 gal of water, with 2.5 gal reacting with the cement, the resulting composition of the mortar after setting is estimated at 48.6 percent by volume of hydrated cement, 42.9 percent sand, 8.0 percent pore water, and 0.5 percent pore air. The specific gravity of the solubles, ρ_1, has been taken at 2.3. The rate of leaching was computed from the volume of water, the area of

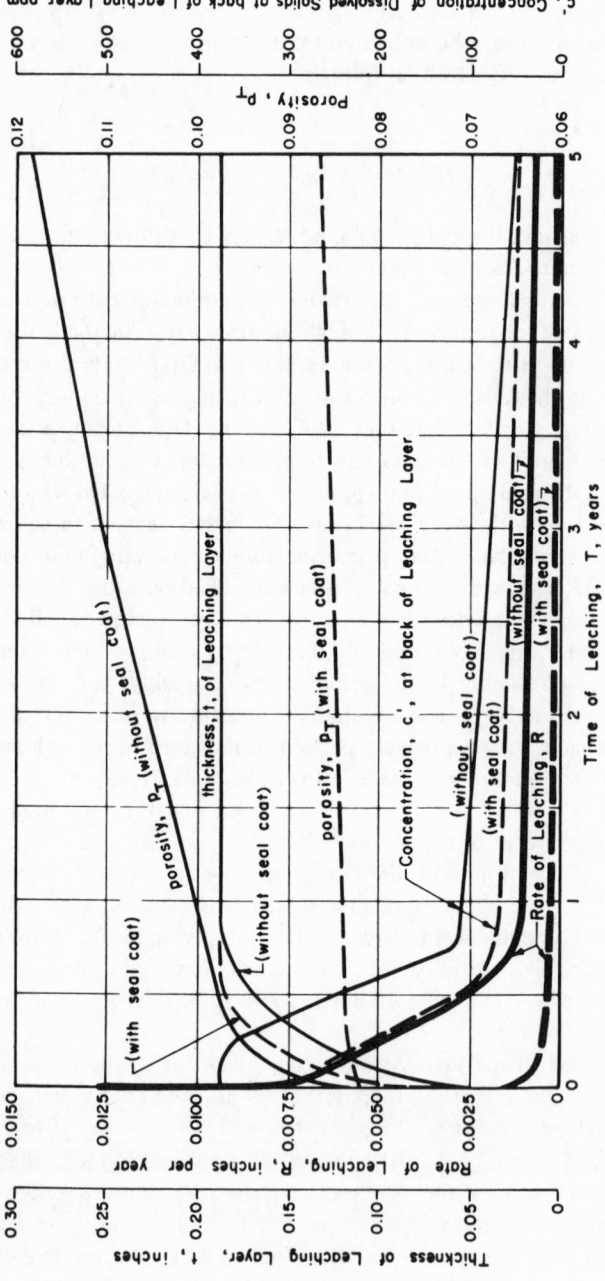

Figure 8-16
Analysis of leaching of $\frac{3}{16}$-in. cement lining from Gregg tests (p_S/t_S of seal coat = 0.155).

lining exposed to the water, and the rate of increase of dissolved solids. The initial porosity, p_i, has been taken at 0.082. By trial-and-error adjustments of Eqs. (8-6) and (8-7), the saturation value of c' was estimated at 373 ppm. The value of p_s/t_s for the seal coat was estimated at 0.155, by means of Eq. (8-9), at the start of leaching, with $\Delta c'$ taken at 360 ppm and R_i at 0.00121 in./year.

It will be noted from Figure 8-16 that the thickness t of the leaching layer extended completely through the $\frac{3}{16}$-in. cement lining in about 10 months without the seal coat and in about 6 months with the seal coat. Without the seal coat, the concentration of dissolved solids in the pores at the back of the leaching layer began to decrease rapidly after about 3 months and dropped to about 100 ppm in about 10 months, at which time leaching had penetrated through the cement lining. Thereafter the concentration decreased slowly to about 50 ppm at the end of 5 years. With the seal coat, however, the concentration c' fell off at once to about 300 ppm within 1 month and then decreased rapidly to about 80 ppm after about 7 months. Thereafter it decreased slowly to about 50 ppm at the end of 5 years. Without the seal coat, the porosity p_T increased rapidly to about 0.096 after about 3 months, and thereafter increased more slowly to about 0.118 after 5 years. With the seal coat, on the other hand, the porosity p_T increased slowly from 0.082 to about 0.087 at the end of 5 years. Without the seal coat, the rate of leaching R_T started at 0.0127 in./year and decreased very rapidly to 0.0075 in less than 1 month; it decreased further to about 0.0012 at the end of 9 months, and thereafter slowly to 0.0006 at the end of 5 years. On the other hand, with the seal coat the rate started at 0.00121 in./year, dropped to 0.005 in about 3 months, and then slowly decreased to 0.00011 in 5 years.

The volume of hydrated cement in the $\frac{3}{16}$-in. lining is equivalent to a thickness of about 0.091 in. ($0.187 \times 0.486 = 0.091$). The equivalent thickness leached in 5 years from the uncoated lining was computed at 0.006819 in., or about 7.5 percent of all the hydrated cement. The rate of leaching at the end of 1 year was estimated at 0.001095 in./year, and it was 0.00061 at the end of 5 years. With the seal coat the equivalent thickness leached from the lining in 5 years was computed at 0.0008732, which is only 12.8 percent of the volume leached without the seal coat. The effectiveness of the seal coat is thus apparent. With the seal coat the rate at the end of 1 year was 0.000153 in./year and at the end of 5 years, 0.000108 in./year; these represent 14 percent and 18 percent, respectively, of the corresponding rates without the seal coat.

The value of the term in Eq. (8-8) involving p_s/t_s was approximately 0.12 for most of the 5-year period of leaching, ranging from 0.14 at the start to 0.117 at the end of the 5 years. For small values of p_s/t_s, 0.15 or less, the value of the term in the denominator of Eq. (8-8) is slightly greater than 1; the rate of leaching is approximately proportional to the value of p_s/t_s for any value of $\Delta c'$. In the authors' opinion, any seal coat with a porosity in excess of 5 percent is of little value. To be effective, the ratio p_s/t_s should preferably be less than 0.1.

A concrete or asbestos–cement pipe or structure that is submerged in, or under, water or groundwater is subject to leaching of cement from the exterior walls. If submerged in water, the rate of leaching may be estimated from Eq. (8-6) for no seal coat and from Eq. (8-8) if a seal coat is used.

If the structure is submerged in groundwater, the soluble products will diffuse into the water in the soil pores and will be carried away at the velocity of the groundwater past the structure. The layer of groundwater into which the dissolved cement diffuses will act in a manner similar to a seal coat to reduce the rate of leaching. The following equation may be used for the rate of leaching into groundwater:

$$R_T = 50 \times 10^{-6} \frac{\Delta c'}{\rho_1} \frac{p_g/t_g}{1 + (p_g/t_g)(t/p_T)} \tag{8-10}$$

where p_g is the porosity and t_g is the thickness in inches of the ground layer into which the dissolved cement diffuses. Equation (8-10) is similar to (8-8) and indicates that the ground is similar to a seal coat. For effective protection p_g/t_g must be low. p_g/t_g might be evaluated from Eq. (8-9) except that R_i is difficult to measure.

Camp has derived an approximate equation for the rate of leaching from the exterior of a concrete pipe submerged completely under groundwater flowing transversely around the pipe, as follows:

$$R_T = 50 \times 10^{-6} \frac{\Delta c'}{\rho_1} \frac{p_T/t}{1 + (p_T/t)(0.465/p_g)\sqrt{D/V}} \tag{8-11}$$

where D is the outside diameter of the pipe in feet and V is the groundwater velocity in feet per day past the pipe in two layers, each of thickness t_g, perpendicular to the direction of flow of the groundwater. The thickness t_g is approximately equal to $0.465\sqrt{D/V}$.

The initial rate of leaching for a 48-in. pipe was computed from Eq. (8-11), assuming a soil porosity of 30 percent, $\Delta c'$ at 300 ppm, p_T at 0.22, ρ_1 at 2.3, and t at 0.01 in. For a groundwater velocity of 0.35 ft/day, the value of R_i was 0.0012 in./year; for V of 10 ft/day R_i was 0.0064 in./year; for V of 100 ft/day, R_i was 0.0183 in./year. The velocity of 0.35 ft/day or less might be expected in very fine sand, glacial till, and mixtures of sand, silt, and clay having a permeability k of about 2×10^{-4} cm/sec. The velocity of 10 ft/day may be expected in fine clean sand with a permeability k of about 6×10^{-3}. For adequate protection against leaching, therefore, the soil permeability should be 2×10^{-4} or lower. If k is higher than about 2×10^{-4}, a thick-walled structure or an effective seal coat should be used.

Prestressed reinforced concrete pipes or tanks with high-tension steel wire on the outside protected only by a mortar coating are particularly vulnerable to damage resulting from leaching in groundwater. Excessive leaching or cracking of the mortar coating will permit corrosive attack of the steel wire and possible failure of the structure.

Thin-walled prestressed reinforced concrete tanks or pipes should, in the authors' opinion, be provided with watertight steel shells if a seal coat is to be applied to a face through which water might otherwise percolate as a result of pressure on the opposite face. Without the watertight steel shell, pressure would build up under the seal coat and rupture it.

Ludwig and Ludwig (*16*) have studied the relative rates of leaching of cement from 4-in. concrete and asbestos-cement pipe in contact with distilled water, sulfuric acid, and sodium sulfate solutions. In distilled water the rate of leaching over a period of 11 days, as measured by the increase in hardness, was about the same for the two materials, despite the fact that the asbestos-cement contained about 3 times as much cement. In sulfuric acid solution with a strength of about 1,600 ppm, the amount leached in the first 11 days, as measured by the hardness (about 60 percent of acidity neutralized), was about the same from the two types of pipe, but it was about nine times the amount with distilled water. After about 20 days with the acid at 1,600 ppm, the amount leached from the asbestos-cement pipe was about 12 percent greater than from the better specimens of concrete pipe, and after about 37 days, it was about 25 percent greater. With weaker solutions of sulfuric acid, the rate of leaching was appreciably reduced. With a 1 percent solution of sodium sulfate, the amount leached in the first 10 days was about 17 percent greater for the asbestos-cement than for the concrete, and for the concrete, it was about 2.2 times the amount in distilled water.

When concrete or asbestos-cement is in contact with a strong acid or sulfate solution, concentration gradients are set up for hydrogen and sulfate ions, which promote diffusion of these ions into the pores of the hydrated cement where reactions take place to remove them. The hydrogen ions react with some of the hydroxyl of the free lime, and the sulfate may precipitate as calcium sulfate or calcium sulfoaluminate. The net effect is to increase the concentration gradient of the dissolved constituents of the hydrated cement and to increase the rate of leaching correspondingly.

Considerable attention has been directed to the corrosion of concrete sewers, flowing partially full, by hydrogen sulfide gas in the sewer air above the water surface (crown corrosion). Such gas is evolved from the anaerobic decomposition of sludge deposits in the sewer or of the organic matter of the sewage itself, if the sewage is stale and devoid of dissolved oxygen. In most cases the inner face of the sewer above the water surface is wet with condensate from the air above the sewage. This water contains oxygen and hydrogen sulfide dissolved from the air, which react with the water to form sulfuric acid. The rapid rate of leaching of cement from this cause has obscured the fact that leaching occurs, albeit at a slower rate, wherever concrete is in contact with water.

High sulfate concentrations in waters in contact with hydrated cement not only increase the rate of leaching from the cement, but they may result in the disintegration of the hydrated cement. As previously explained, calcium sulfate is added before calcining to control the rate of setting. During hydration, the sulfate combines with alumina and lime to form calcium sulfoaluminates. If, after setting, there is an appreciable amount of tricalcium aluminate in the hydrated cement (5 percent or more), the cement does not have *high sulfate resistance*. When a hydrated cement of this character is in contact with water containing about 1,000 ppm or more of sulfates, the sulfates diffuse into the pores to react with tricalcium aluminate to form more calcium sulfoaluminate. Since the crystals of the sulfoaluminates are much larger than the crystals of tricalcium aluminate from which they are formed, expansion of the hydrated cement takes place to disintegrate it. According to Bogue (*13*), the predominating factor in determining the sulfate resistance appears to be the content of tricalcium aluminate.

Manson and Blair (*14*) present experimental evidence that free hydrated lime in the hydrated cement is the predominant factor in determining sulfate resistance. They state that free hydrated lime may react with sulfates to produce calcium sulfate crystals or with sulfates and aluminates to produce ettringite ($3CaO \cdot Al_2O_3 \cdot 3CaSO_4 \cdot 30H_2O$), a calcium sulfoaluminate. Both reactions would produce expansion of the hydrated cement. Both calcium sulfate and ettringite crystals were identified in photomicrographs of normal cured asbestos-cement pipe that had been in contact with high-sulfate groundwater.

Test results presented by Manson and Blair (*14*) showed that normal cured cement, types I, II, and V, without the addition of finely ground silica had 15.5, 14.5, and 13.7 percent free hydrated lime and an expansion in sulfate solution (28 cycles) of 0.16, 0.09, and 0.11 percent, respectively. The tricalcium aluminate contents of these cements were 11, 3, and 0 percent for types I, II, and V, respectively. The same cements with 0.6 parts of finely ground silica to 1 of cement, autoclave cured with saturated steam at 100 psi for 16-hr had only about 0.5 percent free lime and an expansion in sulfate solution of only 0.3 percent.

According to Manson and Blair, the free lime will not be removed by the addition of the finely ground silica if normal low-temperature curing is used. Nevertheless, it is common practice to use portland–pozzolan cement or portland blast-furnace slag cement for normally cured concrete structures to be exposed to seawater (about 2,650 ppm sulfates) or other waters high in sulfates. Both pozzolans and blast-furnace slags are silicious materials that are intimately interground with portland cement clinker. Experience indicates that such cements are more sulfate resistant than ordinary portland cement. Bogue (*13*) presents evidence that some of the constituents of pozzolans and blast-furnace slag do combine with the free lime in the cement during hydration at ordinary temperatures, and that diatomaceous earth and precipitated silica are very reactive with hydrated lime.

REFERENCES

1. T. R. Camp, "Corrosion and Corrosion Research," *Trans. Am. Soc. Civil Engrs., 121,* 791 (1956).
2. J. M. Pearson, *Trans. Electrochem. Soc., 81,* 489 (1942); H. H. Uhlig, *Corrosion Handbook,* New York, John Wiley & Sons, Inc., 1948, p. 483; R. Eliassen and J. C. Lamb, *J. Amer. Water Works Assoc., 45,* 1290 (1953).
3. W. M. Latimer, *Oxidation Potentials,* 2nd ed., Englewood Cliffs, N.J., Prentice-Hall, Inc., 1952.
4. J. W. Gibbs, *The Collected Works of J. Willard Gibbs,* Essex, England, Longman Group Ltd., 1928.
5. T. R. Camp, "Corrosiveness of Water to Metals," *J. New England Water Works Assoc., 60,* 188 (1946).
6. W. M. Latimer, *Oxidation Potentials,* Englewood Cliffs, N.J., Prentice-Hall, Inc., 1938, p. 294.
7. T. E. Larson and R. M. King, "Corrosion by Water at Low Flow Velocity," *J. Amer. Water Works Assoc., 46,* 1 (1954).
8. R. Eliassen and R. T. Skrinde, "Corrosion Control in Potable Water Systems," *Public Works, 91,* no. 7, 92 (1960).
9. See F. L. Futrol and R. S. Ingels, "Copper Catalysis for Manganese Oxidation," *J. Amer. Water Works Assoc., 45,* 804 (1953).
10. Task Group Report, "Cold Water Corrosion of Copper Tubing," *J. Amer. Water Works Assoc., 52,* 1033 (1960).
11. E. W. Moore, "Corrosion of Brass in Water Subjected to pH Correction," *J. New England Water Works Assoc., 48,* 47 (1934).
12. G. B. Hatch and O. Rice, "Influence of Water Composition on the Corrosion of Steel," *J. Amer. Water Works Assoc., 51,* 719 (1959).
13. R. H. Bogue, *The Chemistry of Portland Cement,* New York, Van Nostrand Reinhold Company, 1947, p. 381.
14. P. W. Manson and L. R. Blair, "Sulfate Resistance of Asbestos-Cement Pipe," Johns-Manville Products Corp., June 1961.
15. P. S. Wilson, Discussion of paper, "Solution Effects of Water on Cement and Concrete in Pipe," M. E. Flentje and R. J. Sweitzer, *J. Amer. Water Works Assoc., 47,* 1183 (1955).
16. H. F. Ludwig and R. G. Ludwig, "Corrosion of Cement-bonded Sewer Pipes," *Water & Sewage Works* (Sept. 1951).

CHAPTER 9

Uses and Water Quality Standards of Public Waters

For generations, municipalities and industries in the United States discharged their liquid wastes into public watercourses with little regard for other users of the waters, unless brought to task by these users through claims for damages. Prior to about 1940, the activities of the federal government and the state agencies were limited generally to the giving of advice. During the depression years, the extent to which the spoilage of water resources had been allowed to progress was brought to public attention by the surveys of the National Resources Planning Board. Since then most of the states have enacted laws to regulate pollution, and during 1948 the Eightieth Congress passed a water-pollution control act known as Public Law No. 845.

The first comprehensive federal law, the Federal Water Pollution Control Act, Public Law 660, was enacted in 1956. It was administered by the Public Health Service and, among its many features, provided grants to municipalities for the construction of wastewater treatment plants. In 1961, the Federal Water Pollution Control Act Amendments vested the Public Health Service with authority to enforce the abatement of pollution. The Water Quality Act of 1965 required the states to adopt water quality standards and classify their waters in accordance therewith*; it established the Federal Water Pollution Control Administration

---

*All the states and Puerto Rico, Guam, and the Virgin Islands participated in this effort to set standards. As a typical illustration of the standards that resulted, the water-use classifications and water quality criteria adopted by the Commonwealth of Massachusetts for interstate, coastal, and marine waters, approved by the federal government on August 7, 1967, are given in the appendix.

(FWPCA) in the Department of Health, Education, and Welfare (HEW) to replace the Public Health Service as the pollution control agency. In 1966, the Clean Water Restoration Act authorized increased appropriations for grants for municipal pollution abatement projects, and under a presidential reorganization plan the FWPCA was transferred from HEW to the Department of the Interior. The Water Quality Improvement Act of 1970 increased the powers of the FWPCA in matters of enforcement, and the FWPCA was renamed the Federal Water Quality Administration (FWQA). A presidential reorganization plan in late 1970 established the Environmental Protection Agency (EPA) with a water quality office to which were transferred the functions of the FWQA.

For many years state health departments administered the state water-pollution control statutes. With increased activities in pollution abatement in the 1950s, their duties and responsibilities have gradually been delegated to state water resources commissions or water-pollution control agencies. Most states now have water quality standards and classification systems and effective enforcement powers, and provide grants to supplement the federal grants for the construction of wastewater treatment facilities.

Since 1930 a number of interstate water-pollution-control compacts have been concluded. Among the groups administering these compacts are the Interstate Sanitation Commission, comprising New York, New Jersey, and Connecticut, the Interstate Commission on the Delaware River (now the Delaware River Basin Commission), the Ohio River Valley Water Sanitation Commission, the New England Interstate Water Pollution Control Commission, the Potomac River Commission, and others.

9-1. POLLUTION ABATEMENT POLICY (1)

The primary purpose of water pollution abatement is to reclaim waters or watercourses for uses by the public and the riparian owners. A secondary purpose is to abate atmospheric pollution by noxious gases, emanating from the polluted watercourse or the polluter's premises, which may result in odor nuisances, corrosion, and paint damage over an extended area beyond the immediate confines of the watercourse. Most of the attention of the water-pollution control agencies is directed to the primary purpose of reclamation of the waters for suitable uses. Usually the secondary purpose of eliminating atmospheric pollution will be accomplished if steps are taken to eliminate the polluting substances or if suitable treatment works are provided and properly operated to accomplish the primary purpose.

If the receiving streams are dry during periods of drought, as is the case with many small streams in the eastern United States and with most of the streams in the arid and semiarid regions of the west, the total flow of the stream at the point of pollution will consist of the treated sewage or liquid wastes, and a part or all of this flow may percolate into the groundwater and flow underground. In many cases, industrial wastes and sewage are discharged directly to spreading grounds or leaching fields to permit the liquid to percolate to the groundwaters.

The discharge of wastewaters into the ground instead of into an open watercourse does reduce the pollution of the open watercourse, but it may result in transferring the pollution problem, modified to a considerable extent, to the groundwater. For example, on Long Island many local water supplies are taken from the ground, and most of the wastewaters are discharged back into the sandy soil. The well waters are, for the most part, bacteriologically safe because bacteria and suspended solids are strained out of the wastewaters by the soil. Dissolved substances, however, such as mineral solids, some pesticides, and synthetic detergents, pass right through and accumulate on repeated use of the water.

Substantially the same primary objective applies to pollution abatement and control whether the receiving waters are surface waters or groundwaters; this objective is to reclaim the waters so that they will be suitable for one or more uses. The practical uses of surface waters are, of course, much more numerous and varied than the practical uses of groundwaters. Moreover, the quality of surface waters and the nature of the pollutants are also much more varied than with groundwaters. Hence, the designation of suitable uses, the setting of water quality standards for these uses, and the methods for abating pollution must necessarily vary greatly, not only between these two classes of waters but also from place to place for either class. Nevertheless, it must be kept in mind that there is a continuous interchange between surface and groundwaters, and that both take part in the runoff from rainfall on the watershed.

Most of the efforts of the water-pollution control agencies are now directed to the abatement and control of pollution in surface receiving waters. As a first step in policy making, it is necessary for the pollution control agency to designate the appropriate uses of each public watercourse, or part thereof, and set standards of suitable water quality. Such standards should, of necessity, take into account existing conditions in a stream and be directed toward developing the highest beneficial use that is economically and practically feasible. This procedure is analogous to zoning in the preparation of a city plan. Prior to enactment of the Water Quality Act of 1965 a number of pollution control agencies had already classified their public waters according to use. Since that enactment, all states have complied with the federal government order to adopt classification standards meeting federal approval. This work is extremely important and must be done judiciously if the cost of pollution abatement is to be kept within reasonable bounds. Some watercourses in sparsely inhabited upland areas may readily be reclaimed and controlled for use as drinking-water supplies, whereas other grossly polluted watercourses in highly industrialized regions can never be reclaimed to a high degree of purity without economic waste. The classification policy should be designed to benefit both the public and the riparian owners and should aim at the gradual improvement of all public waters. The policy should be flexible enough to permit changes in classification from time to time as circumstances dictate.

For a single receiving water, the removal of pollutants required at one point on the watershed may be totally different from that required at another point. A high degree of treatment may be necessary at some points of pollution, and no treatment at all at other points. Once a satisfactory standard of water quality is set for a particular public watercourse, any degree of treatment of sewage or industrial wastes in excess of that required to meet the standard will result in economic waste. This matter should be emphasized very strongly. It is wrong to require more treatment than is necessary to produce the desired results, and it is wrong to require any polluter to treat his wastes or sewage unless such treatment is necessary to produce the desired results at the least overall cost.

The problem of pollution abatement for a particular watercourse should be studied as a whole, including the entire watershed, to determine which wastes must be fully treated or eliminated at their source and which may be allowed to go partially treated or untreated for the best economical solution. Pollution control policy should be developed so as to permit the selection of wastes for treatment and of wastes that may be discharged untreated for best overall economy. This proposal is contradictory to the policies that have been followed by some agencies. Under existing law, it is easier for a public agency to enforce a policy that requires treatment by all polluters on a given watercourse and the same degree of treatment for similar wastes regardless of quantity. Such a policy can be unreasonable and wasteful. There are instances, certainly, where *effluent* standards would lead to a more effective and practical enforcement than the general policy implied in maintaining *receiving-water* standards.

Successful abatement and control of water pollution may require (1) prohibition or control of the manufacture and use of some harmful chemical pollutants that are toxic to animal or vegetable life or are chemically or biochemically stable, (2) removal of pollutants from wastewater by changes in the processes that produce the pollutants, or by treatment of the wastewaters and proper dispersion of the treatment-plant effluent in the receiving waters, and (3) allocation of pollution loads among the polluters, and regulation and modification of flow in the receiving waters.

At present, the state and federal governments have inadequate power to prohibit or limit the manufacture and use of harmful new chemicals, and new legislation should be enacted vesting sufficient power and responsibility in the appropriate agencies. At present, the Atomic Energy Commission and the Federal Radiation Council have more power than water-pollution control agencies over the control of pollution by radioactive materials. Similarly, state and federal pollution control agencies must look to the Department of Agriculture, to state agricultural agencies, and to forestry agencies for control of toxic pesticides. Legislation *has* been enacted by which phosphates have been eliminated in the manufacture of detergents, but the environmental improvement to be derived from this measure has yet to be demonstrated.

Pollution abatement involves *processes*. The construction of treatment works is only one of several means to the same end. Successful abatement requires that operation of the treatment works and all other means be given equal attention. The designer of a treatment plant is, in fact, designing processes. The treatment works themselves serve only to house these processes. The processes that occur in the watercourse itself are just as important as those that take place in the treatment plants, for they determine the need and the extent of removal of pollutants. The important factors and processes in the receiving waters are quantity, stages and velocity of flow, dispersion of pollutants, reaeration, deoxygenation, bacterial die-off, uses of receiving waters, and flow regulation. Pollution control agencies should focus their attention on the processes in the receiving waters and on the limiting quantities of polluting materials allowable at each point of pollution.

The control agency should allot to each polluter a definite quantity of polluting substances, not to be exceeded at each point of pollution. The procedure by which the required degree of removal is accomplished should be left to the ingenuity of the polluter. It is wrong to require standard methods of treatment if the polluter can show that he can stay within the allowable pollution loads by other procedures which he considers less expensive and more appropriate for his use. For example, changes in the raw materials or the production methods in an industrial plant or reclamation and reuse (or sale) of the polluting substances are frequently more economical than waste treatment works.

If a pollution control program is to be successful, it is essential to allocate to each polluter allowable pollution loads, which, if complied with, will meet the water quality standards. A single polluter, through his engineer, must estimate an equitable allowable pollution load for himself in order to design his means of pollution abatement and to protect himself from wasteful expenditures for pollution control, but he is powerless to control the pollution by others. It is, therefore, the duty of the pollution control agency to estimate and allocate the allowable pollution loads.

Reduction of pollution by low-flow augmentation is usually a regional matter and requires cooperation of all polluters, as well as of governmental agencies. A master plan should be developed by, or for, the pollution control agency whereby the overall cost is a minimum. Such a plan would include the allocation of allowable pollution loads and the preliminary design of low-flow augmentation works.

9-2. WATER QUALITY STANDARDS

The main uses of public waters in the United States are (1) public water supply, (2) fish or shellfish propagation, (3) recreation and bathing, (4) industrial water supply, (5) agricultural use, (6) electric power production, (7) navigation, and (8) disposal of sewage and industrial wastes. These purposes are listed in the general order of decreasing water quality but not in the order of their economic importance. Standards of quality of water suitable for some of these uses may be defined in fairly definite terms, whereas the quality of water required for other uses is quite variable, and depends upon local circumstances. In any classification program, more than one use must be allocated to most of the watercourses, and the standard of water quality for any public water will be determined by the higher uses. Since the conditions of use vary from one region to another, it is understandable that classifications should also vary. A review of the classifications that have already been made by various pollution control agencies reveals a great diversity of opinions regarding the required quality of waters of the various classes. This is a natural result of the diversified use of the waters. If the classifications and standards are soundly drawn up for two adjoining regions, there is no reason why they should be similar except that agreement should be reached for streams and public waters which are common to both regions.

For most polluted waters, the two most important characteristics that influence their use are their contents of sewage bacteria and dissolved oxygen. The coliform count is an index of the concentration of domestic sewage in the water and of the potential hazards from intestinal disease germs and viruses. This is the only commonly used parameter that bears any direct relation to the public health. The dissolved oxygen content of a public water determines its suitability as a habitat for fish life and indicates whether or not it is on the point of becoming septic with the production of black water and foul odors. Oxygen contents less than the saturation value (about 8 ppm in the summer to 14 ppm in extremely cold weather) indicate that the oxygen is being used up by bacterial decomposition of organic matter in the water or by oxygen-demanding industrial wastes, some of which may be toxic to bacteria and other biota. Some decomposable vegetable organic matter enters streams from swampy regions, but most of the decomposable organic matter emanates from pollution by municipal sewage and industrial wastes. Hence, the oxygen content in the water is determined primarily by the amount of polluting organic matter.

As pointed out in Chapter 6, there is great diversity of opinion among the water-pollution control agencies regarding the minimum concentration of dissolved oxygen for various water uses, and there is even greater diversity of opinion with respect to the allowable coliform count. Some standards have no limit whatever for the coliform count for waters to be used for industrial purposes, because of the misapprehension that the lack of such a limit will make it easier for the industrial users. Such is not the case, of course, because substantially all the coliform bacteria are associated with sanitary sewage. As indicated in Chapter 6, it is the authors' opinion that a coliform limit of some sort should be associated with all use classifications of public waters.

For most of the higher uses of public waters, the water should contain no objectionable odor, oil, scum, floating solids, or debris except from natural sources; it should contain no taste-producing substances in objectionable concentrations and no substances, in harmful concentrations, that are of themselves toxic or become toxic in combination with other substances. Note that these qualities are relative and depend primarily on the concentration of the polluting substances. It is seldom practicable to accomplish the complete elimination of any substance. The amount and character of suspended matter should not be such as to produce objectionable sludge deposits. Inordinate color and turbidity, when introduced with polluting wastes, are of esthetic importance and are particularly noticeable by the general public. Both color and turbidity, so introduced, may be reduced by the treatment methods employed for the removal of organic matter and bacteria, but the extent to which further treatment solely for the removal of color and turbidity can be justified economically is questionable.

Many pollution control agencies require a minimum treatment of primary sedimentation for domestic sewage. This is an unreasonable requirement for small municipalities discharging sewage into large receiving waters, because it puts such municipalities to great expense for sewage treatment and sludge handling with no benefit to anyone. This is also an unreasonable requirement for large municipalities and for industries because it implies that primary settling must be a part of any treatment plant whether there are other less costly processes to accomplish the required results or not. Treatment of wastewaters by a particular polluter for the removal of suspended solids should not be required unless such treatment is essential to meet the water quality standards in the receiving waters. If there are unsightly discharges of sewage from small communities into large receiving waters from visible shore outlets, the unsightliness can be almost completely eliminated by screening and by extending the outlet sewer to discharge submerged at several points. Disinfection should be employed for domestic wastes in any event.

9-3. CRITICAL STREAM FLOWS AND WATER TEMPERATURES

The flow of streams is quite variable, from the low flows during periods of extreme drought to peak flows during major floods, and the range of flows is much greater in some streams than in others, depending on size, the nature of the catchment areas, and the extent of low-flow regulation by storage. It is evident, therefore, that the continuous discharge of a definite quantity of some material into a stream may result in failure to meet water quality standards at some flows, and may not at greater flows. In using the water quality standards, it is necessary to state at what flow in a particular stream, or portion thereof, a limiting standard is to apply. If complete compliance is expected, the flow selected must be the minimum possible at any time. For many streams something short of complete compliance must be accepted. A study of flow duration curves will furnish a basis for the selection of critical flows below which a failure to meet standards must be accepted.

Some states have endeavored to specify the critical stream flow at which the water quality standards should apply, such as, for example, the 1 percent flow, which, on the average, is exceeded 99 percent of the time. In other cases, a 10 percent flow may be recommended, which is normally exceeded 90 percent of the time. The flow will be less than the 10 percent flow only 10 percent of the time, or about 36 days/year. Another critical flow standard is the average flow during a consecutive 7-day period that has the least flow over a period of 1 year or, in some cases, over a period of 10 years.

It is, of course, essential to select a critical flow for a particular stream or portion thereof at which a limiting standard is to apply, but since the flow conditions vary so widely from stream to stream and even between branches of the same stream, it is unwise to adopt the same critical flow for all streams in a state. There are many streams that are widely used by industry and in which the flow is controlled by numerous mill ponds along the streams. In many such cases, water is stored in the mill ponds over the weekend when the mills are not operating and is withdrawn during the work week. The flow just downstream from the mill dams may be zero over the weekend. Such flows may not be reflected in the flow duration data, but they must be taken into account in stream-pollution studies. In some cases it may be more economical to require the release of small quantities of water over the weekend than to construct and operate the treatment works that might otherwise be required.

In many pollution-abatement problems, the possibility of low-flow augmentation from upstream reservoirs should be studied to determine whether such augmentation can reduce the cost of wastewater treatment and, if so, whether the cost of low-flow augmentation is more than offset by the savings in cost of wastewater treatment.

For discharges of wastewaters into large lakes, studies should be made, by means of floats and tracers of various types, to determine the frequency of currents in various directions and at various velocities and the frequency of wind-blown surface currents. These studies should be made by releasing floats and tracers of various types at preselected points in the lake at which a wastewater outlet may be constructed. The purpose of such studies is to determine the rate at which wastewaters will be carried away from the outlet in various directions and the reduction in concentration of the pollutants that will be effected at various distances from the outlet. The location of the outlet with respect to nearby bathing waters or waterworks intakes and the required degree of treatment of the wastewater may thus be estimated.

For wastewater outlets into the ocean or into tidal estuaries, procedures similar to those described for lakes should be undertaken, except that tidal movements and saltwater density currents should also be taken into account.

The temperature of the receiving waters is variable throughout the year. If dissolved oxygen is an important criterion in the quality of the receiving waters, the temperature at which the standards shall apply must be selected. The dissolved oxygen will be at its lowest level when the receiving waters are warmest, and the rate of bacterial utilization of dissolved oxygen will also be most rapid at this time. In many cases the stream flow will be least when the water temperature is greatest. In oxygen-balance studies, therefore, a temperature should be selected that is the highest probable temperature of the water during periods of low stream flow. If there is large use of the receiving waters for cooling purposes, there may be *thermal pollution*, and restrictions may have to be placed on such use in order to maintain some oxygen in the water at all times.

9-4. LIMITING POLLUTION LOADS

After the classification of a public watercourse has been established and the standards of water quality have been set, the limiting quantities of each polluting substance that may be discharged into the watercourse at each point of pollution can be estimated. In making such an estimate, analyses must be made of the quantity and the character of each polluting substance and of the water itself. The most important pollutants, other than coliforms, in municipal sewage and industrial wastewaters are suspended solids and decomposable organic matter. The latter is measured by its BOD (biochemical oxygen demand), which is defined at length in Section 10-7. Concentrations of these pollutants are usually expressed in parts per million, and allowable pollution loads are given in pounds per day.

The stream or tidal estuary must be considered as a dynamic moving entity. Due account must be taken of the dilution available, times of flow, deoxygenation and reaeration rates, sedimentation and benthal decomposition, and bacterial die-off rates; a particular critical regimen of temperature, flow, wind direction, and velocity must be selected to which the water quality standards are to apply. The methods to be used in oxygen balance studies are described in Chapter 11.

After the limiting quantities are determined, the amounts of removal may be estimated by deducting the limiting quantities from the total quantities that are expected to be discharged into the watercourse.

The removals required during the critical flow periods will determine the size and the extent of the treatment works and will fix the capital expenditures required. During cold weather, and for flows greater than the critical flow, a lesser degree of treatment may suffice to meet the water quality standards. Whether such a degree of treatment is permissible is a question of great importance, which depends primarily on the nature and the use of the watercourse. In many streams, organic matter that is not removed by treatment during the winter months deposits in ponds or reservoirs to cause trouble during the following summer.

In any allocation of the pollution-receiving capacity of a watercourse, some allowance for the growth of the municipalities and existing industries must be made, but it is not feasible to predict the demands of riparian owners who may in the future wish to exercise their rights to some of the pollution-receiving capacity of the watercourse. These conditions will surely arise, and reallocation of the allowable pollution loads will have to be made in the future, because the pollution-receiving capacity is limited by nature.

9-5. COMBINED MUNICIPAL SEWERS

Most older cities and towns in the United States have combined sewers carrying both sanitary sewage and storm water. If the combined sewers have adequate capacity for storm-water runoff, their capacities will range from about 50 times the average dry-weather flow of sanitary sewage for densely populated areas to about 200 times the average dry-weather flow for sparsely populated areas. It is usually not economically feasible to intercept the total capacity of the combined sewerage system and to subject this flow to the treatment required by the receiving-water standards. It has long been common practice, therefore, to design interceptors and treatment works with capacities for the peak dry-weather flow, or 2 to 5 times the average dry-weather flow, and to permit overflows of mixed sewage and storm water at the points of interception during and immediately following rainstorms.

For cities in the North and East, with combined sewers, studies over many decades have shown that only about 3 percent of the year's total production of sanitary sewage is discharged without treatment through storm-water overflows to the receiving waters. Since this figure is small compared with the BOD remaining in sewage-treatment plant effluents (10 percent for complete treatment and 70 percent for primary treatment), the effect of the overflows of mixed sewage and storm water upon the receiving waters has been ignored, for the most part, for generations.

In studies by McKee (*2*) in 1946, based on Boston rainfall records, it was shown for the first time that, with interceptors and treatment plants designed for 2 to 3 times the average dry-weather flow, there would be overflows of mixed untreated sewage and storm water on the average of 5 to 6 days of each month. These studies further showed that during rainfalls with intensities of only 0.05 in./hr approximately 60 percent of the sanitary sewage would be discharged overboard into the receiving waters, and that during rainfalls with intensities of about 0.20 in./hr nearly 90 percent of the sanitary sewage would be discharged untreated into the receiving waters. Thus large slugs of pollution of intestinal bacteria and viruses, as well as BOD and suspended solids, will be discharged to the receiving waters nearly every time it rains. Such waters are not safe for either bathing or recreation.

McKee's studies also showed that substantial increases in the capacity of the interceptors and treatment works would serve no useful purpose in reducing the frequency of overflows of raw sewage into the receiving waters unless the capacity was increased to that of the combined sewers. For example, with interceptors having a capacity of 20 times the dry-weather flow, overflows might be expected on about 2 days/month, on the average. Somewhat similar results were found by Palmer (*3*) in 1949 for Detroit conditions and Johnson (*4*) in 1958 for the District of Columbia. The senior author has found similar results for rainfall data in Concord, New Hampshire (*5, 6*) and Bogota, Colombia. Measurements at Northampton, England (*7*), during 1960–1961 indicate that, with interceptors having capacities of three times the average dry-weather flow, overflows will occur on 10 days/month on the average, and with capacities of 45 times the dry-weather flow, overflows will occur on 2 days/month on the average.

If a combined sewer has a capacity when flowing full of 50 to 100 times the average dry-weather flow, the depth of flow during dry weather will be only about 10 percent of the diameter of the sewer if it is a circular sewer, and less if it has a relatively flat bottom. The velocity in a circular sewer during dry weather will average about 30 percent of the velocity flowing full. For a velocity flowing full of 3 ft/sec, the average velocity during dry weather will be only about 0.9 ft/sec. To maintain an average velocity during dry weather in excess of 1.6 ft/sec for adequate scouring of organic solids, the velocity flowing full would have to exceed 5.4 ft/sec. Most combined sewers do not have slopes that are adequate for such a velocity. It is, therefore, evident that a considerable amount of organic solids settles out on the bottoms of combined sewers during dry weather.

Surveys and analyses of a portion of the combined sewerage system at Northampton, England (*7*), 1960–1961, revealed considerable deposition of organic solids and grit in the sewers during dry weather, despite the fact that 80 percent of the system has slopes between $\frac{1}{30}$ and $\frac{1}{160}$ and averaging $\frac{1}{77}$. About 90 percent of the length are egg-shaped brick sewers ranging from 2 × 1.5 ft to

$3 \times 2\frac{1}{3}$ ft, with velocities flowing full estimated to range from about 4.5 to 13.0 ft/sec. It was estimated from the Northampton analyses that an average of about 32 percent of the suspended solids entering the system and about 10 percent of the BOD were deposited in the combined sewers during dry weather.

It has been assumed for many years that the organics that settle out in the sewers during dry weather are picked up in the "first flushings" during a rainstorm and carried out of the combined sewer. Much evidence is accumulating that this is not the case and that scouring of sewer deposits continues for hours during a storm.

In studies of the chlorine demand of the Buffalo sewage during the years 1938 to 1940, Symons et al. (*8*) found that on 126 days when rain occurred the average chlorine demand before the storm was 5.3 ppm; after the first hour it was 6.0 ppm, after the second hour it was 6.8 ppm, and it diminished slowly to 5.8 ppm after 6 hr. These were the average values for 126 days of storm. In cases where rains continued from 8 to 24 hr, the demand was usually less after the first few hours. In cases of sudden showers, the demand usually increased greatly during the first hour and returned to normal within an hour or two. The Symons data indicate a considerable scour of sewer deposits for many hours after the start of a storm.

Studies by Riis-Carstensen (*9*) of gagings and analyses of Buffalo sewage in the Bird Avenue district during August 1936 throw light on the suspended solids discharged in overflows. A rainstorm amounting to 0.55 in. started at 5:30 P.M. and ended at 9:00 P.M. The rainfall intensity averaged about 0.16 in./hr. The maximum gaged rate of flow corresponded to a dilution ratio of only 5.2 times the average dry-weather flow. During runoff from the rainstorm, the rate of flow of suspended solids by weight reached 29.3 times the dry-weather rate, and averaged 11.8 times the dry-weather rate. From the data collected by Symons et al., and from the Riis-Carstensen studies, the senior author has estimated that 35 percent or more of the year's production of suspended solids at Buffalo may be discharged in the overflows of mixed sewage and storm water, despite the fact that only about 3 percent of the sanitary sewage is discharged untreated through the overflows. It is probable that the loss of BOD through the overflows is somewhat less than the loss of suspended solids.

Based on the analyses and flow measurements made over the 2-year period, 1960–1961, at Northampton, England, estimates were made of the yearly discharge of BOD and suspended solids in overflows of mixed sewage and storm water, for various interceptor capacities. The results were presented by Gameson (*7*) in multiples of the daily dry-weather load per year. These results have been recomputed as percentages of the total annual inflow load to the combined sewers and are presented in Table 9-1. It will be noted in Table 9-1 that for interceptors designed with capacities of three times the dry-weather flow about 8.1 percent of the BOD and about 27.4 percent of the suspended

Table 9-1
Estimated Biochemical Oxygen Demand and Suspended Solids Loads in Overflows of
Mixed Sewage and Storm Water for Various Interceptor Capacities, Based on 1960–1961
Studies at Northampton, England

Interceptor Capacity (multiples of average dry-weather flow)	Percentage of Total Annual Inflow Load to Combined Sewers	
	BOD	Suspended Solids
3	8.1	27.4
6	5.2	18.3
9	3.4	15.7
12	2.5	12.9
20	1.22	7.6
30	0.49	3.2
45	0.03	0.22

solids will be lost in the overflows. These percentages decrease with increases in the capacity of the interceptors, but very large interceptors are required for substantial reductions. As previously indicated, the combined sewers in Northampton have relatively high velocities. It may be expected, therefore, that for combined systems with relatively flat slopes the loss of suspended solids and BOD will be much greater. It has been shown in Section 6-5 that effective chlorination of the overflows of mixed sewage and storm water can result in coliform kills of 99.99 percent. In most cases this should be adequate to destroy pathogenic bacteria and viruses.

Holding tanks may be constructed at overflow points to collect all the overflow of mixed sewage and storm water up to the capacity of the tank, thus reducing the frequency of overflows and the amount of BOD and suspended solids lost in the overflows. Following a storm, the tank contents can be pumped back into the intercepting sewer to be treated at the main treatment plant. Holding tanks have been used in a few places for decades, and are now under study (10) in many other places. Chicago has undertaken a system of temporary storage in deep rock tunnels, with storm water being later pumped to the surface, chlorinated, and discharged. A similar system has been proposed for Boston. In some cases, the best solution may be obtained with combinations of partial separation of the combined system, larger interceptors, chlorination of overflows, and holding tanks.

In view of the fact that large fractions of the total annual suspended solids, BOD, and sewage bacteria are discharged untreated from combined systems into the receiving waters through overflows during rainstorms, despite the fact that only about 3 percent of the volume of sanitary sewage is so discharged, solutions of this problem are urgently needed if the water quality standards in the receiving waters are to be met. Refinements in treatment of the dry-weather flow for higher removals are pointless if the overflows of mixed sewage and storm water are not treated, eliminated, or reduced in volume.

REFERENCES

1. See T. R. Camp, "Pollution Abatement Policy," *Trans. Am. Soc. Civil Engrs.*, *116*, 252 (1951).
2. J. E. McKee, "Loss of Sanitary Sewage Through Storm Water Overflows," *J. Boston Soc. Civil Engrs.*, *34*, 55 (1947).
3. C. L. Palmer, "The Pollutional Effects of Storm Water Overflows from Combined Sewers," *Sewage Ind. Wastes*, *22*, 154 (1950).
4. C. F. Johnson, "Nation's Capitol Enlarges Its Sewerage System," *Civil Eng.*, *28*, 428 (1958).
5. T. R. Camp, "Chlorination of Mixed Sewage and Storm Water," *J. San. Eng. Div.*, *Amer. Soc. Civil Engrs.*, Jan. 1961; May 1961; and Sept. 1961.
6. T. R. Camp, "Overflows of Sanitary Sewage from Combined Sewerage Systems," *Sewage Ind. Wastes*, *31*, 381 (1959).
7. A. L. H. Gameson, "Storm Water Investigations at Northampton," London, The Institute of Sewage Purification, June 22, 1962.
8. G. E. Symons, R. S. Simpson, and S. R. Kin, "Variation in the Chlorine Demand of Buffalo Sewage," *Sewage Works J.*, *13*, 249 (1941).
9. E. Riis-Carstensen, "Improving the Efficiency of Existing Interceptors," *Sewage Ind. Wastes*, *27*, 1115 (1955).
10. S. A. Greeley and P. E. Langdon, "Storm Water and Combined Sewage Overflows," *J. San. Eng. Div.*, *Amer. Soc. Civil Engrs.*, *87*, 57 (Jan. 1961).

CHAPTER 10

Bacterial Decomposition
of Organic Matter
in Water

10-1. MECHANISM OF DECOMPOSITION PROCESS

The decomposition of organic matter in water and wastewater is due to the presence of bacteria and other microorganisms. Bacteria use the organic matter for food and break it down to simpler compounds, most of which are rejected as wastes. The *saprophytic* bacteria, which subsist upon dead organic matter and are usually harmless to man, are mainly responsible for this process of degradation. Although *pathogenic* bacteria (disease germs) are often present in sewage and in water polluted with sewage, their proportionate numbers are usually so small that they are negligible factors in the decomposition of dead organic matter. The natural food and environment of the pathogens is the living body of the host. These bacteria, therefore, do not multiply readily in sewage or polluted water. Protozoa and plankton are also present in sewage or surface waters, but since their growth rates and ultimate populations are much less than those of bacteria, their direct influence upon the rate of decomposition of organic matter is small. Their influence, however, is felt indirectly through their depredation of the bacterial population.

The growth and multiplication of bacteria in water depend upon the presence of compounds containing carbon and nitrogen in a form capable of being assimilated by the cells and capable, also, of furnishing energy to the cells.

238

Since the ingestion of food by bacteria is accomplished by the passage of the compounds through the cell walls, it is expedited if the food material is finely divided. Small bacterial cells require food material that is composed of relatively simple molecules dispersed in true solution. Some of the larger bacteria are capable of ingesting more complex molecules, and many protozoa can ingest particles of colloidal size. Bacteria are enveloped whole by the larger protozoa and multicellular fauna.

The energy for the respiration of bacterial cells is supplied by *wet combustion* or oxidation of the chemical compounds broken down by the bacteria. If the oxidation is accomplished by the use of molecular oxygen gas dissolved in the water, the breakdown process is called *aerobiosis;* if it occurs by the transfer of hydrogen from the compound to a hydrogen acceptor other than molecular oxygen, the process is called *anaerobiosis*. Bacteria requiring molecular oxygen for their growth and multiplication are called *obligate aerobes* or *strict aerobes*, and those whose growth requires the complete absence of dissolved oxygen are called *obligate anaerobes* or *strict anaerobes*. Most bacteria are *facultative anaerobes* and obtain their energy for growth either aerobically or anaerobically, depending upon the character of the organic matter available, the amount of dissolved oxygen, and their own specificity toward the compounds available.

Almost any type of organic compound, saturated and unsaturated fatty acids, hydroxy acids, keto acids, di- and tribasic acids, alcohols, carbohydrates, amines, amino acids, amides, and aromatic compounds, may be used as a source of carbon and of energy. Only a small part of the compound destroyed, however, is used for cell building; the remainder is by-product waste as far as the growth of the microorganism is concerned. The part that is synthesized into cell protoplasm is said to be *assimilated*, and the waste portion is said to be *dissimilated*. Nitrogen for cell building may be assimilated in the form of amino compounds, but it is probable that it is usually assimilated as ammonia, since most saprophytic bacteria can grow in media where ammonium salts furnish the only source of nitrogen. The ammonia is usually obtained from sewage by the breakdown of urea and of proteins. Certain types of organisms are able to utilize atmospheric nitrogen, and others use nitrites and nitrates as their source of nitrogen.

The degradation of organic matter by bacterial cells is due to *enzymes*, or *ferments*, secreted by the organisms. These enzymes act as organic catalysts and are specific in the reactions that they promote. Little is known of the nature of the mechanism by which enzymes act. The large number of different reactions produced by some bacteria, for example, *E. coli*, indicates the ability of some organisms to secrete a variety of different enzymes. The first reaction between an enzyme and its substrate appears to be an adsorption phenomenon whereby the molecules of the substrate are oriented and are thus activated.

Quastel (*1*) accounts for specificity on the basis of preferential adsorption of substrates having certain molecular groupings.

Enzymes in general produce colloidal solutions in water. All enzymes contain nitrogen and belong to the class of compounds known as proteins, but many of them have not been isolated in pure form, and the chemical structure has not been determined for a number of them. The rate of the reaction induced by an enzyme is generally proportional to the enzyme concentration, unless the concentration of the *substrate* (the substance being decomposed) is too low, in which case the rate is proportional to the concentration of substrate. The activity of an enzyme is greatly affected by the pH value, since there is an optimum pH for each enzyme. Rise in temperature at first increases the rate of enzymic reactions in the manner common to all chemical reactions, but at comparatively low temperatures an optimum is reached, and at about 70°C the activity of most enzymes is stopped (*2*).

An enzyme is usually designated by affixing the syllable "ase" to the root of the name of the substrate upon which it acts or to the type of reaction that it catalyzes. For example, enzymes that decompose protein are called *proteases* and those that effect oxidations are called *oxidases*. There are two large groups of enzymes active in bacterial decompositions: (1) the *hydrolases*, which break down complex molecules by hydrolysis and which include the carbohydrases, proteases, lipases (attacking fats), esterases, and amidases; and (2) the *desmolases*, whose function is to supply the energy requirements of the cells and which include oxidases and reductases.

The hydrolases are usually *extracellular* enzymes, in which case the reactions they promote take place outside the cell walls, are independent of the presence of the cells, and usually involve small energy changes. The desmolases, on the other hand, are *intracellular* and promote reactions within the cells, involving, in general, large changes in energy. The energy released by extracellular reactions is not available to the cell for growth and respiration. It may, however, serve to raise the temperature of the medium.

The rate of decomposition of organic matter is greatest when the bacteria are undergoing active multiplication. When the number of bacteria in a culture gets large, the accessibility of the food supply to the organisms diminishes because of the interference of the products of decomposition and other causes. A point is soon reached where the rate of adsorption of food supply by the organisms is insufficient to sustain bacterial reproduction. Hence, the rate of decomposition gradually diminishes, and eventually the number of bacteria also decreases. Complete breakdown of organic matter with the liberation of all its available energy is seldom attained. Anaerobic decomposition of carbohydrates is called *fermentation*. Anaerobic decomposition of nitrogenous organic matter is called *putrefaction* or *septic decomposition*.

Ordinary domestic sewage contains from less than $\frac{1}{2}$ million to more than

30 million bacteria per milliliter, and averages about 4 million. The coliform group averages about 1 million per milliliter. According to Harris, Cockburn, and Anderson (*3*), the predominant bacterial flora in the raw sewage at the Shieldhall, Glasgow, sewage-treatment plant belong to the streptococci and intestinal groups. Sixty-one percent of the bacteria of intestinal origin was of the coli-aerogenes type, and the remainder was of the *Bacillus proteus* type. Of the cocci in sewage, *Staphylococcus albus*, which is universally present on the skin and clothing and is the predominating organism in laundry wastes (*4*), is probably the most numerous. *B. subtilis, B. welchii, B. cloacae,* and *Pseudomonas fluorescens* are also found abundantly in sewage. In smaller

Table 10-1

Type of Organism	Energy-Yielding Reaction and Source of Carbon	Source of Nitrogen
Heterotrophs		
(a)	Utilization of carbon compounds more reduced than CO_2, i.e., organic carbon compounds	Array of amino acids and enzymes for synthesis of protoplasm
(b)	Same as (a)	Single amino acid, particularly tryptophane
(c)	Sames as (a)	Ammonia, nitrate, nitrite, N_2
Autotrophs		
(d) Facultative chemo-synthesizing aerobes	Oxidation of thiosulfate, H_2, CO, CH_4, yielding energy for assimilation of CO_2	Ammonia, nitrate
(e) Chemosynthesizing aerobes	Oxidation of NH_3, nitrite, yielding energy for assimilation of CO_2	Ammonia, nitrite
(f) Chemosynthesizing "sulfur" aerobes	Oxidation of H_2S and thiosulfate, yielding energy for assimilation of CO_2	Ammonia
(g) Photosynthesizing "sulfur" anaerobes	$CO_2 + 2H_2S \xrightarrow[\text{energy}]{\text{radiant}} HCHO + 2S + H_2O$	Ammonia
(h) Simplest organisms with chlorophyll or similar photocatalyst	$CO_2 + H_2O \xrightarrow[\text{energy}]{\text{radiant}} HCHO + O_2$	Ammonia
(i) Blue-green algae	$CO_2 + H_2O \xrightarrow[\text{energy}]{\text{radiant}} HCHO + O_2$	Fixation of N_2 gas

numbers, many other types of bacteria and some yeasts and molds are present. In the sewage and sludge treatment processes, the same organisms are present, but their relative numbers are modified by changes in the environment, and other organisms, such as the zoogloea-forming bacteria and the nitrifiers, are introduced from the environment. An important group of bacteria in anaerobic sludge digestion is the methane-producing organisms.

From the standpoint of nutritional requirements, Knight (5) classified bacteria as shown in Table 10-1. The *heterotrophs* require organic compounds for their metabolism, whereas the *autotrophs* are able to subsist on inorganic compounds. The heterotrophs are mainly responsible for the decomposition of sewage organic matter, although in the latter stages of the aerobic processes of sewage treatment and of self-purification of polluted rivers the autotrophic nitrifying bacteria play important roles. It has been shown that the presence of carbon dioxide surrounding the cell is essential to the growth of most bacteria (5), but the part played by CO_2 is unknown except in the case of the autotrophs and a few heterotrophs that use CO_2 as a hydrogen acceptor. CO_2 is an end product in the metabolism of the heterotrophs, and hence is normally present around the cells. There is some evidence that carbon is assimilated in the form of pyruvic acid by heterotrophs (6).

10-2. THERMAL CHEMISTRY

Chemical reactions induced by bacteria usually take place at constant pressure. If the heat evolved in the reaction is written as a part of the equation, the reaction is called a *thermochemical* reaction and the heat evolved is called the *heat of reaction*. For example, in the equation

$$\begin{array}{ccc} 2 & 16 & 18 \\ H_2 + \tfrac{1}{2}O_2 & \longrightarrow & H_2O \quad + 68.317 \text{ kcal} \\ \text{(gas)} \ \text{(gas)} & & \text{(liquid)} \\ H^\circ = 0 \quad\quad 0 & & -68.317 \end{array}$$

(10-1)

in which the pressure, p, is 1 atm, the temperature, T, is $298°K$ ($25°C$), and the heat of reaction, $-\Delta H^\circ = +68.317$ kcal/mole of H_2O. Since the heat of reaction in this case is positive, it represents a loss of heat from the system, and the *change in heat content*, ΔH°, is negative.

When any substrate undergoes bacterial decomposition, heat is necessarily evolved in order to provide the energy for the growth and maintenance of the bacteria. The process as a whole is, therefore, exothermic. Some of the reactions that correspond to steps in the process may be endothermic, however, provided the energy yielded in the rest of the process is more than sufficient to overcome the deficit.

The heat of reaction in the preparation of a compound from its elements is called the *heat of formation* of the compound. In Eq. (10-1) the heat evolved, +68.317 kcal, is the heat of formation of liquid water from the elements. The *heat content, $H°$*, of each of the elements is taken as zero in the state in which the element usually occurs and at a pressure of 1 atm. Since the reaction involves a loss of heat from the system, the heat content of water is −68.317 kcal.

Thermochemical reactions may be treated as simultaneous equations, and additions and subtractions may be made freely of reactants, products, and heat contents. This is in accord with the law of Hess,* which states that the heat evolved in a given chemical process is the same whether the process takes place in one or several steps. It is often possible to compute the heat of reaction from the heat contents of the reactants and products, or to compute a heat content from the heat of reaction.

The state of the reactants and resultants should be given in an equation, since most reactions involve a change of state. If, for example, in Eq. (10-1) part of the water is water vapor, the heat of reaction will be less by the latent heat of vaporization, which for 1 g molecule of water at 25°C would be about 10.519 kcal. In enzymic reactions, the water formed is always liquid.

For precision the temperature at which a reaction takes place should be stated, since the heat of reaction may vary with the temperature. The heat of reaction for Eq. (10-1) at 10°C, for example, is about 300 g cal less than the heat of reaction at 50°C, because the specific heat of the gaseous reactants is different from that of water. Since the range of temperatures within which an enzymic reaction may take place is usually less than the range in the above example, the variation in the heat of reaction with the temperature is small enough to neglect. Hence, temperatures are frequently not stated.

The *heat of combustion* of a substance is the heat of reaction of the substance with molecular oxygen at constant pressure, usually 1 atm. The products formed in wet combustion are liquid water and oxides, such as CO_2 and nitrates. In Eq. (10-1), for example, the heat of reaction is the heat of combustion of hydrogen gas. If hydrogen is burned in air, it combines with gaseous oxygen to produce water vapor, liberating a certain amount of heat. If the process takes place in a calorimeter without heat loss, the temperature of the products will be raised, and the amount of heat evolved may be computed from the final temperature of the products. The heat evolved is the amount that must be withdrawn from the resultant water vapor to condense it and lower its temperature to the original temperature of the reactants.

When a substance is added to a solvent, heat may be evolved, and the amount

---

*Germain Henri Hess, Swiss–Russian chemist (1802–1850).

of heat per mole of solute changes with the amount of solvent until a dilution is reached beyond which no further heat change takes place. The *heat of solution* of a substance is the heat evolved in the solution of 1 mole of the substance in a volume of solvent so great that no further change is produced by the addition of solvent. For bacterial reactions the solvent is usually water. The heats of solution of strong acids are appreciable, but those of some organic compounds are negligible. It is unsafe to neglect the heats of solution in writing bacterial reactions in aqueous solution if precision is required.

The heat of formation of water from its ions is +13.36 kcal as follows:

$$
\begin{array}{ccc}
1 & 17 & 18 \\
H^+ + & OH^- \longrightarrow & H_2O \quad + 13.36 \text{ kcal} \\
H^\circ = 0 & -54.957 & -68.317
\end{array}
\qquad (10\text{-}2)
$$

If the heat content of H^+ is taken as zero, it follows that the heat content of OH^- is -54.957 kcal, and its heat of formation is $+54.957$ kcal. The *heat of ions* is of importance in reactions taking place in water.*

10-3. RESPIRATION OF CELLS

The energy required for the work and growth of living cells is obtained by the cells through the decomposition of foodstuff molecules. The maximum amount of energy which can be obtained from organic compounds is that resulting from their combustion in air at high temperatures. Hydrogen is oxidized to water and carbon to carbon dioxide, the energy being given off as heat. A spontaneous reaction of this type, which will run by itself, is of no use to the cell. The cell can use only enzyme-catalyzed reactions which take place in such a way that the energy released is not lost as heat, but is utilized by the cell. The process by which the cell obtains its energy is called respiration.

Since the products of aerobic respiration include CO_2 and H_2O, the process has been considered as an oxidation phenomenon similar to combustion. According to Szent-Györgyi (7), however, life knows but one fuel, hydrogen. Practically all the expired CO_2 is formed by *decarboxylation*, a splitting off of CO_2 from the food molecule through the action of enzymes. This process is

---

*The heats of formation, heats of solution, and free energies of formation of a number of substances of importance in enzymic reactions are given by F. R. Bichowsky and F. D. Rossini (*Thermochemistry of Chemical Substances*, New York, Van Nostrand Reinhold, 1936); by G. S. Parks and H. M. Huffman (*The Free Energies of Some Organic Compounds*, New York, Van Nostrand Reinhold, 1932). See also W. M. Latimer, *Oxidation Potentials*, Englewood Cliffs, N.J., Prentice-Hall, Inc., 1952.

independent of the presence of atmospheric oxygen, and it explains the production of CO_2 in both aerobiosis and anaerobiosis.

The carbon frame of the food molecule serves to carry a number of H atoms, which are plucked off by the living cell through the action of enzymes known as *dehydrogenases*. This process occurs in both aerobiosis and anaerobiosis. In aerobiosis, the H atom is eventually handed over to O_2 to be oxidized to water; the energy released, $\Delta H°$ of about 68 kcal, is utilized by the cell. In anaerobiosis the H atom is handed to some H acceptor other than O_2, and the energy yield is considerably less than in aerobiosis. Even in aerobiosis, all the energy obtained by the oxidation of hydrogen is not secured by the cell in one step. The H is passed from one H carrier to another, and in each step, according to Szent-Györgyi, approximately 11 kcal is obtained by the cell. The initial steps in the process appear to be the same for both aerobic and anaerobic decomposition, as will be shown later for the sugar molecule.

The oxidation of a food molecule may be considered as a replacement of hydrogen by water, as illustrated by Figure 10-9. Such a reaction, however, does not yield energy. It is necessary to have an H acceptor, such as O_2, that will combine with the H for the cell to obtain its energy. In order for the reaction to take place, both the food molecule and the H acceptor or H carrier must be *activated*. The food molecule is activated by the dehydrogenases, and the H acceptors are activated by specific proteins usually carrying a metal, such as iron or copper. These metals are capable of being alternately oxidized and reduced by giving up or accepting an electron. It appears, therefore, that the oxidation process consists of a transfer of electrons from the H to the metals of the H carriers and, thence, to the final H acceptor.

10-4. GROWTH OF BACTERIA AND ENERGY RELATIONS

Under optimum conditions of growth, bacteria divide about every half-hour. The conditions affecting the growth rate are the presence of water, the size of the cells, the temperature, the concentration and availability of food, the pH value, and the presence of poisons and predatory organisms in the medium. Of less importance is the content and composition of mineral salts in the medium, unless the metals are germicidal or take part in enzyme action. In the case of obligate aerobes, the presence or absence of an adequate supply of dissolved oxygen in the medium is an important factor, which determines whether growth will take place; in the case of facultative anaerobes, the presence or absence of oxygen determines the rate of growth.

The effect of temperature upon the generation time of *E. coli* in an environment suitable for reproduction is shown in Figure 10-1 (*6*, p. 93). This organism, it will be noted, grows best at the temperature of the human body,

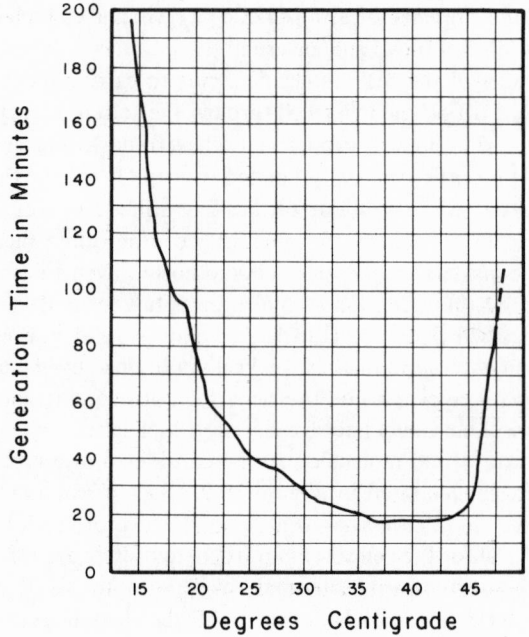

Figure 10-1
Growth rate of *E. coli*. [After Barker (*11*).]

37°C, which appears to be slightly above the optimum temperature for the group of sewage bacteria that are responsible for the anaerobic decomposition of sewage sludge. Another group, called *thermophilic* bacteria, which is also developed anaerobically in sewage sludge, grows best at temperatures between 50 and 60°C. *Escherichia coli* is able to reproduce in a pH range between 5 and 9, but the optimum pH is between 7 and 8 (*6*, p. 119). The same pH appears to be optimum for the thermophilic flora in sewage (*8*).

Penfold and Norris (*9*), in studies of the rate of growth of *B. typhosis* at 37°C in various concentrations of peptone ranging from 0.0125 to 1.25 percent, found that the generation time was inversely proportional to the concentration or that growth rate was directly proportional to food concentration. Studies by Weston and Eckenfelder (*10*) of experimental data on the maximum growth rates of activated sludge, in various concentrations of yeast nutrients spent beer, indicated that the growth rate increased, approximately, as the two thirds power of the concentration of substrate.

Barker (*11*) has identified four different types of methane-producing bacteria that are doubtless active in sewage sludge digestion. These organisms are strict anaerobes, permanently immotile, and do not form spores. They are *Methanosarcina methanica*, large spherical cells characteristically grouped in cubical sarcina packets; *Methanococcus mazei* n.sp., small spherical cells that occur singly, in small groups, or in large irregular aggregates; *Methanobacterium söhngenii* n.sp., rod-shaped cells joined into long threads that lie parallel to one another in bundles; and *Methanobacterium omelianskii*, n.sp., thin, frequently bent rods. The first three organisms produce methane from acetic and butyric acids, but not from alcohols. The other organism ferments ethyl alcohol to acetic acid and butyl alcohol to butyric acid, thereby producing methane. The same organism probably ferments butyric to acetic acid, but does not attack acetic acid. According to Heukelekian and Heinemann (*12*), the methane-producing bacteria grow best at a pH of 7.0, in a temperature range from 28 to 35°C, and with a substrate concentration of 0.75 to 1.0 percent. They grow very slowly, the generation time under optimum conditions being from 15 to 20 h. According to these investigators, methane production in the thermophilic temperature range is probably due to other organisms.

Butterfield (*13*) has shown that the aerobiosis of sewage solids in sewage-treatment processes is due primarily to zoogloea-forming bacteria, probably *Zoogloea ramigera* or other organisms of the same type. This bacterium is an obligate aerobe, but it is not killed when subjected to anaerobic conditions for as long as 7 days. It does not form spores, and shows a marked tendency to grow in flocs or zoogloeal masses in liquid media when the food supply is adequate. When the food supply runs short, the organisms separate themselves from the zoogloeal matrix and disperse themselves in the liquid, each bacterium being propelled by a single flagellum. The bacterium is rod-shaped, averaging about 1.5 by 3 μm in size. It grows at temperatures as low as 4°C, but most vigorous multiplication occurs between 20 and 37°C with an optimum temperature at 28 to 30°C. Growth takes place over a pH range from 5.6 to 8.5, and the optimum pH is between 7.0 and 7.4. The decomposition products produced by the organism invariably increase the alkalinity of the medium, raising the pH as high as 8.8. The organism does not oxidize ammonia or nitrites.

Under strict aerobic conditions, ammonia may be oxidized to nitrite and nitrate through the action of the *nitrifying* bacteria, *Nitrosomonas* and *Nitrobacter*. These organisms are obligate autotrophs, *Nitrosomonas* oxidizing only ammonia and *Nitrobacter* only nitrites. The energy obtained from these reactions is used for the assimilation of carbon, which is obtained from carbon dioxide or bicarbonate, not from organic carbon compounds. These bacteria have been studied extensively by the Winogradskys, Meyerhof, and Boltjes (*5*, p. 28, and *6*, p. 239). The optimum pH value is from 8.5 to 8.8 for *Nitrosomonas* and from 8.5 to 9.3 for *Nitrobacter*. The presence of organic matter, particularly amino compounds, in excessive concentrations, inhibits the growth and respiration of both organisms. For example, urea in concentrations exceeding 1,500 ppm inhibits the respiration of *Nitrosomonas*, and in concentrations exceeding 2,400 ppm it inhibits growth. *Nitrobacter* is not so sensitive, since

9,000 ppm of urea is required to inhibit growth and respiration. Asparagine in concentrations as low as 600 ppm inhibits the growth of *Nitrosomonas*, although a concentration of 22,500 ppm is required for perceptible inhibition of the growth of *Nitrobacter*. The ratio of the weight of nitrogen in the ammonia oxidized by *Nitrosomonas* to the weight of carbon assimilated has been estimated at 35 to 1; a similar ratio for oxidation of nitrite by *Nitrobacter* has been estimated at 135 to 1.

Nitrification is an important contributory factor in the depletion of oxygen from sewage and polluted water, but its operation is usually delayed for several weeks pending the development of a suitable environment for the organisms. In many activated sludge and trickling filter plants, nitrification is very active.

When a pure culture of bacteria is inoculated into a suitable culture medium, it is found that multiplication of the viable organisms does not proceed at the geometric rate that would be indicated by a constant generation time. Buchanan (*14*) distinguishes seven phases of growth, as shown in Figure 10-2. There is a

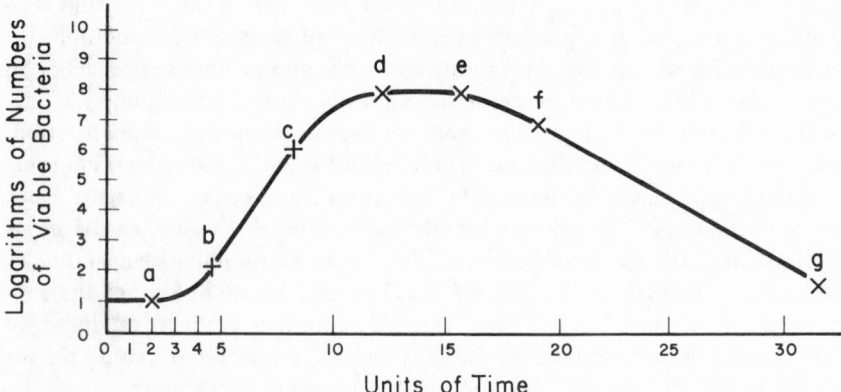

Figure 10-2
Growth phases of bacteria. [After Buchanan and Fulmer (*14*).]

short period of no growth, followed by a short lag phase during which there is a positive acceleration in the rate of multiplication. The third period is the logarithmic growth phase, indicated by a constant generation time. This is the maximum rate of growth. This period is followed by one of negative growth acceleration, and then by a period of substantially constant numbers as the limiting bacterial population is reached. The concentration of accessible food in this period is just sufficient to sustain life, but not adequate for reproduction. There follows then a phase of accelerating death rate, and a final phase with a logarithmic decrease in numbers resulting from death. The latter phases are probably associated with exhaustion of the accessible food and with alterations in the medium due to pH changes and increase in concentration of inhibitory products. The maximum rate of decomposition of organic matter is associated with the latter part of the logarithmic growth phase and the beginning of the period of growth deceleration. The maximum rate of decomposition occurs with the greatest number of actively growing bacteria.

Butterfield (15) has shown that the limiting count of microorganisms in pure cultures in liquid media increases with the original concentration of food material in the medium, provided the concentration is sufficient to support growth. For *Aerobacter aerogenes* growing aerobically at 20°C in a medium containing equal parts of dextrose and peptone and with the pH maintained between 6 and 7.1, it was found that the following empirical equation gave the relation between the limiting bacterial population and the food concentration:

$$y = 2.29x^{0.818} \tag{10-3}$$

where y is the bacteria in millions per milliliter, and x is the initial concentration of each food constituent, dextrose and peptone, in parts per million.

It was also indicated by Butterfield's experiments that the limiting population in a given medium depends upon the size of the microorganism. The total volume occupied by the number of cells in the limiting population appears to be of the same order of magnitude, regardless of the size of the cell. For example, the volume occupied by the limiting number of cells of a small sewage coccus was about the same as that occupied by *A. aerogenes*, although the volume of the individual cell was only about one fourth that of *A. aerogenes*. Similarly, the protozoan *Colpidium*, whose cell volume is about 32,000 times that of *A. aerogenes*, grew to a limiting total volume about equal to that occupied by *A. aerogenes* in the same medium.

Heukelekian and Heinemann (16) have found that the limiting number of methane-producing bacteria in sewage sludge digesting anaerobically is from 10 million to 1 billion per milliliter, the ethyl alcohol fermenters constituting the largest number.

One *A. aerogenes* of average size (0.8 × 2 μm) has a volume of about 1 μ^3 or 10^{-12} cm^3 and weighs about 10^{-12} g. According to Anderson (2, p. 49), the water content of most bacteria is between 75 and 85 percent, of yeasts about 75 percent, and of molds about 85 percent. Spores contain only 40 to 50 percent water. The carbon content of all three types of microorganisms is 45 to 55 percent of their dry weights, and the nitrogen content is from 2 to 14 percent; the latter is quite variable in organisms of the same species and depends upon the condition of the cell and the environment.

If the water content is 75 percent, 1 g dry weight of *A. aerogenes* contains about 4×10^{12} bacterial cells. A count of 1 million *A. aerogenes* per milliliter corresponds to a dry weight of about 0.25 ppm, and the volume occupied by the cells is approximately 0.0001 percent of the total volume of the liquid medium. One zoogloea-forming organism has about 5 times the volume of *A. aerogenes.* Hence, a count of 1 million per milliliter of these bacteria corresponds to about 0.0005 percent of the volume and to a dry weight of about 1.25 ppm, exclusive of the gelatinous matrix synthesized by the cells.

The dry cell material of bacteria may be represented approximately by the empirical formula $C_5H_7NO_2$, suggested by Hoover and later used by Weston and Eckenfelder *(10).* The heat of combustion of bacteria, according to Tangl *(6,* p. 12), is from 4 to 5 kcal/g dry weight of cells. This is equivalent to a heat of reaction of 452 to 565 kcal/mole for a molecular weight of 113. The combustion reaction is approximately as follows:

$$
\begin{array}{ccccc}
113 & 184 & 220 & 63 & 14 \\
\end{array}
$$

$$
C_5H_7NO_2 \;+\; \tfrac{23}{4}O_2 \longrightarrow \; 5CO_2 \;+\; \tfrac{7}{2}H_2O \;+\; \tfrac{1}{2}N_2 \;+\; 452 \text{ to } 565 \text{ kcal} \quad (10\text{-}4)
$$

$$
\begin{array}{ccccc}
\text{(solid)} & \text{(gas)} & \text{(gas)} & \text{(gas)} & \text{(gas)} \\
\end{array}
$$

$$
H^\circ = -107 \text{ to } -220 \quad 0 \quad\quad -470.2 \quad -202.3 \quad 0
$$

Based on Tangl's values for the heat of combustion, this heat balance indicates a heat of formation of bacterial cell material of 107 to 220 kcal/mole or 0.95 to 1.95 kcal/g dry weight of cells. This is an average of about 1.5 kcal/g dry weight of cells. Numerous measurements *(10)* indicate that the heat of combustion of activated sludge (which includes inert organic solids as well as living cells) averages about 10,000 Btu/lb of volatile solids (5.56 kcal/g). The corresponding heat of formation of the hypothetical cell material represented in Eq. (10-4) is about 0.40 kcal/g dry weight of volatile solids.

A study of experiments by Butterfield and Wattie *(17)* on the aerobic growth of *A. aerogenes* in a dilute solution of peptone and dextrose in water at 20°C reveals that the generation time at which growth was most rapid was from 60 to 100 min. Therefore, the generation time of *A. aerogenes* appears to be about the same as that of *E. coli,* which, according to Figure 10-1, is 80 min at 20°C. Based upon these figures, the heat required for cell building of *A. aerogenes,* for most rapid growth, in dilute substrates at 20°C is about 1.2 kcal/hr/g dry weight of cells.

In contrast with this figure, the energy required for *maintenance* or *endogenous respiration* in the absence of cell multiplication is much less. After the limiting bacterial population is reached, the substrate will continue to be decomposed at a decreasing rate without change in the bacterial count until the accessible food is depleted to the point just sufficient to maintain life. After this stage is reached, the number of organisms will decrease by death, the number of living bacteria remaining at any time being limited by the residual utilizable food. Terroine and Wurmser *(6,* p. 17) estimated that the cost of maintenance of the mold *Aspergillus niger* was only 0.041 kcal/hr/g dry weight of cells. Butterfield, Purdy, and Theriault (*18*) have shown that the limiting population of *A. aerogenes* cultures grown aerobically in liquid media at 20°C is sustained for 30 or 40 days and that the oxygen requirement for maintenance is less than 0.01 ppm daily per million bacteria. This oxygen demand corresponds with an energy requirement for the maintenance of *A. aerogenes* of only about 0.005 kcal/hr/g dry weight of cells. Hoover, Jasewicz, and Porges (*19*) have shown endogenous respiration rates at 22°C for various bacteria, which range from 6 to 23 mg of O_2/hr/g of cell. The corresponding energy required is 0.02 to 0.08 kcal/g dry weight of cells.

Butterfield, Purdy, and Theriault (*18*) found that the minimum generation time of the protozoan *Colpidium* was about 1 day, when the cells were grown aerobically at 20°C in a bacteria-free liquid medium containing 5,000 ppm each of peptone and dextrose. Assuming the same heat of formation for the cells of *Colpidium* as for *A. aerogenes*, the heat required for the most rapid growth of *Colpidium* at 20°C is only about 0.07 kcal/hr/g dry weight of cells or about $\frac{1}{18}$ of the heat required for the most rapid growth of *A. aerogenes*. It is evident from these figures that the rate of decomposition of organic matter by protozoa is negligible as compared to the rate of breakdown by bacteria. It is of interest to note that these rates are about in proportion to the surface area of the organisms per unit of volume. The energy required for maintenance of *Colpidium*, based upon the oxygen used after the limiting population was attained, was about 0.015 kcal/hr/g dry weight of cells.

The 15- to 20-hr generation time of the anaerobic methane-producing bacteria, as found by Heukelekian and Heinemann (*12*), appears to be due not to the size of the organisms but to the small energy content available anaerobically from the food supply of these bacteria. These microorganisms are of the same order of size as *A. aerogenes*. The energy content per gram of food for the anaerobes is about 0.15 to 0.4 kcal, whereas the aerobic destruction of dextrose and peptone yields about 4 kcal. The growth rate of the organisms is, therefore, about proportional to the energy content of the food supply.

It has been shown by Butterfield, Purdy, and Theriault (*18*) that good growth of *A. aerogenes* at 20°C is obtained in liquid media containing as little as 2.5 ppm each of dextrose and peptone, whereas a bacteria-free culture of the protozoan *Colpidium* required a minimum concentration of 500 ppm each of dextrose and peptone. The protozoan *Paramecium* could not be grown at all in a bacteria-free medium. Nevertheless, good growth of *Colpidium* was obtained together with *A. aerogenes* in a medium containing only 5 ppm each of peptone and dextrose. It is evident that in dilute solutions protozoa must feed upon the bacteria which serve as concentrators of the protozoan food. The effect of the protozoa upon the bacterial population is to keep it below the limiting number and, thereby, to stimulate bacterial growth.

It has been shown that the rate of decomposition of organic matter by pure cultures of protozoa is very small as compared with the corresponding rate by pure bacterial cultures. Nevertheless, the presence of protozoa with bacteria increases the decomposition rate markedly through the protozoa influence in stimulating bacterial multiplication. It was shown by Butterfield, Purdy, and Theriault (*18*) that the 5-day oxygen demand of a mixed culture of *Colpidium* and *A. aerogenes* was about 60 percent greater than the 5-day demand for a pure culture of *A. aerogenes* in the same medium. Similarly, the 5-day demand of a mixed culture of the bacteria and plankton of river water was more than twice the demand for *A. aerogenes* in the same medium. These relations are shown in Figure 10-3.

The cause of formation of spores has not been definitely ascertained. Buchner (*6*, p. 105), in studying sporulation of the aerobic organism *B. anthrax*, concluded that exhaustion of food supply was the prime cause of spore formation. This organism does not form spores in the absence of molecular oxygen.

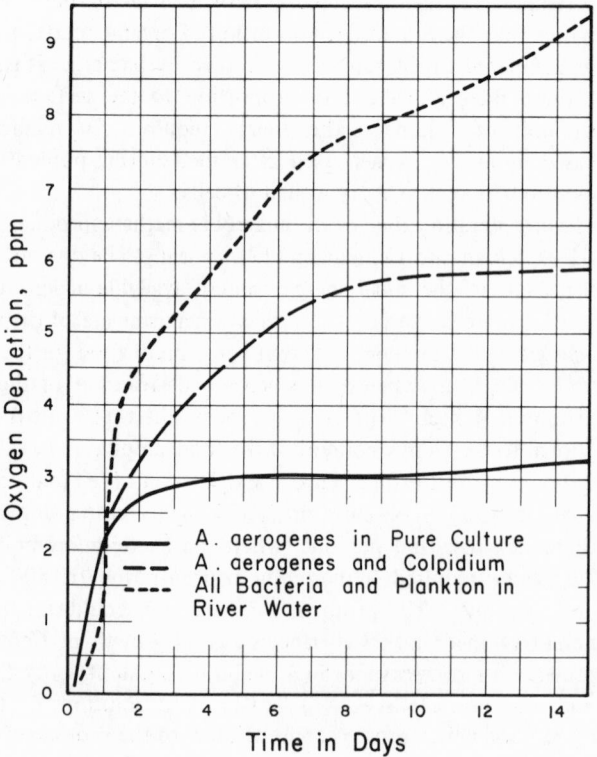

Figure 10-3
Oxygen depletions at 20°C in medium containing 5 ppm each of dextrose and peptone. [After Butterfield, Purdy, and Theriault (*18*).]

Buswell (*20*) also concludes that spore formation is due to lack of sufficient food, as a result of his studies on the cultivation of *B. subtilis* in peptone broth. Although spore formation sometimes occurs under what appear to be optimum conditions, it does not occur during logarithmic growth. Sporulation may develop as a result of accumulation of substances in the medium, or because of a diminution of inhibiting nutrients (*21*). Some regard sporulation as a normal part of the life cycle of certain bacteria.

The preceding discussion deals with the energy requirements for growth of bacteria. It is, of course, obvious that the available food is used for both the energy for growth and the material used in synthesis of the bacterial cells. Based on an average heat of formation of 1.5 kcal/g dry weight of cells synthesized, aerobiosis with dextrose, according to reaction (10-5), should require 0.42 g of dextrose for energy and 1.0 g of dextrose for cell material. In other words, each gram dry weight of cell synthesized should require 1.42 g of dextrose, 30 percent of which is oxidized for energy requiring 0.45 g of oxygen. Aerobiosis with a protein, according to reaction (10-6), should require 0.30 g of protein for energy and 1.0 g for synthesis. In other words, each gram dry weight of cell synthesized should require 1.30 g of protein, 23 percent of which is oxidized for energy requiring 0.45 g of oxygen. Various experimenters have shown that glucose is synthesized to the extent of 44 to 77 percent of the amount used, with the remainder being oxidized for energy.

The reactions involved in the synthesis of cell protoplasm are highly complex and little is known about them. Claims have been made (*10, 22*) that all cell synthesis reactions are oxidations and are exothermic, supplying energy to the cell. This cannot be true, because such a reaction would not only form the protoplasm, but would also release enough heat to burn the cell. Moreover, the energy produced by oxidation of the substrate to CO_2 and water would not be needed and would further increase the temperature of the cell. A heat balance based on the synthesis of cell protoplasm by oxidation of glucose in a single thermochemical reaction indicates a heat production excess of about 1.8 kcal/g dry weight of cell formed. This excess heat would increase the temperature of the cell by about 1,800°C. Since the energy-producing reactions and the synthesis reactions are both intracellular, it is evident that the synthesis reactions must be endothermic to utilize all the energy produced by the decomposition of the food molecules without changing the temperature of the cell. Small rises in temperature may be produced in bacterial cultures, but they occur in the medium as a result of extracellular hydrolytic reactions to break down the food particles so that they may pass through the cell walls.

10-5. AEROBIOSIS AND ANAEROBIOSIS

In the complete aerobiosis of sugar dissolved in water, the end products are carbon dioxide and water. Most of the carbon dioxide will be liberated as gas, but a part of it will remain in solution as carbonic acid. Since carbonic acid is partially ionized, the decomposition of sugar tends to lower the pH value of the medium. Neglecting the carbonic acid, the complete aerobiosis of sugar may be represented by the following hypothetical reaction:

$$
\begin{array}{ccccc}
180 & 192 & 264 & 108 \\
C_6H_{12}O_6 & + \; 6O_2 \longrightarrow & 6CO_2 & + \; 6H_2O + 649.8 \, kcal & (10\text{-}5) \\
\text{(aq. dextrose)} & \text{(aq.)} & \text{(gas)} & \text{(liquid)} \\
H° = \quad -301.5 & -22.8 & -564.3 & -409.8
\end{array}
$$

The heat of reaction is 649.8 kcal/mole or 3.60 kcal/g of dextrose. Since, from the preceding article, the heat required for cell building is about 1.5 kcal/g dry weight of cells, the amount of sugar required to supply this energy aerobically is about 0.42 g/g dry weight of cells. The oxygen demand is about 0.45 g/g dry weight of cells.

In the complete aerobiosis of proteins by heterotrophic bacteria, the end products are carbon dioxide, ammonia, water, and sulfates. As in the case of sugars, most of the carbon dioxide will be evolved as gas, but a small amount of it will remain as carbonic acid and tend to lower the pH value. The ammonia, on the other hand, is highly soluble; a saturated solution in pure water at 20°C contains approximately 25 percent NH_3. Most of the ammonia combines with hydrogen ions to form ammonium ions, NH_4^+, thus tending to raise the pH value. At pH 7 practically all the ammonia is present as NH_4^+, at pH 8 only about 5 percent is present as NH_3, and at pH 9 about 35 percent is present as NH_3. Ammonia gas is, therefore, not evolved in the decomposition process unless the pH value of the medium is alkaline and the concentration of ammonia is great.

For convenience in writing the hypothetical thermochemical reactions, an average protein consisting of 53 percent carbon, 7 percent hydrogen, 23 percent oxygen, 16 percent nitrogen, and 1 percent sulfur will be used. The molecular weights of proteins are known to be very high, but for convenience in writing the equations, the molecular weight will be taken arbitrarily as 100. The heats of combustion of 19 different proteins, as given by Buchanan and Fulmer (*14*, p. 389), range from 5.358 to 5.916 kcal/g and average 5.64 kcal/g dry weight. The average heat of formation of proteins computed from this average heat of combustion is about 0.55 kcal/g. The complete aerobiosis of this hypothetical protein by heterotrophs may be represented approximately by the following reaction. (It is assumed in the equation that a molecule of water is shared by a molecule of NH_3 and a molecule of CO_2 to produce NH_4^+ and HCO_3^-.)

$$100 \qquad\qquad 148.5 \qquad 20.6 \quad 69.8 \qquad 144$$

$$C_{53/12}H_7O_{23/16}N_{16/14}S_{1/32} + 4.63O_2 \longrightarrow \tfrac{16}{14}NH_4^+ + \tfrac{16}{14}HCO_3^- + 3.273CO_2$$

$$\text{(protein)} \qquad\qquad \text{(aq.)} \qquad \text{(aq.)} \quad \text{(aq.)} \qquad \text{(gas)}$$

$$H° = \qquad -55 \qquad\qquad\quad -17.7 \qquad -36 \quad -188.4 \quad -309$$

$$11.0 \qquad 3.1$$

$$+\, 0.611H_2O + \tfrac{1}{32}H_2SO_4 + 509.2 \text{ kcal} \quad (10\text{-}6)$$

$$\text{(liquid)} \qquad \text{(aq.)}$$

$$-41.8 \qquad -6.7$$

The heat of reaction in this case is about 5.09 kcal/g of protein, and the amount of protein required to supply the energy for cell growth is about 0.30 g/g dry weight of cells. The corresponding oxygen demand is about 0.45 g/g of cells.

Reactions (10-5) and (10-6) are typical of the *first stage* in the *deoxygenation* of polluted waters. The *nitrification* stage is due to the autotrophic aerobes, the nitrifying bacteria. *Nitrosomonas* oxidizes ammonia to nitrate as follows:

$$18 \qquad 48 \qquad\quad 46 \quad 2 \qquad 18$$

$$NH_4^+ + 1\tfrac{1}{2}O_2 \longrightarrow NO_2^- + 2H^+ +\ H_2O\ + 56.3 \text{ kcal} \qquad (10\text{-}7)$$

$$\text{(aq.)} \quad \text{(aq.)} \qquad \text{(aq.)} \qquad\quad \text{(liquid)}$$

$$H° = -31.7 \quad -5.7 \qquad -25.4 \qquad\quad -68.3$$

And *Nitrobacter* oxidizes nitrite to nitrate as follows:

$$46 \qquad 16 \qquad\quad 62$$

$$NO_2^- + \tfrac{1}{2}O_2 \longrightarrow NO_3^- + 22.1 \text{ kcal} \qquad (10\text{-}8)$$

$$\text{(aq.)} \quad \text{(aq.)} \qquad \text{(aq.)}$$

$$H° = -25.4 \quad -1.9 \qquad -49.4$$

The heat of the reactions in nitrification is about 3.3 kcal/g of NH_3 in the first stage, reaction (10-7), and an additional 1.3 kcal/g of NH_3 is obtained for the growth of *Nitrobacter*. The oxygen requirements are 2.8 and 1.0 g/g of NH_3 for reactions (10-7) and (10-8), respectively.

In the complete decomposition of sugar by anaerobiosis, the end products are carbon dioxide and methane. As in the case of aerobic breakdown, most of the carbon dioxide is liberated as gas, but some remains in solution as carbonic acid and tends to lower the pH value. Methane is very slightly soluble in water, and practically all of it is evolved as gas. The complete anaerobiosis of dextrose, neglecting carbonic acid, is represented by the following hypothetical equation:

$$
\begin{array}{ccc}
180 & 132 & 48
\end{array}
$$

$$
\underset{\text{(aq. dextrose)}}{C_6H_{12}O_6} \longrightarrow \underset{\text{(gas)}}{3CO_2} + \underset{\text{(gas)}}{3CH_4} + 34.4\,\text{kcal} \tag{10-9}
$$

$$
\begin{array}{cccc}
H° = & -301.5 & -282.2 & -53.7
\end{array}
$$

The heat liberated in this reaction is about 0.19 kcal/g of sugar, or only about 5.3 percent of the energy evolved by aerobiosis of the sugar.

In the complete anaerobiosis of proteins by heterotrophs, the end products usually formed are carbon dioxide, methane, ammonia, and hydrogen sulfide. A variety of other products, such as hydrogen gas, nitrogen gas, acetic acid, and mercaptans, are frequently present and may be regarded as end products for certain types of bacteria. For the purpose of this discussion, however, the hypothetical protein will be assumed to yield the usual end products. As in the case of aerobiosis, the ammonia is present as NH_4^+, and an equivalent amount of HCO_3^- will be assumed in the yield. The hypothetical reaction is as follows:

$$
\begin{array}{ccccc}
100 & 71.4 & 20.6 & 69.8 & 43.4
\end{array}
$$

$$
\underset{\text{(protein)}}{C_{53/12}H_7O_{23/16}N_{16/14}S_{1/32}} + \underset{\text{(liquid)}}{3.966H_2O} \longrightarrow \underset{\text{(aq.)}}{\tfrac{16}{14}NH_4^+} + \underset{\text{(aq.)}}{\tfrac{16}{14}HCO_3^-} + \underset{\text{(gas)}}{0.986CO_2}
$$

$$
\begin{array}{cccccc}
H° = & -55 & -271 & -36 & -188.4 & -97.2
\end{array}
$$

$$
\begin{array}{cc}
36.6 & 1.0
\end{array}
$$

$$
+ \underset{\text{(gas)}}{2.287CH_4} + \underset{\text{(gas)}}{\tfrac{1}{32}H_2S} + 36.8\,\text{kcal} \tag{10-10}
$$

$$
\begin{array}{cc}
-41 & -0.2
\end{array}
$$

Reaction (10-10) is exothermic to the extent of about 0.37 kcal/g of protein. The energy available is only about 7.2 percent of that evolved by aerobiosis.

It is obvious from the preceding equations that it is much more economical, with regard to food supply, for bacteria to obtain their energy aerobically than anaerobically. The same food supply will support a much larger bacterial flora by aerobiosis than by anaerobiosis. Hence, aerobic decomposition of organic matter is much more rapid than anaerobic breakdown. Aerobiosis in the activated sludge process of sewage treatment is substantially complete in 6 to 8 hr, whereas the septic digestion of sewage sludge requires about 60 days at ordinary temperature. The great difference in the time required is indicated roughly by the difference in the heats of reaction.

In the absence of dissolved oxygen, the bacteria turn to other sources of oxygen which may yield more energy that can be obtained without oxygen. Both nitrates and sulfates are used by bacteria as a source of oxygen. The energy yield is high when nitrates are used to oxidize the food molecule, but with sulfates the energy yield is little better than with anaerobic decomposition without an oxygen source.

The decomposition of dextrose with nitrates proceeds, approximately, as follows:

$$\underset{\substack{180 \\ C_6H_{12}O_6 \\ \text{(aq. dextrose)}}}{} \quad \underset{\substack{4.8 \\ +\frac{24}{5}H^+}}{} \underset{\substack{297.6 \\ +\frac{24}{5}NO_3^- \\ \text{(aq.)}}}{} \longrightarrow \underset{\substack{264 \\ 6CO_2 \\ \text{(gas)}}}{} \underset{\substack{151.2 \\ +\frac{42}{5}H_2O \\ \text{(liquid)}}}{} \underset{\substack{67.2 \\ +\frac{12}{5}N_2 \\ \text{(gas)}}}{} + 599.7 \text{ kcal} \quad (10\text{-}11)$$

$$H^\circ = \quad -301.5 \qquad 0 \quad -237 \qquad -564.3 \quad -573.9 \quad 0$$

The heat of reaction is 599.7 kcal/mole or 3.33 kcal/g of dextrose. This is about 92 percent of the energy produced by dissolved oxygen. Nitrogen gas, which is evolved in this reaction, has been observed in final settling tanks of the activated sludge process accompanying rising of settled sludge when the dissolved oxygen is exhausted. Sodium nitrate may be used as a supplementary source of oxygen in polluted streams to prevent or reduce odor nuisance.

The decomposition of dextrose with sulfates to produce hydrogen sulfide occurs approximately as follows:

$$\underset{\substack{180 \\ C_6H_{12}O_6 \\ \text{(aq. dextrose)}}}{} \quad \underset{\substack{6 \\ +6H^+}}{} \underset{\substack{288 \\ +3SO_4^{2-} \\ \text{(aq.)}}}{} \longrightarrow \underset{\substack{264 \\ 6CO_2 \\ \text{(gas)}}}{} \underset{\substack{108 \\ +6H_2O \\ \text{(liquid)}}}{} \underset{\substack{102 \\ +3H_2S \\ \text{(gas)}}}{} + 36.4 \text{ kcal} \quad (10\text{-}12)$$

$$H^\circ = \quad -301.5 \qquad 0 \quad -650.7 \qquad -564.3 \quad -409.8 \quad -14.5$$

The heat of reaction is 36.4 kcal/mole or 0.20 kcal/g of dextrose.

A similar reaction with sulfates to produce elemental sulfur probably takes place in the following manner:

$$\underset{\substack{180 \\ C_6H_{12}O_6 \\ \text{(aq. dextrose)}}}{} \quad \underset{\substack{8 \\ +8H^+}}{} \underset{\substack{384 \\ +4SO_4^{2-} \\ \text{(aq.)}}}{} \longrightarrow \underset{\substack{264 \\ 6CO_2 \\ \text{(gas)}}}{} \underset{\substack{180 \\ +10H_2O \\ \text{(liquid)}}}{} \underset{\substack{128 \\ +4S \\ \text{(solid)}}}{} + 78.4 \text{ kcal} \quad (10\text{-}13)$$

$$H^\circ = \quad -301.5 \qquad 0 \quad -867.6 \qquad -564.3 \quad -683.2 \quad 0$$

The heat of reaction is 78.4 kcal/mole or 0.43 kcal/g of dextrose. The energy yield of this reaction is about twice as great as the energy yield in the anaerobic decomposition of sugar with no oxygen source, reaction (10-9).

Sulfates are known to be decomposed in sewage sludge digestion tanks, which, when operating properly, produce negligible amounts of H_2S gas. Reactions similar to (10-13) are indicated for this condition. When the pH of the digesting sludge is low, H_2S gas may be produced in reactions similar to (10-12).

The concentration of sulfates in water is an important factor in determining whether H_2S will be evolved in reactions similar to (10-12). Where H_2S is evolved in the anaerobic decomposition of bottom deposits in a polluted stream or estuary, it will usually be oxidized chemically to sulfates if there is sufficient dissolved oxygen in the overlying waters, and if the H_2S gas does not rise in large gas bubbles along with CO_2. When the tide rises and covers flats in a polluted tidal estuary, polluted water high in sulfates will seep into the sands of the flats. When the dissolved oxygen in the water in the pores is exhausted by bacterial decomposition of the organic pollutants, the sulfates in the pore water are then broken down in reactions similar to (10-12) and (10-13). When the tide recedes, the H_2S gas in the pores of the exposed sand flats is released to the atmosphere, causing odor nuisances. If the incoming tidal waters contain sufficient dissolved oxygen, the odors will be reduced or abated at high tide.

10-6. COMPOSITION AND BREAKDOWN OF SEWAGE SOLIDS

Sewage from sanitary sewers consists of the used water from the water supply of the community, together with extraneous matter that has been added to the water. Substances originally present in the water are, therefore, present in the sewage, and the dissolved mineral solids in sewage from hard-water regions are correspondingly higher than in sewage from soft-water regions. The material added to the water consists principally of urine, feces, paper, and some garbage from waterclosets; soap, detergents, grease, extracts of meat and vegetables, and waste food from kitchen sinks; soap, detergents, starch, grease, dirt, skin excretions, and tissue from baths and laundries. Most of these wastes contain a myriad of bacteria, both living and dead. Sewage usually contains some grease, oil, and mineral suspended matter from service stations and garages that have improperly trapped drains. Sewage from combined sewers includes, in addition, the suspended load of soil and grit from surface drainage, together with a small amount of organic and bacterial pollution. Sewage from either class of sewers may include industrial wastes.

The studies of Wolff and Lehmann (23) with a mixed population consisting of 37,610 men, 34,630 women, 14,060 boys, and 13,700 girls indicates that an average daily excrement of 1,170 g of urine and 90 g of feces per person may be expected. The average composition of the fresh excreta is estimated as shown in Table 10-2 (23, 24).

All the solids in feces are shown in the table as suspended matter, although much of this material is doubtless in solution in the entrained moisture. The

Table 10-2
Average Composition of Human Excrements (grams per capita per day)

| | Dissolved Solids | | | | | | | Suspended Solids | | | | | | |
| | | | Organic | | | | | | | Organic | | | | |
	Inorganic	Total	N	C	H	O	S	Inorganic	Total	N	C	H	O	S
Urine (1,170 g, 96.3% water)														
Urea		24.30	11.34	4.87	1.62	6.47								
Creatinine		1.45	0.54	0.62	0.09	0.20								
Ammonia (NH_3)		0.60	0.49	–	0.11	–								
Uric acid		0.55	0.18	0.20	0.01	0.16								
Hippuric acid		0.51	0.04	0.31	0.02	0.14	+							
Indican		0.01	+	+	+	+								
Allantoin		0.005	+	+	+	+								
Glucose		0.70	–	0.28	0.05	0.37								
Oxy acids		0.05	–	0.02	+	0.02								
Oxalic acid		0.015	–	+	+	0.01								
Acetone		0.01	–	+	+	+								
NaCl	9.3													
H_3PO_4	1.9													
SO_4^{2+}	1.8													
K^+	1.5													
Ca^{2+}	0.2													
Mg^{2+}	0.2													
Fe^{2+}	0.005													
Subtotal	14.9	28.2	12.59	6.30	1.90	7.37								
Feces (90 g, 77.2% water)									17.8	1.8	9.4	1.4	5.0	0.2
H_3PO_4								1.0						
K^+								0.2						
Ca^{2+}								0.5						
Mg^{2+}								0.3						
Total inorganic								2.0						

dry solids in feces contain a good deal of indigestible substance, such as crude cellulose. The number of bacteria (*24*) in feces is enormous. About one fourth the dried organic matter, or 4 to 5 g, consists of bacterial bodies, both alive and dead; and about one half the nitrogen is bacterial nitrogen. The total number of bacterial bodies in fresh feces is estimated at about 5×10^{12} per person daily, of which about one half, or 26 billion per gram of moist feces, are alive. From 10 to 20 percent of the dry solids of feces is fatty material (*25*), nearly all of which is within the bacterial bodies or other cellular structure.

The gases of the lower human intestine are due to anaerobic bacterial decomposition of the feces within the intestine. Analyses of human intestinal gases by Ruge (in *26*) have shown the composition by volume as illustrated in Table 10-3.

Table 10-3

	CO_2 (%)	H_2 (%)	CH_4 (%)	N_2 (%)
Vegetable diet	21–34	1.5–4	44–55	10–19
Meat diet	8–13	0.7–3	26–37	45–64
Milk diet	9–16	43–54	0.9	36–38

According to Neave and Buswell (*24*), the total nitrogen in domestic sewage can be almost entirely accounted for in the excreta of the population. The average total nitrogen excretion, from Table 10-2, is about 14.4 g/capita/day. From Table 10-4, which gives the organic constituents of a number of typical municipal sewages, it may be seen that the total nitrogen ranges from 6.8 to 31.9 g/capita/day, with an average of about 15. As can be seen from Table 10-2, nearly 80 percent of this nitrogen is excreted as urea ($NH_2 \cdot CO \cdot NH_2$), and about half the remainder is in the feces. Urea hydrolyzes, as shown later, producing NH_3 and CO_2. Since ammonia is oxidized only by the nitrifiers that are strict aerobes and are, therefore, not present in fresh sewage, the urea is not associated with the oxygen demand of fresh sewage. The nitrogen in the feces, about 10 percent of the whole, is associated with about 97 percent of the readily oxidizable matter in the sewage. This oxidizable matter consists mainly of proteins, unabsorbed carbohydrates, and decomposition products of these substances.

Since fresh sewage is made up of the spent water supply, it contains plenty of dissolved oxygen. The bacterial decomposition process, which was anaerobic in the intestine, becomes aerobic in the sewers. The process will remain aerobic until the dissolved oxygen is exhausted. In most systems the period of flow in the sewers is not long enough to completely exhaust the dissolved oxygen.

Table 10-4
Organic Constituents of Typical American Sewages
(grams per capita per day)

Municipality:	Roch-ester	Akron	Glov-ersville	Worces-ter	Fitch-burg	Clin-ton	Paw-tucket	Allen-town	Brock-ton	Day-ton	Rome	Percentage of Total Volatile Solids	
												With Ind. Wastes	Without Ind. Wastes
Type of Sewage:	Large Comb.	Comb. IW	Comb. IW	Comb. IW	Comb. No IW	Sep. IW	Sep. IW	Sep. No IW	Sep. No IW	Sep. No IW	Sep. No IW		
Total volatile solids	158	336	269	144	86	420	161	144.7	65	69	138	100	100
Suspended volatile solids	53	72	143	72	46	240	35	45.8	30.4	29	54	21–57	32–53
Dissolved volatile solids	105	264	126	72	40	180	126	98.9	34.6	40	84	43–79	47–68
Fats	–	–	268.5	–	–	169.5	–	–	–	9.7	–	–	–
5-day 20°C BOD	98.7	57.1	92.5	78.5	56	–	62	71.2	–	37.0	54.0	–	–
Total BOD, $L(k_1 = 0.1)$	144	83.5	135	114.6	81.9	–	90.6	104	–	54	79	–	–
Ammonia N	5.8	3.3	9.6	6.5	4.4	14.4	5.8	5.8	6.3	4.3	3.2	1–4.5	2.3–10
Organic N	–	7.6	19.9	6.0	–	17.5	6.0	2.8	–	2.5	5.8	2.3–7.4	2–4.2
Nitrite N	–	–	–	0.04	0.03	–	–	0.11	0.03	–	–	–	–
Nitrate N	0.14	–	–	0.33	0.32	–	–	0.09	0.02	–	–	–	–
Ratio $\frac{L}{TVS}$	0.91	0.25	0.50	0.80	0.95	–	0.56	0.72	–	0.78	0.57	0.25–0.80	0.57–0.95
Ratio $\frac{\text{5-day BOD}}{NH_3 \text{ nitrogen}}$	17	17.3	9.6	12.1	12.7	–	10.7	12.3	–	8.6	16.8	9.6–17.3	8.6–16.8

10-7. BIOCHEMICAL OXYGEN DEMAND

The rate at which dissolved oxygen is used up in fresh sewage or sewage-polluted waters has been studied extensively by Phelps, Streeter (*27*), and many others. The biochemical oxidation has been found to proceed in a manner similar to a *unimolecular chemical reaction;* that is, the rate is approximately proportional to the remaining concentration of unoxidized organic matter, measured in terms of oxidizability, and it is a function of the temperature. These relations are shown in Figures 10-4 (*28*) and 10-5. The amount of oxygen

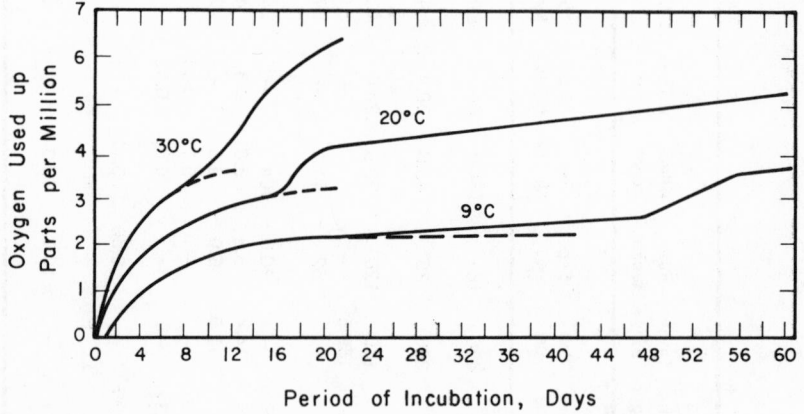

Period of Incubation, Days

Figure 10-4
General course of the deoxygenation curve.

used up is called the *biochemical oxygen demand* or *BOD* for a particular time and water temperature. BOD tests have become standard practice and are now widely used. The standard procedures for the test are described in ref. *29*. The 5-day 20°C BOD test is used in routine analyses.

The BOD of a sample of sewage, industrial wastes, or polluted water is a measure of the concentration of decomposable organic matter in that sample. It has become the most useful single determination in the routine examination of sewages, industrial wastewaters, and treatment plant effluents and, especially, in the examination of the receiving waters in pollution studies. The BOD concept involves not only the amount of organic material that is decomposable by bacteria, but also the time rate at which it will decompose aerobically. The BOD test may, therefore, be used as a laboratory model of the deoxygenation process in the receiving waters.

The concentration of decomposable organic matter in wastewater or polluted water is not directly measurable and is of little importance, except that it controls the demand for dissolved oxygen. The associated dissolved oxygen

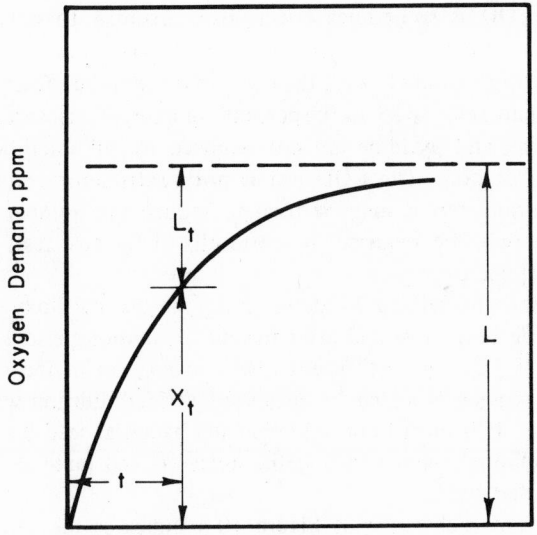

Period of Incubation, Days

Figure 10-5
Biochemical-oxygen-demand curve [see Eqs. (10-15a) and (10-15b)].

demand, on the other hand, is of major importance, and this is measured by the BOD test. Decomposable organic matter may be reduced in concentration by a number of different processes in treatment plants or polluted waters. Since the amount of the associated oxygen demand, rather than the amount of organic matter, is the parameter of interest, it has become common practice to use the term "BOD" interchangeably with the term "decomposable organic matter" in discussions of pollution loads and changes in amounts of organic matter. It is important to note that the quantities are of oxygen and not organic matter.

The BOD test has been subjected to adverse criticism because of the long period required for its completion, 5 days or more, and substitute chemical tests have been sought. Chemical tests may be used to determine the concentration of chemically oxidizable organic matter, but the result is not the same as the biochemically oxidizable organic matter, and it throws no light on the biochemical time rate.

Potassium permanganate was first used as a reagent to determine the *oxygen consumed*, OC, by organic matter. In (29), this reagent has now been superseded by potassium dichromate, and the values obtained are known as the *dichromate oxygen consumed*, DOC, or the *chemical oxygen demand*, COD. Since the carbon and hydrogen, and not the nitrogen, are oxidized by the chemical oxidants, the COD is sometimes erroneously assumed to indicate the total carbonaceous organic matter present. For most organic compounds, oxidation is 95 to 100 percent complete with the dichromate method. Oxidation of short-chain alcohols and acids is 85 to 95 percent complete if a silver catalyst is used. Benzene, toluene, and pyridine are not oxidized by this method whether the catalyst is used or not. The COD test is not a satisfactory substitute for the BOD determination, but it may be used to reduce the number of BOD tests that might otherwise be required, if the results of the two tests are correlated on occasional replicate samples.

The BOD test is performed by determining the dissolved-oxygen content of the sample at the beginning, and after various incubation periods at the desired temperature. If there is insufficient dissolved oxygen in the sample for the complete test, oxygen is added by means of aerated dilution water before incubation begins. It is most important that the bacterial seed be the same as in the waste or polluted water and that the nutrients and minerals be representative of the polluted water.

The first part of each curve of Figure 10-4 indicates the aerobic oxidation of the organic matter other than NH_3 by the normal bacterial flora in the sewage. This part of the decomposition process is known as the *first stage of deoxygenation*. The latter part of each curve is associated with the growth of the nitrifiers and the oxidation of NH_3 to nitrites and, subsequently, to nitrates. This stage is known as the *second stage* or *nitrification stage*. The end products in the first stage include NH_3, as shown by reaction (10-6). In sewage and polluted water, the two stages do not ordinarily occur together, and the first stage is substantially complete before nitrification becomes appreciable.

A convenient mathematical formulation for the first stage oxidation (Figure 10-5) is as follows: At any time, t days, the rate of oxidation at a particular water temperature is

$$\frac{dX_t}{dt} = \frac{d(L - L_t)}{dt} = -\frac{dL_t}{dt} = K_1 L_t \text{ or } 2.3 k_1 L_t \qquad (10\text{-}14)$$

in which X_t is the O_2 used in t days, that is, the biochemical oxygen demand (BOD) of the sample of polluted water or sewage for t days; L is the total first stage O_2 demand; L_t is the first-stage O_2 demand remaining after t days; and K_1 or $2.3 k_1$ is a proportionality constant (where $K_1 = 2.3 k_1$).

The integration of Eq. (10-14) yields

$$L_t = L \cdot e^{-K_1 t} = L \cdot 10^{-k_1 t} \qquad (10\text{-}15\text{a})$$

or

$$X_t = L(1 - e^{-K_1 t}) = L(1 - 10^{-k_1 t}) \qquad (10\text{-}15\text{b})$$

in which K_1 and k_1 are the deoxygenation constants for natural and common logs, respectively.

The value of the first-stage deoxygenation constant k_1 at $20°C$ was first determined by Phelps (*30*) in 1909 in an analytical study of 2,600 tests on the *relative stability* of effluents of several sewage filters. The relative stability is defined in reference *29* as the percentage ratio of oxygen available (dissolved oxygen, nitrite, and nitrate oxygen) to the total oxygen required to satisfy the BOD. The test is made by adding a methylene blue solution and noting the time required for decolorization when the mixture is incubated at $20°C$. Phelps presented in 1909 a table of relative-stability numbers for incubation periods up to 20 days, which is still used. The *relative stability number* is the ratio as a percentage of X_t/L in Eq. (10-15b), and it is, therefore, determined by the value of k_1. The relative-stability numbers presented by Phelps correspond with a value of k_1 at $20°C$ of approximately 0.1. Phelps (*30*, p. 66) describes early work done for the British Royal Commission on Sewage Disposal on *oxygen absorption* at $65°F$, in which the oxygen absorbed in 1 and 2 days was 31 and 52 percent, respectively, of the 5-day absorption. These ratios indicate a k_1 value of about 0.1 at $18.3°C$, in rough confirmation of Phelps relative-stability numbers.

Extensive experiments by the Cincinnati Laboratory (*31*) of the U.S. Public Health Service in connection with pollution studies of the Ohio River indicated that k_1 has a value of 0.1 at $20°C$ for all sewage and polluted waters. This belief was widely held until the 1930s when evidence to the contrary began to accumulate. It is now known that the value of k_1 at $20°C$ may range from a low of about 0.01 for slowly oxidizable organic matter, including some industrial wastes and some well-oxidized treatment plant effluents, to about 0.30 for some fresh, readily oxidizable organic wastes. Sawyer et al. (*32*) have shown that the presence of toxic substances, such as Cu^{2+}, CrO_4^{2-}, Pb^{2+}, $HgCl_2$, and chloramine, substantially reduce the 5-day BOD of several pure organic substances, and that a limited seed or the absence of NH_3-N (ammonia nitrogen) or P also reduces the BOD. For reliable studies of pollution, it is necessary to measure the BOD at the temperature desired for periods from 1 to 10 days, or longer, and to compute the k_1 value from the BOD test results.

Phelps and Streeter (*27*) have found experimentally that the value of the deoxygenation constant k_1 varies with temperature according to the relation

$$\frac{k_1(T')}{k_1(T)} = \theta^{(T'-T)} \tag{10-16}$$

where $k_1(T')$ and $k_1(T)$ are values of k_1 at temperatures T' and T, respectively, and θ is a thermal constant with a value of 1.047 for temperatures from 10 to $37°C$.

Theriault (28) confirmed this value of θ experimentally and has shown an analogy between Eq. (10-16) and the van't Hoff–Arrhenius equation:

$$\log \frac{k_1(T')}{k_1(T)} = \frac{\mu(T' - T)}{2.3R(T' + 273)(T + 273)} \qquad (10\text{-}17)$$

where μ is the heat of activation, which ranges from 7,400 to 8,500 g cal for all experimental values of k_1 noted by Theriault, and R is the gas constant (1.985 g cal/mole).

The Arrhenius equation in its differential form is

$$\frac{d(\ln k)}{dT_a} = -\frac{\mu}{RT_a^2} \qquad (10\text{-}17a)$$

where T_a is the absolute temperature in $°K$ $(T + 273)$. It is analogous to the Lewis–Randall (33) equation for the effect of temperature on the equilibrium constant of chemical reactions at constant pressure, in which k_1 is replaced by the equilibrium constant and μ is replaced by the heat of the reaction. Fair and Moore (34) show a number of values of the heat of activation, μ, for typical chemical and biochemical reactions. For biochemical reactions the values of μ, and hence of θ, are constant over relatively narrow temperature ranges.

Moore (35) conducted long-time BOD experiments on 2 percent dilutions of settled domestic sewage in both bicarbonate and phosphate dilution waters and on undiluted samples from two streams containing moderate pollution of remote origin. The 5-day 20°C BOD of the sewage was 80 to 160 ppm and of the river water was 0.8 to 2 ppm. The samples were incubated at temperatures ranging from 0.5 to 20°C inclusive. The values of k_1 computed from the results of these experiments are plotted in Figure 10-6, together with straight-line fits of the data on semilog paper. It will be noted that the k_1 values for a particular temperature vary widely, nearly 2 to 1 at some temperatures. These variations indicate the danger of using a formula, such as Eq. (10-16), to estimate the value of k_1 at one temperature from its value at another temperature, unless the formula was derived from experimental data on the organic wastes being studied. The values of θ obtained by Moore, 1.145 and 1.065 for the sewage and 1.026 for the river water, not only vary widely from each other but also from the Phelps–Streeter value of 1.047, which further indicates the unreliability of using a formula to estimate the effect of temperature on the value of k_1 in the absence of experimental data.

It will be noted from Figure 10-4 that the first-stage BOD, L, is greater for higher temperatures. Theriault (28) has formulated an empirical relation to account for this variation as follows:

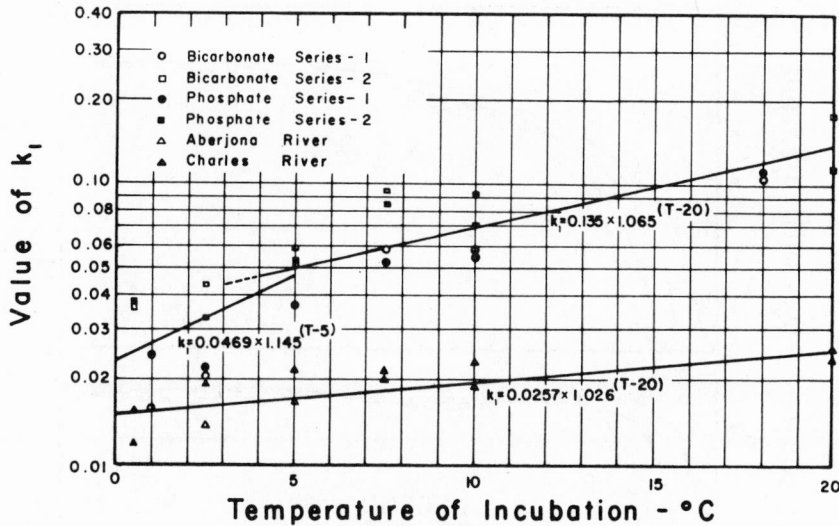

Figure 10-6
Effect of temperature on the velocity constant k_1. [After Moore (*35*).]

$$L_T = L_{20}[1 + 0.02(T - 20)] \qquad (10\text{-}18)$$

where L_T and L_{20} are the total first stage O_2 demand at $T°$ and $20°C$, respectively. This indicates an increase in the value of L of 2 percent for each degree above $20°C$, and similar decreases for lower temperatures. Moore's (*35*) experiments showed changes in L, based on values at $20°C$, of 0.8 and 2 percent per degree for the two series on sewage, and 2.8 percent per degree for the river samples.

Butterfield and Wattie (*36*) showed with pure cultures of *A. aerogenes* grown aerobically at $30°C$ in a dilute liquid medium that the total count and total oxygen utilized after 5 days was substantially the same, regardless of the initial bacterial count (see Figure 10-7). Initial inoculations of 3,200 to 3,200,000 bacteria per milliliter were used. The maximum growth rate for the lowest initial count corresponded to a generation time of about 72 min, which, in turn, corresponds with about 375 mg of O_2/hr/g dry weight of cells. At the end of 1 day, it will be noted, neither the bacterial count nor the oxygen utilization for the smaller initial inoculation was equal to those of the larger initial counts. These data indicate that if only the 5-day $20°C$ BOD is wanted, the initial bacterial count is immaterial, but if the 1-day BOD is desired the initial number will affect the results. Figure 10-7 also shows that the maximum growth rate decreases as the initial count increases.

It will be noted from Figure 10-7 that the oxidation rate after the first day

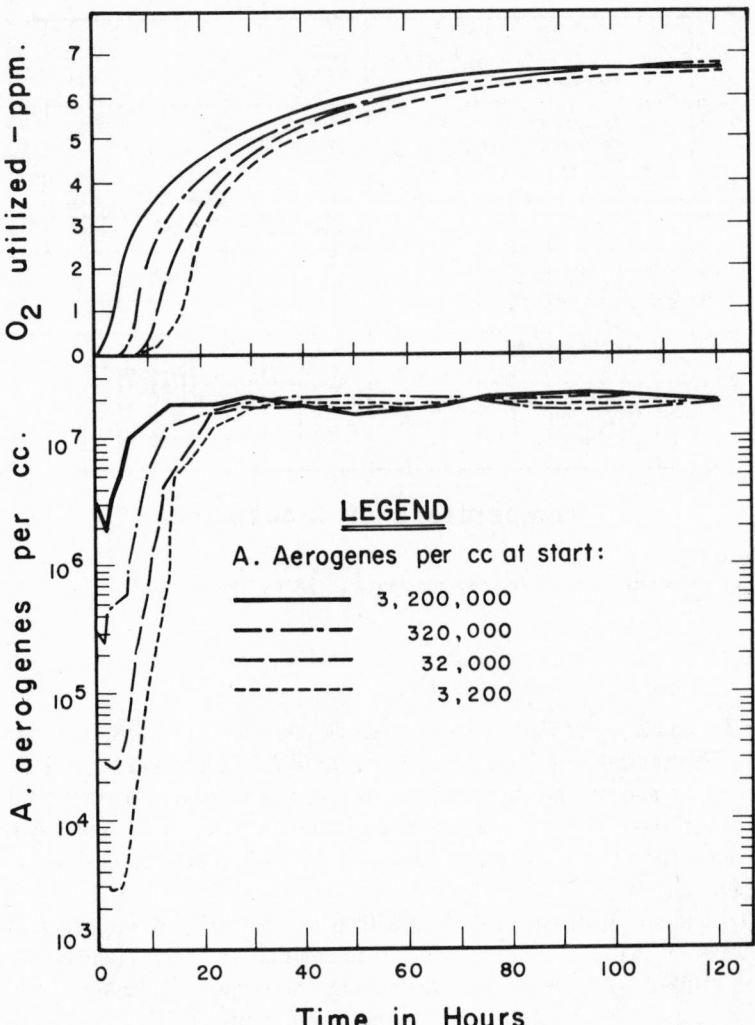

Figure 10-7
Biochemical-oxygen-demand relations with pure cultures of *A. aerogenes*. [After Butterfield and Wattie (*36*).]

corresponds with endogenous respiration. In the series with the highest initial seed, the limiting count of about 20 million per milliliter was reached in about 14 hr. Of the total 5-day 20°C BOD of about 6.7 ppm, approximately 40 percent was for endogenous respiration at a rate varying from 200 mg O_2/hr/g of cells, at 14 hr, down to about 1 mg O_2/hr/g, at the end of 5 days, and averaging 4.5 mg O_2/hr/g.

Hoover et al. (*19*) cite experiments by others in which the viable bacteria in BOD bottles were found to reach the maximum count on the second day and diminish rapidly thereafter. Experiments by Hoover and co-workers on the oxidation of skim-milk solids with a 5-day BOD of 500 to 800 ppm indicate that the maximum count was usually reached in 1 day, with a decrease up to 40 percent in 5 days.

The maximum count of 20 million per milliliter of *A. aerogenes* in the Butterfield and Wattie experiments represents a suspended solids increase of not more than about 5 ppm, if it is assumed that the bacteria which are colloidal in size flocculate to become suspended solids. This process, known as *bioflocculation*, is of major importance in the biological processes for treatment of sewage and organic industrial wastes, and may result in the production of large quantities of sludge, which must be removed and disposed of.

Bioflocculation in a polluted stream may result in the formation of sludge banks and bottom deposits of organic matter, even if all settleable solids are removed by primary settling of wastewaters. Sludge banks may, of course, be formed through the settling out of coarse suspended organic matter without bioflocculation, if the concentration of suspended organic matter in the stream at the point of pollution is appreciable and if the velocity is low enough.

It has been found by Theriault and McNamee (in *27*) that, when sewage solids which have undergone anaerobic decomposition are brought into contact with dissolved oxygen, there is an *immediate* O_2 demand as distinguished from the normal BOD. The rate of this demand at 20°C is roughly 16 times the normal rate, so that the immediate demand is satisfied in less than 1 day. The immediate demand also includes the chemical demand to oxidize ferrous iron, sulfides, and other chemicals in the reduced state. To compute the value of k_1, the immediate demand should be determined and deducted from the total first-stage demand.

Streeter (*27*) has shown that the normal rate of oxidation in certain polluted streams in the *second* or *nitrification stage* is about one third the first-stage rate at 20°C. The corresponding value of the second-stage deoxygenation constant, k_n, is about 0.03. Moore (*35*) found k_n for sewage to be about 0.06 at 20°C and about 0.035 at 10°C.

10-8. OXIDATION BY ACTIVATED SLUDGE

Numerous studies have been made of the rate of oxidation of sewage substrates by activated sludge. Butterfield, Ruchhoft, and McNamee (*37*) studied the rates of oxidation of sterile and synthetic sewages by pure culture activated sludges. These sludges, which did not include nitrifiers, oxidized about 50 percent of the 5-day BOD of the sewages in 5 hr and about 80 percent in 24 hr.

The velocity constant at 20°C in the first 3 to 5 hr of aeration was found to be 6.8 to 12.6 with the time expressed in days, which corresponds with an initial O_2 utilization rate of 65 to 120 ppm/hr/100 ppm of total first-stage demand. The rapid rate was found to break after 3 to 5 hr. Ruchhoft, McNamee, and Butterfield (*38*) found that, with actively nitrifying activated sludge, the oxygen utilized in 5 hr corresponded with about 120 percent of the 5-day BOD and in 10 hr with about 160 percent. They further concluded that there is a great variation in oxidizing ability of activated sludges, and that the rate cannot be expressed satisfactorily in terms of a unimolecular reaction. The high oxidizing rates are due to enormous numbers of bacteria in the mixed liquor, estimated by Butterfield and Wattie (*36*) at about 10 billion per milliliter in a mixed liquor containing 1,000 ppm of suspended solids.

Sawyer and Nichols (*39*) found at least three types of O_2 utilization curves for Wisconsin activated sludges. High initial oxygen utilization rates sustained for nearly 2 hours were found with nitrifying sludges. The base rate of O_2 demand was found to be proportional roughly to the suspended solids of the mixed liquor, and to range from 1.85 to 9.8 mg/hr/g of dry solids in the sludge. Grant, Hurwitz, and Mohlman (*40*) found that the activated sludge at the Chicago North Side plant used about 7 mg of O_2/hr/g of dry sludge. The oxidation rates of these sludges indicate endogeneous respiration for most of the aeration period.

Sawyer and Rohlich (*41*) have investigated the effects of temperature upon the initial oxygen utilization rates of activated sludge. A study of their results shows that the thermal constant, θ, has a value of about 1.10 between 10 and 25°C for both nitrifying and nonnitrifying sludges. Sawyer (*42*) found that dissimilar activated sludges gained similar characteristics in about 3 weeks, as measured by the deoxygenation curves, when fed on the same sewage. The character of the sludge thus depends principally upon the diet. Nitrifying ability requires a diet high in ammonia. Sawyer found that the ratio of 5-day BOD to ammonia nitrogen ranged from 8.2 to 21 for the different sewages studied, the sewages with the lower ratios producing the greatest nitrification with activated sludge.

It has long been known from laboratory studies that nearly all the first-stage BOD, L, that is removed in an aeration tank is transferred from the substrate (sewage or wastewater) to the activated sludge solids in less than 1 hr. Similarly, in a one-pass trickling filter, all the removal of BOD takes place while the substrate trickles through the filter in 15 to 30 min. The BOD removed from the substrate is absorbed by the activated sludge or by the bacterial slimes on the trickling filter medium. It is also well known that the rate of O_2 utilization is highest when the activated sludge and the substrate are first brought together, and rapidly decreases during a period of about 1 hr, after which the decrease is relatively small for the remainder of the aeration period.

These observations indicate that sludge growth (cell synthesis) usually takes place in the first hour or two in a manner characterized by the logarithmic rate, b-c, and the post logarithmic rate, c-d, of Figure 10-2, and the remainder of the aeration period, and all the settling period, is characterized by endogenous respiration and some bacterial die-off.

The logarithmic growth rate of activated sludge cell material is proportional to the concentration of living cells at any time. Since it is not feasible to measure the concentration of viable cells in the mixed liquor suspended solids during the period of active growth, the concentration of volatile suspended solids is used as a measure of the concentration of living cells. The logarithmic growth rate is expressed as follows:

$$\frac{dS}{dt} = 2.3k'S \tag{10-19}$$

where S is the mixed liquor volatile suspended solids in parts per million at time t hours and k' is the growth rate per hour.

The integration of Eq. (10-19) yields

$$\ln\frac{S_2}{S_1} = 2.3k'(t_2 - t_1) \tag{10-20a}$$

or

$$\log\frac{S_2}{S_1} = k'(t_2 - t_1) \tag{10-20b}$$

The generation time, g, is the time interval required for S to double, or

$$g = t_2 - t_1 = \frac{\log 2}{k'} = \frac{0.301}{k'} \tag{10-21}$$

Laboratory studies by Garrett and Sawyer (*43*) of the growth of mixed cultures in 3-normal peptone (5-day BOD, about 432 ppm) indicated a minimum generation time of about 2.7 hr at 30°C and 3.2 hr at 20°C. The concentration of viable bacteria was not measured. At both temperatures, however, the maximum rate of oxygen utilization of about 45 ppm/hr was reached after about 4 hr. If, as indicated in Sec. 10.4, each gram dry weight of cell synthesized should require about 0.45 g of oxygen, 45 ppm/hr of oxygen would correspond with an increase of S at the rate of 100 ppm/hr. From Eq. (10-19), the mixed liquor suspended solids can be estimated at about 400 ppm at the end of the log-growth phase.

Since the mixed liquor suspended solids carried in most activated sludge plants range from about 500 to 5,000 ppm and the oxygen demand is highest at the start of aeration, it is evident from the preceding discussion that the log-growth phase does not consume much of the aeration period in most activated sludge plants.

Weston and Eckenfelder (*10*) have presented graphs of experimental data by Gellman and Heukelekian (*44*) on yeast nutrients spent beer that indicate logarithmic growth for periods of 4 to 8 hr in a total aeration time of 24 hr. The generation times range from about 7.7 hr, with an initial value of S of about 450 ppm, to 20 hr, with an initial value of S of about 3,600 ppm. The initial 5-day BOD of the substrate was 1,380 ppm. The value of S reached during endogeneous respiration was not the same for all four experiments in this series, as might be expected from Figure 10-7. The limiting value of S was only about 1,100 ppm for an initial suspended solids of 450, and was about 1,600, 2,500, and 4,300 ppm for initial suspended solids of 900, 1,800, and 3,600 ppm, respectively.

Another series of experiments on yeast nutrients spent beer by Gellman and Heukelekian (*44*) was shown by Weston and Eckenfelder (*10*) to have generation times ranging from 10 hr for an initial BOD of 2,760 ppm to 20 hr for an initial BOD of 690 ppm. The initial value of S was 1,800 ppm in all three experiments of the series. In this series, the final suspended solids in endogenous respiration ranged from about 2,200 ppm for the small BOD load to 3,200 for the large BOD load.

As a result of their studies of the experiments of Gellman and Heukelekian and of others, Weston and Eckenfelder conclude that the log-growth rate is proportional to some power (about 0.7) of the ratio of the L value of the BOD load to the initial value of S.

The BOD reduction during cell growth represents the food required for cell growth, a part of which becomes a part of the cell. The BOD equivalent, b, of each gram of food synthesized to cell material varies with the food; it is approximately 1.067 g for dextrose [Eq. (10-5)], 1.485 g for a typical protein [Eq. (10-6)], and ranges from 0.7, or less, for intermediate products to nearly 3 for long-chain fatty acids. An approximate average value of b for most readily decomposable organic wastes is 1.3.

During each cycle of the activated sludge process, aeration period plus time of settling and returning sludge, the sludge synthesized may be represented as follows:

$$\Delta S = \frac{a}{b} L_r \tag{10-22}$$

where ΔS is the sludge growth in parts per million per cycle, L_r is the first stage BOD removed in parts per million, a is the fractional part of the BOD synthesized into bacterial cells, and b is the BOD equivalent of the food material. Equation (10-22) is representative of a complete cycle, and is useful in laboratory and pilot-plant studies where determinations are limited to mixed liquor suspended solids and to BOD removal at the beginning and end of each run. Equation (10-22) gives no information, however, with which to determine the portion of the aeration period during which endogeneous respiration takes place, the generation time during sludge growth, and the portion of the BOD oxidized during endogeneous respiration. Samples should be collected at intervals throughout the aeration period to determine when sludge growth stopped and how much BOD had been removed at that time.

It is evident from Eq. (10-22) that ΔS should be zero if treatment is wholly endogeneous. Under such conditions, there would be no sludge disposal problem for excess activated sludge. The detention period would have to be long, and the mixed liquor suspended solids, high. On the other hand, with a low initial value of S, a shorter aeration period may be used; but ΔS will be great with a costly sludge disposal problem.

Experimental studies by Helmers et al. (45) on activated sludge treatment of domestic sewage mixed with various quantities of cotton kiering, rag rope, and brewery wastes indicated that the sludge growth per cycle, ΔS, from an initial value of S of 1,500 ppm was directly proportional to the 5-day BOD of the feed mixture and, hence, to L_r. The cycle was 8 hr with a 7-hr aeration period for all runs. With an assumed value of b of 1.3, the value of a estimated by means of Eq. (10-22) was about 0.26 at $10°C$ and 0.32 at $20°C$ for L_r of 135 ppm, and about 0.32 at $10°C$ and 0.43 at $30°C$ for L_r of 810 ppm. The value of g computed from Eqs. (10-19) and (10-21), on the assumption of log-growth throughout the aeration period, was 27 hr at $10°C$ and 22 hr at $20°C$ for L_r of 135 ppm, and about 3.6 hr at $10°C$ and 2.7 hr at $30°C$ for L_r of 810 ppm. The computed g values tend to indicate endogenous respiration for most of the aeration period at the low BOD loading and log-growth for most of the aeration period at the high loading. Sludge production was 8 times as great at the higher loading.

The value of a/b in the experiments by Helmers (45) and co-workers for the same mixed liquor suspended solids content, 1,500 ppm, ranged from a low of 0.2, with low values of L_r, to a high of 0.33, with high values of L_r. In many conventional activated sludge plants, about 0.5 lb of excess activated sludge dry solids is produced per pound of 5-day BOD removed in the aeration tanks and final settling tanks, representing an a/b value of about 0.34. The Helmers experiments indicate that the value of a/b for a particular plant may be reduced to almost any low value, approaching zero for endogenous respiration, by minimizing the withdrawal of excess sludge and allowing the mixed liquor suspended solids to seek its maximum concentration, provided, however, that the sludge will settle rapidly and will not be lost in the effluent, and that the BOD removal will not be reduced appreciably. In pilot-plant studies under the senior author's direction, an a/b value of 0.15 has been maintained with an aeration period of 12 hr without impairing the BOD removal.

The experiments of Helmers et al. *(45)* indicate that the maximum nitrogen requirement for stabilization of the three industrial wastes was 5 to 6 lb/100 lb of 5-day BOD, which is equivalent to a ratio of BOD removed to nitrogen utilized of 20 and 17, respectively. The maximum phosphorus requirement was 0.8 to 1.0 lb/100 lb of 5-day BOD, or a ratio of BOD removed to phosphorus utilized of 125 and 100, respectively.

10-9. TYPICAL AMERICAN SEWAGES

Table 10-4 *(46)* shows the organic constituents of a number of typical American sewages. From Eqs. (10-5) and (10-6), it will be noted that the O_2 requirement for sugar is 1.067 and for proteins is about 1.485 g/g of substrate decomposed. This would indicate a requirement of between 1 and 1.5 parts of O_2 per part of sewage organic solids for complete aerobic decomposition by a bacterial flora not including nitrifiers. The actual requirement, however, as measured by the ratio of the total O_2 demand, L, to the total volatile solids, ranges from 0.25 to 0.95 for the sewages of Table 10-4. In other words, from about one third to two thirds of the volatile solids in sewage are ordinarily not readily decomposed by bacteria. The residue is largely the crude suspended material that makes up the body of fully digested sewage sludge. The total oxygen demand for the nitrifiers, from Eqs. (10-7) and (10-8), is 4.6 g/g of ammonia–nitrogen oxidized. The total theoretical demand for nitrification of the sewages in Table 10-4 is, therefore, from 31 to 147 g of O_2/capita/day. The actual demand would be less because of the residual nitrogen in the sludge.

The information contained in Table 10-4 is representative of American sewages prior to 1935. Since that time, sewages have become stronger because of the increased discharge of ground garbage into sewage and the increased quantities of organic industrial wastes. The average 5-day BOD of the sewages in Table 10-4 is about 0.15 lb/capita/day, and the average volatile suspended solids is about 0.16 lb/capita/day. For design of sewage-treatment works in the absence of reliable analytical data on raw sewage, it has been customary to use 0.17 lb/capita/day for BOD and 0.2 lb/capita/day for total suspended solids. More recently, with increased use of household garbage grinders, 0.2 to 0.25 lb/day has been used for BOD and 0.25 to 0.3 lb/day for total suspended solids. Syndets are now present in sewage in increasing concentrations, from 5 to 15 ppm in normal municipal sewage, and higher where large quantities of industrial wastes containing syndets are involved. This concentration occurred as non-biodegradable ABS (alkyl benzene sulfonate) prior to 1965 and as biodegradable LAS (linear alkyl sulfonate) after 1965.

10-10. ANAEROBIC DECOMPOSITION

The anaerobic decomposition of organic matter in sewage, as used in sewage-treatment plants, is usually limited to sludges, scum, and screenings. Such materials contain very little of the urine or dissolved nitrogenous matter and carbohydrates originally present in the sewage. The dissolved solids remain in the supernatant liquid of the tanks and are decomposed aerobically in the liquid or pass out in the plant effluent. According to Buswell and Neave, (47) the fats or grease (ether-soluble matter) in American municipal sewage amounts to about 20 g/capita/day in the absence of fatty industrial wastes. Much of the grease remains with the settled solids, and grease recovered by skimming is frequently added to the sludge digestion tanks. Sewage grease is composed principally of a mixture of glycerides and soaps of the higher fatty acids. This picture has been changed considerably by the widespread use of synthetic detergents since World War II.

The average analysis by Buswell and Neave (47) of 39 samples of fresh sewage solids and a similar number of samples of sewage solids following anaerobic digestion showed that 59.4 percent of the fresh solids and 37.8 percent of the digested solids were organic. This indicates an average decomposition of about 36 percent of the organic solids by anaerobiosis. Ordinarily, from 30 to 50 percent reduction in volatile solids may be expected by anaerobic digestion. The average composition of the organic solids in sewage sludge, as determined by Buswell and Neave, is shown in Table 10-5.

Table 10-5

	Fresh Organic Solids (% of fresh)	Digested Organic Solids (% of fresh)
Protein (6.25 × N)	32.7	21.0
Fats (ether soluble)	42.4	14.5
Crude fiber	18.2	16.5
Humic acids (by pyridine)	6.7	11.6
	100.0	63.6

The anaerobic decomposition of sewage solids results in the production of gases, of which about two thirds by volume is methane, CH_4, and the remainder is principally CO_2. Some N_2 and H_2 are usually present and, occasionally, H_2S is produced in very small quantity. Since these are the same gases found to be present in the human intestine, as shown in Table 10-3, it is evident that the same bacteria are responsible. The total gas yield is about 20 percent greater by weight than the reduction in volatile solids. This is explained by

Buswell and Boruff (*48*) as being due to the entrance of water into the decomposition reaction.   The manner in which this takes place is indicated in the typical reaction (10-10), and is shown in more detail on the following pages. The greases, fats, and soaps furnish a gas with 62 to 75 percent methane by volume. Cellulose digests to gas having only 45 to 50 percent CH_4.

The rate at which sewage sludges of various types are digested anaerobically has been studied by Fair and Moore (*49, 34*).   Figure 10-8 is a typical curve

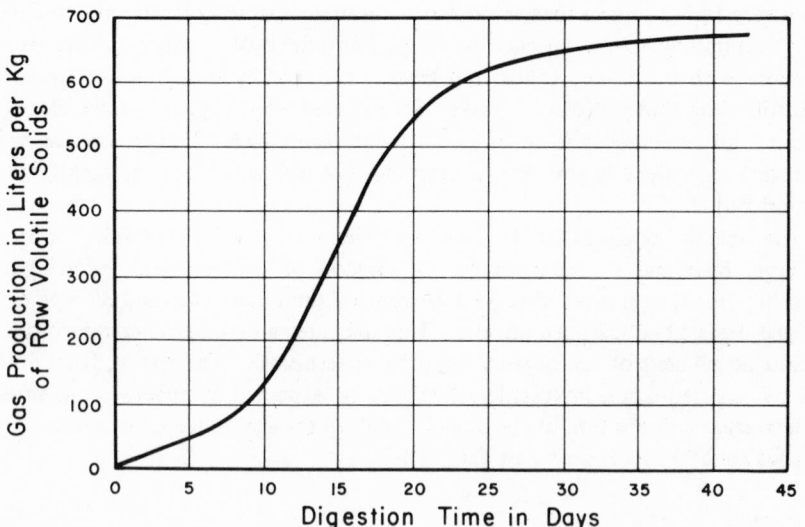

Figure 10-8
Course of anaerobic digestion of sewage sludge in batches. [After Fair and Moore (*34*).]

showing the course of digestion of seeded sludge as measured by the volume of gas produced per kilogram of raw volatile solids. It will be noted that the curve has an S shape in contrast with the typical first-stage BOD curve. The shape of the first part of the curve is doubtless due to the increasing rate of destruction of the gas-producing compounds, and it is probably accompanied by an increasing rate of growth of the responsible bacteria. The curve is similar in shape to the lower of the O_2 utilization curves of Figure 10-7; the reason for the increasing slope in both cases is probably that the decomposition rate in the early stages is proportional to the number of bacteria present, and not to the quantity of readily available food. The rate is so rapid with the O_2 utilization curves, however, that the shape of the initial part of the curve is usually obscured by lack of data.

Fair and Moore have presented two mathematical formulations of the course of gas production: (1) an autocatalytic equation, or equation of a unimolecular reaction accelerated by the catalytic activity of the products of reaction, and (2) a two-stage formulation involving a discontinuous equation of constant rate of increase, followed by a discontinuous unimolecular reaction. The two-stage formulation is as follows:

$$\text{First stage:} \qquad \frac{dG}{dt} = K_1 G \qquad\qquad (10\text{-}23)$$

$$\text{Second stage:} \qquad \frac{dL_t}{dt} = -K_2 L_t \qquad\qquad (10\text{-}24)$$

Integration and transformation to common logarithms yield

$$\text{First stage:} \qquad \log G = \log \frac{L}{2} - k_1(t_1 - t) \qquad\qquad (10\text{-}25)$$

$$\text{Second stage:} \qquad \log L_t = \log \frac{L}{2} - k_2(t - t_1) \qquad\qquad (10\text{-}26)$$

in which G is the amount of gas produced in time t days during the first stage; L is the total amount of gas produced in both stages; L_t is $L - G$ or the amount of gas remaining to be produced at time t days during the second stage; k_1 is the first-stage gasification constant; k_2 is the second-stage gasification constant; and t_1 is the time in days to the transition from first to second stage. It is assumed that at this point

$$G = L_t = \frac{L}{2}$$

A number of determinations of the values of k_1, k_2, and the weighted mean gasification constant k were made by Fair and Moore (*34*) for sludge derived from plain sedimentation and for excess activated sludge. The data indicate that, at 20°C, k should vary from about 0.04 to 0.06 for fresh sludge and from 0.008 to 0.012 for activated sludge, the higher values being obtained where seeding is best. In studying the effect of temperature, it was found that the heat of activation, μ, was about 5,000 for fresh sludge from plain sedimentation, and about 10,000 for activated sludge in the temperature range from 25 to 40°C. The corresponding values of θ in Eq. (10-16) are 1.027 for the digestion of fresh primary tank sludge and 1.055 for the digestion of activated sludge. These studies indicate that at the usual temperatures employed in sludge digestion, 85 to 95°F, 50 percent of the gas is produced in 10 to 17 days for primary sludge and in 16 to 36 days for activated sludge, and 90 percent of the gas is produced in 22 to 30 days for primary, and 61 to 78 days for activated sludge.

Ninety percent gas production is the accepted definition of "complete" digestion.

It should be noted here that the experiments of Fair and Moore were batch incubations in which the sludge to be digested was first mixed with the seeding sludge in volatile solids ratios of 0.8 to 2.2. In normally operating sludge digestion tanks, the fresh sludge is added daily to a volume of the digesting sludge 20 to 50 times the volume of the daily increment. In batch experiments, the bacteria must grow in numbers until the maximum count is reached. In a well-operated digester, the maximum bacterial numbers should be constantly present. If this is the case, the first stage of gas production, as noted by Fair and Moore, should be absent in a digester, and the initial rate of gas production should be very high. Hatfield, Buswell, and others (*50*) report that 50 percent of the gas was produced in the first 24 hr in continuous digestion experiments with fresh sludge, and that 90 percent of the gas was recovered in 5 to 7 days. The data of Fair and Moore indicate that 50 percent of the gas should come off in about 5 days if the initial stage of digestion is absent. The slow rate obtained by Fair and Moore is probably due to a poor inoculum or seeding material, as pointed out by Hatfield, Buswell, and others (*50*). On the other hand, the high rate of digestion noted by Hatfield, Buswell, and others indicates a bacterial flora too ideal to be realized in practical sludge digestion.

In later experiments of high-rate digestion of primary sludge from the Nut Island Sewage Treatment Plant at Boston, Sawyer and Roy (*51*) found that 91.5 percent as much gas was produced in 6 days and 95.6 percent as much in 10 days, as in 20 days. The gas contained 55.2, 56.8, and 57.2 percent methane for the 6, 10, and 20-day periods, respectively. The digestion units were operated at 33 to 40°C and were fed twice daily to give theoretical detention periods of 6, 8, 10, 15, and 20 days. All units were started by using a digested sewage sludge with 4 percent total solids. Feedings were started on a 20-day detention basis and gradually increased at weekly intervals until all units were operated at the desired detention periods. Continuous mixing was accomplished by recirculating the gas. The raw sludge contained 68 to 82 percent volatile solids and 13.4 to 16.8 percent grease. In the 20-day period, 58.2 percent of the volatile matter and 71.4 percent of the grease were destroyed, whereas in 6 days 52.2 percent of the volatile matter and only 53 percent of the grease were destroyed.

The studies of Fair and Moore indicate that the total gas yield, L, increases with the temperature as follows:

$$L_T = L_{20}[1 + a(T - 20)] \tag{10-27}$$

where L_T and L_{20} are the total gas yields at $T°$ and 20°C, respectively, and a is a constant whose value was observed to be about 0.0072 for primary sludge and 0.024 for activated sludge.

10-11. BREAKDOWN OF STARCH

Starch is a polysaccharide, having, like cellulose, the general empirical formula $(C_6H_{10}O_5)_n$, and occurring in plant grains of concentric structure, such as corn, rice, and potatoes. Starches are generally insoluble in water, but may form colloidal dispersions.

A large number of bacteria, including *B. subtilis* and *Clostridium sporogenes*, have the power to break down starch through action of a bacterial *diastase*, an exocellular hydrolase produced by the bacteria that operates independently of the presence of living cells. The breakdown of the starch is an hydrolysis, which, when carried to completion, yields a hexose:

$$\underset{\text{(starch)}}{\overset{162n}{(C_6H_{10}O_5)_n}} + \underset{\text{(liquid)}}{\overset{18n}{nH_2O}} \longrightarrow \underset{\text{(aq. hexose)}}{\overset{180n}{nC_6H_{12}O_6}} + 2.1n \text{ kcal} \qquad (10\text{-}28)$$

$$H^\circ = \quad -231.0n \qquad -68.4n \qquad -301.5n$$

The first products formed in the hydrolysis are of high molecular weight and have a variety of names, such as soluble starch, amylodextrin, and dextrin. The first well-defined product is maltose, which is produced through the action of the diastase. Maltose or malt sugar is a disaccharide, which yields two molecules of *d*-glucose (also known as grape sugar and as dextrose) on hydrolysis, as follows:

$$\underset{\text{(maltose)}}{\overset{342}{C_{12}H_{23}O_{11}}} + \underset{\text{(liquid)}}{\overset{18}{H_2O}} \longrightarrow \underset{\text{(aq. dextrose)}}{\overset{2 \times 180}{2C_6H_{12}O_6}} + 0.0 \text{ kcal} \qquad (10\text{-}29)$$

$$H^\circ = \quad -534.6 \qquad -68.4 \qquad -603.0$$

This reaction takes place through the activity of the enzyme *maltase*, also called glucase, which is secreted by yeasts and a number of bacteria.

The above reactions, it may be seen, involve very little energy change. The decomposition of glucose will be considered later.

10-12. BREAKDOWN OF CELLULOSE

Pure cellulose is a polysaccharide of the empirical formula $(C_6H_{10}O_5)_n$, which composes the bulk of the supporting structure of plants. It is an important component of wood fiber. Cotton fiber and filter paper are almost pure cellulose. Cellulose molecules are fibers of variable length consisting of

100 to 500 glucose molecules. Cellulose is insoluble in water and in all organic solvents.

Cellulose is hydrolyzed in two stages (6, p. 131). The first stage involves its hydrolysis to *cellobiose*, a disaccharide with the structural formula

by means of the enzyme *cellulase*. This is analogous to the hydrolysis of starch to maltose. Cellulase is secreted by many bacteria, both aerobic and anaerobic, and by some molds. It is not secreted by yeasts. The first stage reaction is

$$162n \qquad 18\frac{n}{2} \qquad 342\frac{n}{2}$$

$$(C_6H_{10}O_5)_n + \frac{n}{2}H_2O \longrightarrow \frac{n}{2}C_{12}H_{22}O_{11} + 3.4n \text{ kcal} \qquad (10\text{-}30)$$

$$\text{(cellulose)} \quad \text{(liquid)} \qquad \text{(cellobiose)}$$

$$H° = \quad -230.0n \qquad -34.2n \qquad \quad -267.6n$$

In the second stage, cellobiose is hydrolyzed to glucose through the action of the enzyme *cellobiase*, a reaction similar to the breakdown of maltose. Cellobiase and cellulase occur together in the same bacteria. Above 67°C, cellobiase is destroyed but cellulase is still active. Below 20°C, decomposition of glucose is arrested but cellobiase is still active. The second reaction proceeds as follows:

$$342 \qquad\qquad 18 \qquad\qquad 2 \times 180$$

$$C_{12}H_{22}O_{11} + H_2O \longrightarrow 2C_6H_{12}O_6 \quad - 0.6 \text{ kcal} \qquad (10\text{-}31)$$

$$\text{(cellobiose)} \quad \text{(liquid)} \qquad \text{(aq. dextrose)}$$

$$H° = \quad -535.2 \qquad -68.4 \qquad , \qquad -603.0$$

10-13. BREAKDOWN OF HEMICELLULOSES, PECTINS, AND LIGNIN

The hemicelluloses are polysaccharides that form, with cellulose, the walls of plant cells. They also act as reserve carbohydrates for the metabolism of the cells. They are easily hydrolyzed by dilute acid and yield different products than cellulose. The principal hemicelluloses are

1. The *mannans*, which hydrolyze to 1 part *d*-fructose and 20 parts *d*-mannose. Cell walls of yeasts are composed largely of mannans.
2. The *galactans*, which hydrolyze to *d*-galactose.
3. The *pentosans*, which hydrolyze to pentoses: xylan to *l*-xylose and araban to *l*-arabinose. Araban is an important constituent of wood, bran, and straw.

The decomposition of hemicelluloses is important in soils. It is carried on by molds rather than by bacteria.

Pectins are polysaccharides of gelatinous consistency that form the middle lamellar layer of plant tissues. They are composed of tetragalacturonic acid combined with araban and *d*-galactose, and hydrolyze to galacturonic acid, *l*-arabinose and *d*-galactose.

The bacterial breakdown of pectins (*6*, p. 136) through the action of the enzyme *pectinase* is an anaerobic process. It occurs in retting of flax and hemp and in rotting of vegetables like kale, turnips, and carrots. In retting, pectin is dissolved, but cellulose fibers are left intact. Pectinase appears to be exocellular. Aerobic retting takes place to some extent through the action of molds.

Lignin in woody tissues is present, apparently, to give strength and rigidity to the cell wall. According to Gortner (*52*), lignin is extremely resistant to bacterial action, but Buswell (*53*) reports some gasification of lignin in anaerobic digestion of cornstalks.

10-14. BREAKDOWN OF HEXOSES

The simple sugars or monosaccharides all have the empirical formula $C_6H_{12}O_6$ and, hence, are called hexoses. They are characterized by a chain of six carbon atoms with several hydroxyl groups and, usually, not more than one carbonyl group. Thus they combine the chemical properties of alcohols with those of ketones or aldehydes, depending upon the position of the carbonyl group. They are, then, divided into two groups, the *aldohexoses* and the *ketohexoses*.

The principal aldohexoses are *d-glucose* (grape sugar or dextrose) and *d-galactose*. The latter is obtained by the hydrolysis of *lactose* or milk sugar, a disaccharide that yields one molecule of *d*-glucose and one molecule of *d*-galactose. These sugars are represented by the following formulas:

$$
\begin{array}{c}
CH_2OH \\
| \\
HO-C-H \\
| \\
HO-C-H \\
| \\
H-C-OH \\
| \\
HO-C-H \\
| \\
O=C-H
\end{array}
\qquad\qquad
\begin{array}{c}
CH_2OH \\
| \\
HO-C-H \\
| \\
H-C-OH \\
| \\
H-C-OH \\
| \\
HO-C-H \\
| \\
O=C-H
\end{array}
$$

d-glucose   *d-galactose*

The most important ketohexose is *d-fructose* (fruit sugar or levulose). It is represented by the formula

$$
\begin{array}{c}
CH_2OH \\
| \\
HO-C-H \\
| \\
HO-C-H \\
| \\
H-C-OH \\
| \\
C=O \\
| \\
CH_2OH
\end{array}
$$

d-fructose

Among the important disaccharides, which have the empirical formula $C_{12}H_{22}O_{11}$, are *sucrose* (cane sugar), *lactose*, and *maltose*. All the disaccharides hydrolyze by the addition of one molecule of water to form two molecules of hexose; the reactions are similar to reaction (10-31) and involve very little energy change. Sucrose yields one molecule each of *d*-glucose and *d*-fructose, and maltose yields two molecules of *d*-glucose. Sucrose is hydrolyzed by the enzyme *invertase*, secreted by yeasts, and the resulting mixture of *d*-fructose and *d*-glucose is called *invert sugar*. Honey is an example of natural invert sugar.

The decomposition of hexose occurs both anaerobically and aerobically; the products formed depend upon the organisms present, the configuration of the sugar molecule, the presence or absence of dissolved oxygen, and other environmental conditions. In general, the breakdown of hexose furnishes both energy and carbon for the living cell. Knowledge concerning the bacterial degradation of hexose is quite incomplete and is speculative at many points.

The first step in the breakdown of hexose under either aerobic or anaerobic conditions is, according to Szent-Györgyi (7), the *phosphorylation* of the hexose molecule, which is subsequently split into two three-carbon molecules of triose phosphate. The phosphate appears to operate in a closed cycle, and is given up by the triose after its oxidation. According to Szent-Györgyi, the triose molecule is oxidized to pyruvic acid in both oxidation and fermentation through the action of the enzyme *cozymase*, as shown in Figure 10-9 for the

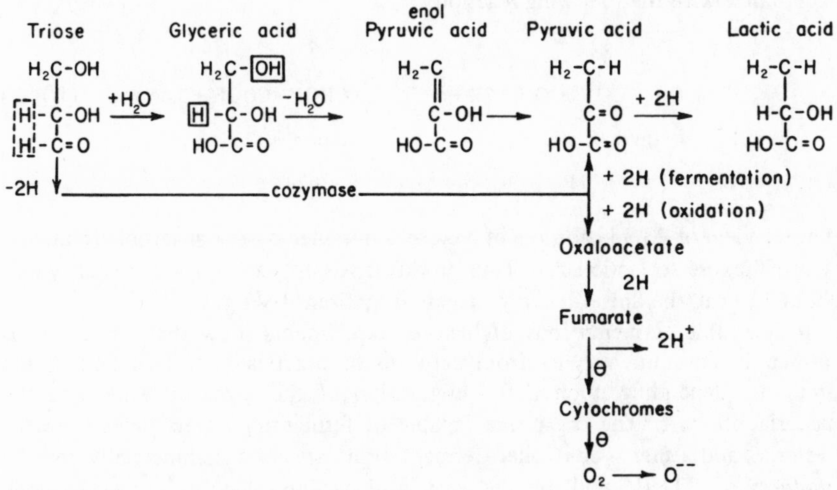

Figure 10-9
Breakdown of hexose through triose. [After Szent-Györgyi (7).]

respiration of muscle. The first step in this process is the removal of two H atoms by the cozymase, which are replaced by water, changing the triose to glyceric acid. In the next step the H and OH in the squares are taken from the glyceric acid as H_2O, yielding enol pyruvic acid. The enol pyruvic acid, by a rearrangement of one H atom, gives up a double bond and becomes pyruvic acid.

Pyruvic acid is very unstable. It can be oxidized, reduced, decarboxylated, or polymerized. It is probably one of the basic constituents for cell building, for it produces the necessary carbon. In order to be utilized for this purpose, however, energy is required. The production of pyruvic acid from hexose is endothermic, as may be seen from the following reaction:

$$\underset{\substack{\text{(aq. hexose)}}}{\overset{180}{C_6H_{12}O_6}} \xrightarrow{\text{cozymase}} \underset{\substack{\text{(pyruvic acid)}}}{\overset{4}{4H} + \overset{2 \times 88}{2CH_3 \cdot CO \cdot COOH}} - 19.7 \text{ kcal} \qquad (10\text{-}32)$$

$$H° = \quad -301.5 \qquad\qquad\qquad -281.8$$

To obtain the necessary energy for cell life and cell building, a suitable acceptor must be available for the H atoms taken by the cozymase.

In oxidation, the two H atoms taken over by the cozymase are eventually handed to dissolved O_2 after the energy is extracted by a piecemeal oxidation process. In fermentation, there is no hydrogen acceptor available. The two H atoms are, therefore, given back to the pyruvic acid to form lactic acid, in accordance with the following reaction:

$$88 \qquad\qquad 2 \qquad\qquad\qquad 90$$

$$CH_3 \cdot CO \cdot COOH + 2H \longrightarrow CH_3 \cdot CHOH \cdot COOH + 21.5 \text{ kcal} \qquad (10\text{-}33)$$

(pyruvic acid) (lactic acid)

$$H^\circ = \qquad -140.9 \qquad\qquad\qquad -162.4$$

A total yield of 23.3 kcal/mole of hexose is obtained by the anaerobic fermentation of hexose to lactic acid. Thus, in this first step, each triose molecule yields about 11 kcal, the unit of energy suggested by Szent-Györgyi.

In anaerobic fermentations of hexose, experiments show that lactic acid is formed in amounts varying from zero up to practically 100 percent of the products, depending upon the configuration of the sugar molecule and the bacteria present. The organisms capable of fermenting sugars include yeasts, bacteria, and fungi. Controlled fermentations are used commercially for the production of lactic acid, by the lactic acid bacillus; alcohol, by yeasts; acetic acid, by acetic bacteria; as well as acetone, *n*-butyl alcohol, amyl alcohol, various aliphatic aldehydes, formic acid, and citric acid, all by specific organisms. It has been shown recently that when methane bacteria are present in mixed cultures, such as in sewage sludge, complete decomposition of carbohydrates to CH_4 and CO_2 is obtained for much of the substrate.

Lactic acid fermentation and alcoholic fermentation are nearly identical processes, according to Szent-Györgyi. In alcoholic fermentation, the pyruvic acid is *decarboxylated* by means of the enzyme *carboxylase*, leaving acetaldehyde, which accepts the two H atoms from the cozymase and is reduced to ethyl alcohol as follows:

$$88 \qquad\qquad\qquad 44 \qquad\qquad 44$$

$$CH_3 \cdot CO \cdot COOH \xrightarrow{\text{carboxylase}} CO_2 + CH_3 \cdot CHO + 7.7 \text{ kcal} \qquad (10\text{-}34)$$

(pyruvic acid) (gas) (aq. acetaldehyde)

$$H^\circ = \qquad -140.9 \qquad\qquad -94.4 \qquad -54.2$$

$$44 \qquad\qquad 2 \qquad\qquad 46$$

$$CH_3 \cdot CHO + 2H \longrightarrow CH_3 \cdot CH_2OH + 15.6 \text{ kcal} \qquad (10\text{-}35)$$

(aq. acetaldehyde) (aq. alcohol)

$$H^\circ = \qquad -54.2 \qquad\qquad\qquad -69.8$$

Reactions (10-32), (10-34), and (10-35) give a total yield of 26.9 kcal/mole of hexose by alcoholic fermentation.

The enzyme carboxylase has been shown to be composed of thiamin pyrophosphate together with a specific protein. Thiamine is the name suggested by Williams (*54*) for *vitamin B₁*. It is essential in the human diet for the prevention of polyneuritis (beriberi), a disease that arises from a deficient decarboxylation in carbohydrate metabolism. The structure of the thiamin molecule has been determined, and thiamin chloride is being prepared synthetically. Thiamin enters into several enzyme systems in carbohydrate metabolisms, and is secreted by a number of bacteria, including *E. coli* and *A. aerogenes*.

In anaerobic fermentation, the dissimilation of pyruvic acid to formic and acetic acids may also occur by hydrolysis, as follows:

$$
\begin{array}{cccc}
88 & 18 & 46 & 60 \\
CH_3 \cdot CO \cdot COOH + & H_2O \longrightarrow & HCOOH & + \quad CH_3 \cdot COOH \\
\text{(pyruvic acid)} & \text{(liquid)} & \text{(aq. formic acid)} & \text{(aq. acetic acid)}
\end{array}
$$

$$H° = \qquad -140.9 \qquad -68.4 \qquad\quad -100.2 \qquad\qquad -118.1$$

$$+ 9.0 \text{ kcal} \quad (10\text{-}36)$$

Buswell (*55*) has shown that a mixed flora from sewage sludge, including methane bacteria, accomplished almost complete destruction of a large number of different carbohydrates fed consecutively without apparent specificity. These pure substances included pyruvic acid, lactic acid, methyl, ethyl, and higher alcohols, formic and acetic acids. They were found to react with water to produce CH_4 and CO_2 in accordance with the following general equations:

$$
C_nH_aO_b + \left(n - \frac{a}{4} - \frac{b}{2}\right) H_2O \longrightarrow \left(\frac{n}{2} - \frac{a}{8} + \frac{b}{4}\right) CO_2 + \left(\frac{n}{2} + \frac{a}{8} - \frac{b}{4}\right) CH_4 \quad (10\text{-}37)
$$

The anaerobiosis of a number of the carbohydrate derivatives, in accordance with Eq. (10-37), is as follows:

$$
\begin{array}{cccc}
4 \times 88 & 2 \times 18 & 7 \times 44 & 5 \times 16
\end{array}
$$

$$4CH_3 \cdot CO \cdot COOH + 2H_2O \longrightarrow 7CO_2 + 5CH_4 + 51.4 \text{ kcal} \quad (10\text{-}38)$$

$$
\begin{array}{cccc}
\text{(pyruvic acid)} & \text{(liquid)} & \text{(gas)} & \text{(gas)}
\end{array}
$$

$$H° = \qquad -563.6 \qquad\quad -136.8 \qquad -660.8 \quad -91.0$$

$$
\begin{array}{ccc}
2 \times 90 & 3 \times 44 & 3 \times 16
\end{array}
$$

$$2CH_3 \cdot CHOH \cdot COOH \longrightarrow 3CO_2 + 3CH_4 + 13.0 \text{ kcal} \quad (10\text{-}39)$$

$$
\begin{array}{ccc}
\text{(lactic acid)} & \text{(gas)} & \text{(gas)}
\end{array}
$$

$$H° = \qquad -324.8 \qquad\quad -283.2 \quad -54.6$$

$$4 \times 32 \qquad\qquad 44 \quad 3 \times 16 \quad 2 \times 18$$

$$4CH_3OH \longrightarrow CO_2 + 3CH_4 + 2H_2O + 47.0\,kcal \qquad (10\text{-}40)$$

(methyl alcohol)   (gas)   (gas)   (liquid)

$$H° = \qquad -238.8 \qquad\qquad -94.4 \quad -54.6 \quad -136.8$$

$$2 \times 46 \qquad\qquad 44 \quad 3 \times 16$$

$$2CH_3 \cdot CH_2OH \longrightarrow CO_2 + 3CH_4 + 9.4\,kcal \qquad (10\text{-}41)$$

(ethyl alcohol)   (gas)   (gas)

$$H° = \qquad -139.6 \qquad\qquad -94.4 \quad -54.6$$

$$4 \times 46 \qquad\quad 3 \times 44 \quad 16 \quad 2 \times 18$$

$$4HCOOH \longrightarrow 3CO_2 + CH_4 + 2H_2O + 37.4\,kcal \qquad (10\text{-}42)$$

(formic acid)   (gas)   (gas)   (liquid)

$$H° = \quad -400.8 \qquad\quad -283.2 \quad -18.2 \quad -136.8$$

$$60 \qquad\qquad 44 \quad 16$$

$$CH_3 \cdot COOH \longrightarrow CO_2 + CH_4 - 5.5\,kcal \qquad (10\text{-}43)$$

(acetic acid)   (gas)   (gas)

$$H° = \quad -118.1 \qquad\qquad -94.4 \quad -18.2$$

Reactions (10-38) to (10-43), it will be noted, are all exothermic with respect to the heats of reaction, except (10-43) which involves acetic acid. The anaerobic breakdown of acetic acid and higher fatty acids, as will be shown later, is endothermic with respect to the heat of reaction.

There is an appreciable amount of H_2 in intestinal gas (see Table 10-3) and it is also present at times in the anaerobic digestion of sewage sludge. Hoppe-Seyler (in 6) has shown that formic acid is the source of the hydrogen gas appearing in anaerobic fermentations, as follows:

$$46 \qquad\quad 2 \quad 44$$

$$HCOOH \longrightarrow H_2 + CO_2 - 5.8\,kcal \qquad (10\text{-}44)$$

(formic acid)   (gas)   (gas)

$$H° = \quad -100.2 \qquad\quad 0 \quad -94.4$$

Here again the reaction is endothermic with regard to the heat of reaction.

With methane bacteria present, H_2 and CO_2 are readily converted into methane and water, as has been shown by Söhngen (55) and others and verified by Buswell (55). The reaction is as follows:

$$4 \times 2 \quad 44 \qquad 16 \quad 2 \times 18$$

$$4H_2 + CO_2 \longrightarrow CH_4 + 2H_2O + 60.6 \text{ kcal} \qquad (10\text{-}45)$$

(gas) (gas) (gas) (liquid)

$$H^\circ = \quad 0 \qquad -94.4 \qquad -18.2 \quad -136.8$$

Stephenson and Stickland (5, p. 48) isolated an organism from river mud that uses formate as its sole source of carbon and energy in accordance with reaction (10-42). They found, however, that the decomposition proceeds in two stages, the first stage producing H_2 and CO_2, according to Eq. (10-44), and the second stage producing CH_4 and water, in accordance with Eq. (10-45).

In the aerobiosis of sugar by muscle tissue, referring again to Figure 10-9, the H given up by the triose is accepted by oxaloacetate in the cell tissue. Oxaloacetic acid ($COOH \cdot CH_2 \cdot CO \cdot COOH$), it will be noted, is carboxy-pyruvic acid, that is, pyruvic acid with its methyl group carboxylated. The pyruvic acid appears to be held in the tissue by the carboxyl group, $-COOH$. The existence of a second carboxyl group in oxaloacetic acid explains its greater affinity for the enzyme in the cell tissue. According to Szent-Györgyi (7), if both pyruvate and oxaloacetate are present in the tissue, the latter will take over the H atom because of its greater affinity for the enzyme. In taking over the H, oxaloacetic acid is reduced to malic acid through the action of the enzyme *malic dehydrogenase* as follows:

132   134

$$COOH \cdot CH_2 \cdot CO \cdot COOH + 2H \xrightarrow[\text{dehydrogenase}]{\text{malic}} COOH \cdot CH_2 \cdot CHOH \cdot COOH \quad (10\text{-}46)$$

(oxaloacetic acid)   (malic acid)

If O_2 is present, the malic acid passes on its H atoms to the next H transmitter, and is itself oxidized back to oxaloacetic acid. Hence, if O_2 is present, oxaloacetic acid is recovered and respiration proceeds aerobically. No lactic acid is produced.

The H atoms are received from the malic acid by a flavoprotein, consisting of vitamin B_2 and a specific protein, and are transmitted to fumaric acid in the cell tissue. The fumaric acid, which is activated by the succinicdehydrogenase, accepts the H and is thereby reduced to succinic acid, as follows:

116   118

$$COOH \cdot CH \cdot CH \cdot COOH + 2H \xrightarrow[\text{dehydrogenase}]{\text{succinic}} COOH \cdot CH_2 \cdot CH_2 \cdot COOH \quad (10\text{-}47)$$

(fumaric acid)   (succinic acid)

The succinic acid transmits the two H atoms to cytochrome b, and is thus oxidized back to fumarate. The succinate and oxaloacetate in muscle tissue act as catalyzers. The presence of these substances in the tissue may arise through carboxylation of three-carbon molecules. Succinic acid is found in some bacterial fermentations of sugar.

The cytochromes (7) are remarkable substances that are linked to the protein of the cell. They are so constituted that their color and spectrum fade out when the cytochrome is oxidized and reappear again when it is reduced. The cytochromes consist of three different entities, cytochromes a, b, and c. Cytochrome c has been extracted in pure condition and its analysis is known. The most essential part of the molecule is an iron atom, which is alternately oxidized and reduced as the cytochrome functions.

The metals of the cytochromes act as *electron transmitters*, being alternately reduced and oxidized. Thus cytochrome b accepts electrons from the H atoms, changing them to H ions. The electrons are passed from cytochrome b to c and, thence, to cytochrome a. Cytochrome a passes the electrons to the metal-protein *cytochrome oxidase*, which also activates the oxygen to receive the electrons. All the energy is, thus, obtained by the cell in the oxidation of the H atoms to water.

In the aerobic decomposition of sugar by bacteria, the production of pyruvic acid in accordance with Figure 10-9 is probably general. The pyruvic acid, if not assimilated, may be completely oxidized to CO_2 and H_2O, as follows:

$$\overset{88}{CH_3 \cdot CO \cdot COOH} + \overset{5 \times 16}{\tfrac{5}{2}O_2} \longrightarrow \overset{3 \times 44}{3CO_2} + \overset{2 \times 18}{2H_2O} + 269.6 \text{ kcal} \qquad (10\text{-}48)$$

	(pyruvic acid)	(aq.)	(gas)	(liquid)
$H° =$	-140.9	-9.5	-283.2	-136.8

The extent to which this reaction is completed by various organisms and the intermediate products involved have been little studied. With a mixed flora, such as is present in sewage and polluted water, the reaction doubtless goes to substantial completion in short order because of the enormous energy yield.

All the intermediate products formed in sugar fermentation, such as lactic acid, alcohol, formic acid, and acetic acid, are readily oxidized by suitable organisms to CO_2 and H_2O with large energy yields. One partial oxidation used commercially is the production of acetic acid from alcohol by the *vinegar bacteria*, as follows:

$$\overset{46}{CH_3 \cdot CH_2OH} + \overset{32}{O_2} \longrightarrow \overset{60}{CH_3COOH} + \overset{18}{H_2O} + 112.9 \text{ kcal} \qquad (10\text{-}49)$$

	(alcohol)	(aq.)	(acetic acid)	(liquid)
$H° =$	-69.8	-3.8	-118.1	-68.4

In the absence of sufficient dissolved O_2, nitrates and other similar substances may be used as hydrogen acceptors, as shown by reactions (10-11), (10-12), and (10-13) for hexose.

10-15. BREAKDOWN OF NITROGENOUS COMPOUNDS

The nitrogenous compounds in sewage consist of both animal and vegetable proteins, urea, and numerous decomposition products of the proteins, such as proteoses and higher polypeptides, amino acids, amines, amides, and ammonia.

Proteins are substances of high molecular weight that make up the principal portion of animal tissues and of the protoplasm of both animal and vegetable cells. By boiling with dilute acid, they may be hydrolyzed almost quantitatively into a mixture of about 20 or more amino acids. Almost all these acids are produced by the hydrolysis of every protein, but the proportions of each vary widely. According to various authorities, the chemical composition of proteins in terms of the elements varies within narrow limits as follows:

		C	H	O	N	S
	from	50	6.7	19.0	15.2	0.3
Percentage						
	to	55	7.3	24.0	19.3	2.4

Proteins, proteoses, and higher polypeptides must be broken down by bacterial *proteolytic* enzymes, similar to pepsin and trypsin of animal digestive tracts, before these substances can be utilized by bacteria and yeasts. Proteolytic enzymes are exocellular and are produced by *B. proteus*, *B. subtilis*, and most anaerobes. Proteolytic exoenzymes do not break down proteins completely but only enough to permit the fragments to enter the cell walls, after which more complete decomposition is effected by proteolytic endoenzymes. Some simple source of nitrogen must be available in small quantities at the start, even for such active producers of proteolytic exoenzymes as *B. proteus*, in order to stimulate the growth of the cell while it is producing a sufficient quantity of the exoenzyme to effect the breakdown of proteins. The optimum pH for the production of proteolytic enzyme by *B. proteus* is about 8.0. Good aeration and the presence of calcium and magnesium salts favor the production of the exoenzyme.

Such organisms as *E. coli*, which is nonproteolytic, and the staphylococci must use simpler nitrogen compounds than proteins. *Peptone*, a protein partially hydrolyzed by mild acid, is a suitable source of nitrogen for *E. coli*.

The structural formulas of the amino acids usually met with as components of proteins are as follows:

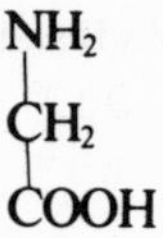

NH₂
CH₂
COOH
glycine

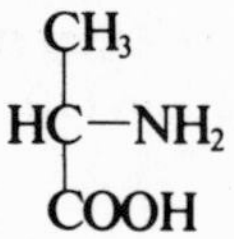

CH₃
HC—NH₂
COOH
alanine

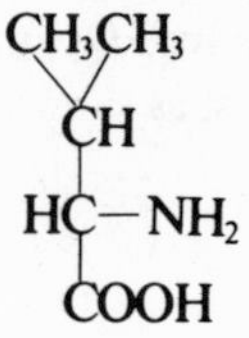

CH₃CH₃
CH
HC—NH₂
COOH
valine

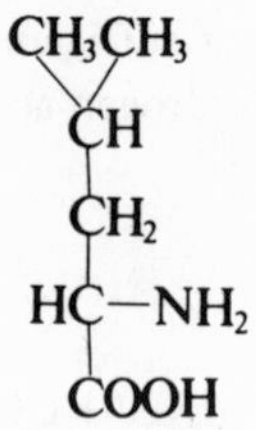

CH₃CH₃
CH
CH₂
HC—NH₂
COOH
leucine

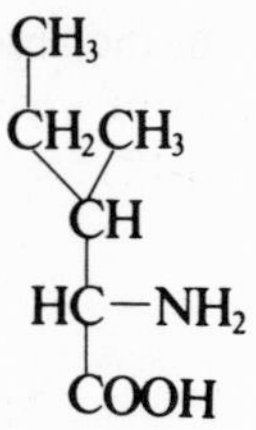

CH₃
CH₂CH₃
CH
HC—NH₂
COOH
isoleucine

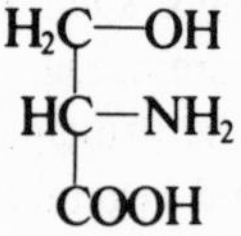

H₂C—OH
HC—NH₂
COOH
serine

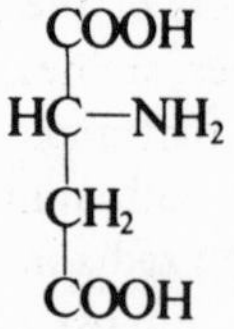

COOH
HC—NH₂
CH₂
COOH
aspartic acid

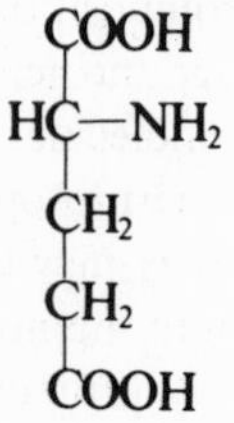

COOH
HC—NH₂
CH₂
CH₂
COOH
glutamic acid

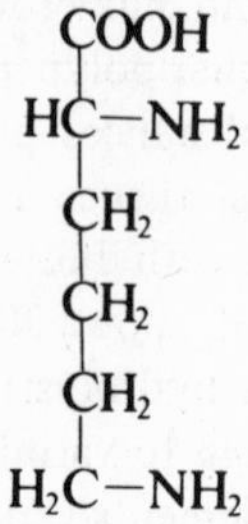

COOH
HC—NH₂
CH₂
CH₂
CH₂
H₂C—NH₂
lysine

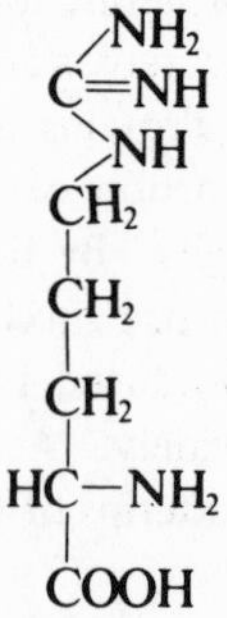

NH₂
C=NH
NH
CH₂
CH₂
CH₂
HC—NH₂
COOH
arginine

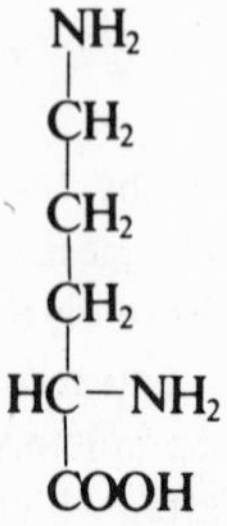

NH₂
CH₂
CH₂
CH₂
HC—NH₂
COOH
ornithine

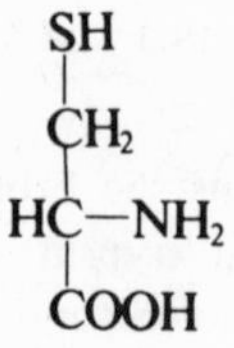

SH
CH₂
HC—NH₂
COOH
cysteine

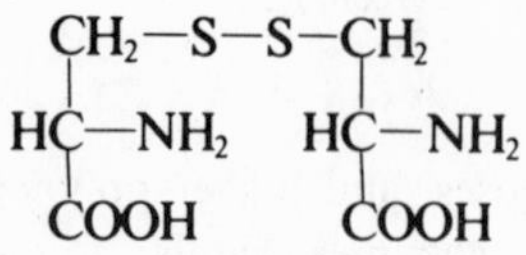

CH₂—S—S—CH₂
HC—NH₂   HC—NH₂
COOH   COOH
cystine

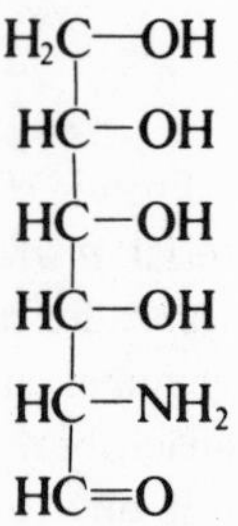

H₂C—OH
HC—OH
HC—OH
HC—OH
HC—NH₂
HC=O
glucosamine

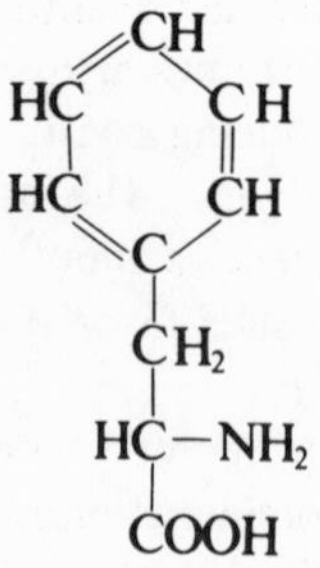

CH
HC   CH
HC   CH
C
CH₂
HC—NH₂
COOH
phenylalanine

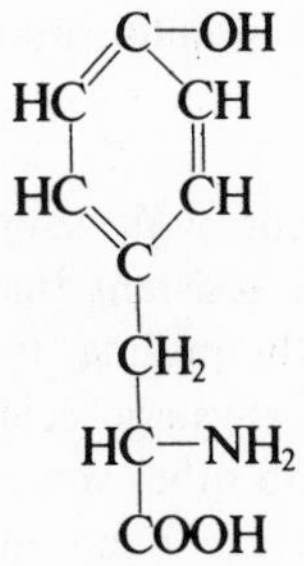

tyrosine

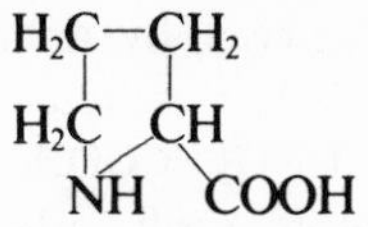

proline

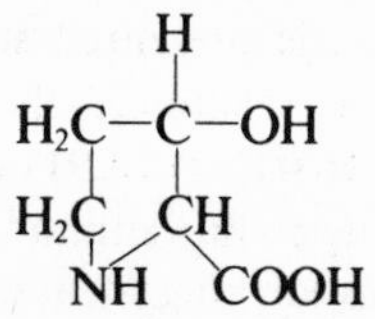

hydroxyproline

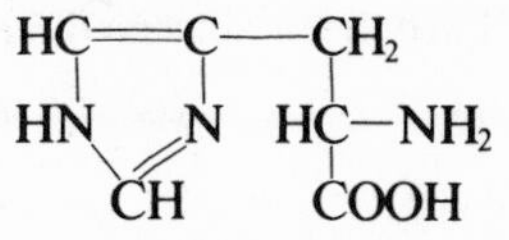

histidine

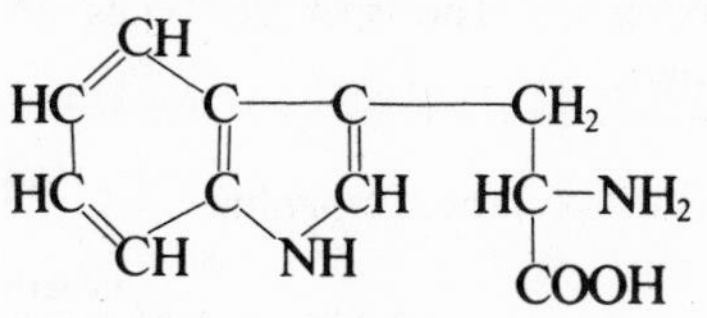

tryptophan

In polypeptide chains and proteins, the amino acids are linked together by an amide bond, formed by the reaction of the carboxyl group of one amino acid with the amino group of another acid and the splitting off of H_2O. Thus, it is by hydrolysis that the proteins and polypeptide chains are broken down into their constituent amino acids. Amino acids are very soluble in water, and exist in solution principally in the form of salt-like double ions (zwitterions), $^+NH_3 \cdot R \cdot COO^-$. The group R for any amino acid is obvious from its structural formula. Amino acids are amphoteric, the carboxyl group reacting with H^+ to produce $^+NH_3 \cdot R \cdot COOH$ and the amino group reacting with OH^- to produce $NH_2 \cdot R \cdot COO^-$ and water. They thus exert strong buffer action in aqueous solution.

The vast majority of the amino acids in the human body are derived from plant proteins, where they are produced from their inorganic constituents by the action of photosynthesis. Man and other vertebrates appear to be unable to synthesize amino acids, but may synthesize proteins from amino acids received from plants.

The breakdown of amino acids appears to be largely endocellular and is the mechanism by which NH_3 is released from the amino acids for assimilation by bacteria. As has been previously stated, however, amino acid breakdown may also furnish carbon and energy to the cells. Amino acids decompose in a number of different ways. The course of the decomposition reactions, as given below, is presented substantially as given by Anderson (*2*). For simplification of the reactions, the amino acids are represented by the general formula $R \cdot CHNH_2 \cdot COOH$ or $R \cdot CH_2 \cdot CHNH_2 \cdot COOH$.

Long-chain amino acids are more easily attacked than acids with short chains or ring structures. Alanine, for example, is much less resistant than glycine. Putrefactive organisms, such as *B. proteus*, can open the proline ring to produce ω-aminovaleric acid, $NH_2 \cdot (CH_2)_4 \cdot COOH$, and *n*-valeric acid, $CH_3 \cdot (CH_2)_3 \cdot COOH$. Molds can open any ring structure if no other source of nitrogen is available. *Escherichia coli* and *Ps. fluorescens* can open the histidine and tryptophan rings, but *B. proteus* cannot. Anaerobic conditions in the medium are accompanied by reductions and the accumulation of saturated acids and hydrocarbons as end products; aerobic conditions yield other types of products. The specific types of reactions that have been observed are as follows:

A. Decarboxylation to amine

$$R \cdot CHNH_2 \cdot COOH \xrightarrow{\text{decarboxylase}} R \cdot CH_2NH_2 + CO_2 \qquad (10\text{-}50)$$

This type of reaction is produced only by bacteria and is favored by anaerobic conditions. For example, *Ps. fluorescens* breaks down glycine to methylamine:

$$\overset{75}{CH_2 \cdot NH_2 \cdot COOH} \xrightarrow{\text{decarboxylase}} \overset{31}{CH_3 \cdot NH_2} + \overset{44}{CO_2} - 18.5 \text{ kcal} \quad (10\text{-}51)$$

$$\text{(glycine)} \qquad\qquad \text{(methylamine)} \quad \text{(gas)}$$

$$H^\circ = \quad -122.2 \qquad\qquad\qquad -9.3 \qquad -94.4$$

and mixed cultures can break down lysine to cadaverine.

B. Deamination

(1) Reductive to saturated acids:

$$R \cdot CHNH_2 \cdot COOH \xrightarrow{+H_2} R \cdot CH_2 \cdot COOH + NH_3 \qquad (10\text{-}52)$$

Bacteria are particularly active in promoting this reaction, but yeasts and molds are also capable of it. Anaerobic conditions are essential. Almost all amino acids are subject to reductive deamination, and the products are usually substituted propionic acids. For example, tryptophan yields indolepropionic acid. It is probable that reduction is continued in most cases, yielding methane and substituted acetic acids, for indole acetic acid is also produced in the putrefaction of tryptophan. For example, glycine yields acetic acid and alanine yields propionic acid through putrefaction by *B. putrificus* as follows:

$$\overset{75}{CH_2 \cdot NH_2 \cdot COOH} + \overset{2}{H_2} + \overset{18}{H_2O} \longrightarrow \overset{60}{CH_3COOH} + \overset{18}{NH_4^+} + \overset{17}{OH^-}$$

      (glycine)   (liquid)   (acetic acid)

$H° =$ -122.2 -68.4 -118.1 -31.5 -54.6

$$+13.6 \text{ kcal} \quad (10\text{-}53)$$

$$\overset{89}{CH_3CHNH_2 \cdot COOH} + \overset{2}{H_2} + \overset{18}{H_2O} \longrightarrow \overset{74}{CH_3CH_2COOH} + \overset{18}{NH_4^+} + \overset{17}{OH^-}$$

   (alanine)   (liquid)   (propionic acid)

$H° =$ -135.1 -68.4 -124.0 -31.5 -54.6

$$+6.6 \text{ kcal} \quad (10\text{-}54)$$

(2) Hydrolytic to α-hydroxy acids:

$$R \cdot CHCH_2 \cdot COOH \xrightarrow{+H_2O} R \cdot CHOH \cdot COOH + NH_3 \quad (10\text{-}55)$$

The products are substituted lactic acids. For example, tryptophan yields indole lactic acid and phenylalanine yields phenyl lactic acid.

(3) Desaturative to unsaturated acids:

$$R \cdot CH_2 \cdot CHNH_2 \cdot COOH \longrightarrow R \cdot CH=CH \cdot COOH + NH_3 \quad (10\text{-}56)$$

The α-β linkage is attacked with the formation of substituted acrylic acids. The reaction is produced by the coli-typhoid group. For example, aspartic acid yields fumaric acid, and histidine yields urocanic acid, the latter as follows:

      (histidine)   (urocanic acid)

(4) Oxidative to α-keto acids:

$$R \cdot CH_2 \cdot CHNH_2 \cdot COOH \xrightarrow{+O_2} R \cdot CH_2 \cdot CO \cdot COOH + NH_3 \quad (10\text{-}58)$$

This reaction is favored by aerobic conditions, but may be accomplished by means of a suitable hydrogen acceptor. The products are substituted pyruvic acids which are unstable and undergo further decomposition. For example, alanine yields pyruvic acid through the action of *B. proteus* as follows:

$$
\begin{array}{ccccccc}
89 & & 16 & 18 & 88 & 18 & 17
\end{array}
$$

$$CH_3 \cdot CHNH_2 \cdot COOH + \tfrac{1}{2}O_2 + H_2O \longrightarrow CH_3 \cdot CO \cdot COOH + NH_4^+ + OH^-$$

$$
\begin{array}{cccc}
\text{(alanine)} & \text{(aq.)} & \text{(liquid)} & \text{(pyruvic acid)}
\end{array}
$$

$$
\begin{array}{cccccc}
H^\circ = & -135.1 & -1.9 & -68.4 & -140.9 & -31.5 \quad -54.6
\end{array}
$$

$$+ \, 21.6 \, kcal \quad (10\text{-}59)$$

Tryptophan yields indole pyruvic acid by *E. coli* and putrefactive organisms as follows:

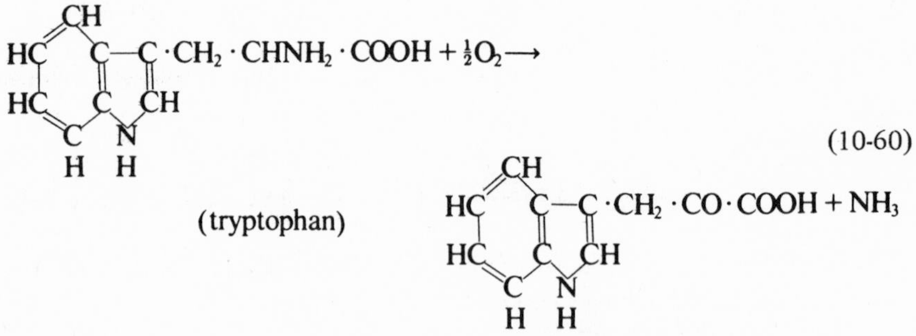

$$(10\text{-}60)$$

(tryptophan)

(indole pyruvic acid)

C. Breakdown of sulfur amino acids

Little is known about the decomposition of cystine and cysteine, except that they produce H_2S and possibly methyl mercaptan and ethyl sulfide. It appears that the production of methyl mercaptan, $CH_3 \cdot SH$, depends upon the presence of carbohydrate. These amino acids are responsible for some of the disagreeable odors produced in putrefaction.

D. Further breakdown of products

(1) Oxidation and hydrolysis of unsaturated acids:

$$R \cdot CH{=}CH \cdot COOH \xrightarrow{+O_2} R \cdot CO \cdot CH_2 \cdot COOH \xrightarrow{+H_2O} R \cdot COOH + CH_3 \cdot COOH$$

$$(10\text{-}61)$$

This reaction takes place under aerobic conditions, and the products are acetic acid and substituted formic acids. For example, *E. coli* produces *p*-hydroxybenzoic acid from tyrosine and indole carboxylic acid from tryptophan as follows:

$$O_2 + H_2O \quad\text{(10-62)}$$

(tyrosine) → (p-hydroxy-benzoic acid) $+ NH_3 + CH_3COOH$

(tryptophan) $\xrightarrow{O_2 + H_2O}$ (indole carboxylic acid) $+ NH_3 + CH_3 COOH$ (10-63)

(2) Decarboxylation

Saturated acids produced by reductive deamination, reaction (10-52), are decarboxylated to hydrocarbons under anaerobic conditions through the agency of the enzyme decarboxylase as follows:

$$R \cdot CH_2COOH \xrightarrow{\text{decarboxylase}} R \cdot CH_3 + CO_2 \qquad (10\text{-}64)$$

For example, glycine yields methane through acetic acid by means of reaction (10-43). Buswell (55) and others have found that propionic acid and higher fatty acids also break down in anaerobic digestion by hydrolysis to methane and carbon dioxide, as will be shown later. It is possible in some cases that decarboxylation may precede deamination, or that the two reactions may take place together.

Substituted acetic acids with ring structure undergo anaerobic decarboxylation with putrefactive organisms to produce hydrocarbons, some of which are malodorous. For example, tyrosine yields *p*-cresol through *p*-hydroxyphenylacetic acid, and tryptophan yields skatole through indole acetic acid as follows:

$$\text{tyrosine} \longrightarrow \underset{\text{OH}}{\overset{\text{CH}_2\text{COOH}}{\bigcirc}} \xrightarrow{\text{decarboxylase}} \underset{\text{OH}}{\overset{\text{CH}_3}{\bigcirc}} + CO_2 \qquad (10\text{-}65)$$

$$\left(\begin{array}{c}p\text{-hydroxy-}\\\text{phenylacetic acid}\end{array}\right) \qquad (p\text{-cresol})$$

$$\text{tryptophan} \longrightarrow \underset{\underset{\text{H}}{\text{N}}}{\bigcirc\!\!\!\bigcirc}\!\!\text{CH}_2\text{COOH} \xrightarrow{\text{decarboxylase}}$$

$$\left(\begin{array}{c}\text{indole}\\\text{acetic acid}\end{array}\right)$$

$$\underset{\underset{\text{H}}{\text{N}}}{\bigcirc\!\!\!\bigcirc}\!\!\text{CH}_3 + CO_2$$

$$(10\text{-}66)$$

$$(\text{skatole})$$

These reactions are produced by *E. coli* and many other bacteria.

Substituted formic acids with ring structure, as produced by reaction (10-61), are decarboxylated by *E. coli* under aerobic conditions. For example, tyrosine produces phenol through *p*-hydroxybenzoic acid and tryptophan produces indole through indole carboxylic acid as follows:

$$\text{tyrosine} \longrightarrow \underset{\text{OH}}{\overset{\text{COOH}}{\bigcirc}} \xrightarrow{\text{decarboxylase}} \underset{\text{OH}}{\bigcirc} + CO_2$$

$$(10\text{-}67)$$

$$\left(\begin{array}{c}p\text{-hydroxy-}\\\text{benzoic acid}\end{array}\right) \qquad (\text{phenol})$$

$$\text{tryptophan} \longrightarrow \underset{\underset{\text{H}}{\text{N}}}{\bigcirc\!\!\!\bigcirc}\!\!\text{COOH} \xrightarrow{\text{decarboxylase}} \underset{\underset{\text{H}}{\text{N}}}{\bigcirc\!\!\!\bigcirc} + CO_2$$

$$(10\text{-}68)$$

$$\left(\begin{array}{c}\text{indole}\\\text{carboxylic acid}\end{array}\right) \qquad (\text{indole})$$

α-keto acids produced by reaction (10-58) are decarboxylated to substituted aldehydes as follows:

$$R \cdot CH_2 \cdot CO \cdot COOH \xrightarrow{\text{decarboxylase}} RCH_2CHO + CO_2 \qquad (10\text{-}69)$$

The aldehydes are unstable and undergo further decomposition. Under aerobic conditions, they are oxidized by bacteria to substituted acetic acids as follows:

$$R \cdot CH_2 \cdot CHO \xrightarrow{+O_2} RCH_2COOH \qquad (10\text{-}70)$$

For example, phenylalanine produces phenylacetic acid through phenylpyruvic acid and phenylacetaldehyde as follows:

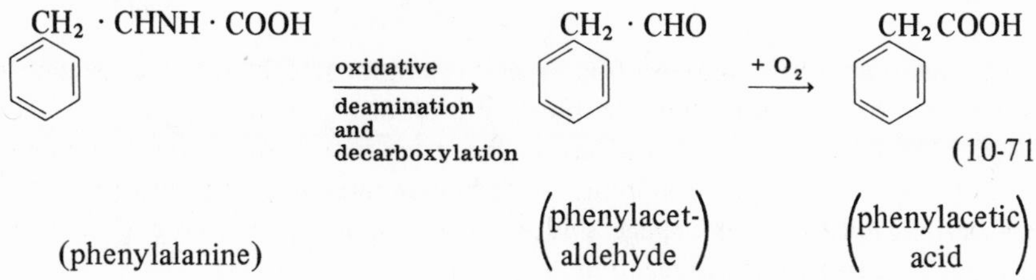

(phenylalanine) — oxidative deamination and decarboxylation → (phenylacetaldehyde) + O_2 → (phenylacetic acid) $\qquad (10\text{-}71)$

Under anaerobic conditions, the aldehydes are reduced to alcohols as follows:

$$R \cdot CH_2 \cdot CHO \xrightarrow{+H_2} R \cdot CH_2CH_2OH \qquad (10\text{-}72)$$

This reaction is common with yeasts, but it is also brought about by molds, *B. proteus*, and lactic acid bacteria. For example, by this reaction leucine produces indole ethanol, tyrosine produces *p*-hydroxyphenylethanol, and tryptophan yields tryptophol.

It will be noted that nitrogen metabolism does not produce urea by bacterial breakdown as it does in human digestion. Urea is produced, however, in the hydrolysis of arginine by *arginase* secreted by the sac fungus, *Aspergillus niger*, the denitrifying bacteria *Ps. fluorescens*, and other organisms (*14*, Vol. 111, p. 90). The reaction may be represented as follows:

$$NH_2 \cdot CNH \cdot NH \cdot (CH_2)_3 \cdot CHNH_2 \cdot COOH + H_2O \xrightarrow{\text{arginase}}$$

(arginine)

$$NH_2CONH_2 + NH_2 \cdot (CH_2)_3 \cdot CHNH_2 \cdot COOH \quad (10\text{-}73)$$

(urea)   (ornithine)

Urea is readily utilized as a source of nitrogen by many organisms, as has been noted. It seems that urea cannot be utilized as a source of both nitrogen and carbon, however. The breakdown of urea is accomplished through the agency of *urease*, a hydrolase that seems to be an endoenzyme. According to Buchanan and Fulmer (*14*, Vol. 11, p. 501; Vol. 111, p. 391), urease is produced by lactic acid bacteria, *Ps. fluorescens, B. ammoniagenes, A. aerogenes, B. proteus, Staph. aureus*, and others. The reaction, which proceeds as follows, is accelerated by the presence of utilizable carbohydrates:

$$
\overset{60}{NH_2CONH_2} + \overset{3 \times 18}{3H_2O} \xrightarrow{urease} \overset{2 \times 18}{2NH_4^+} + \overset{2 \times 17}{2OH^-} + \overset{44}{CO_2} - 14.2\,kcal \quad (10\text{-}74)
$$

$$\text{(aq. urea)} \quad \text{(liquid)} \quad\quad\quad\quad\quad \text{(gas)}$$

$$H° = \quad -75.6 \quad\quad -205.2 \quad\quad -63.0 \quad -109.2 \quad -94.4$$

It will be noted that the reaction is endothermic, which probably accounts for the fact that it is not a suitable source of both nitrogen and carbon.

Reductive deamination of urea has not been reported. Such a reaction might be the source of the small amount of carbon monoxide sometimes produced in anaerobic digestion of sewage sludge. The reaction would be endothermic to about the same extent as reaction (10-74).

Under strictly aerobic conditions, ammonia may be oxidized to nitrite and nitrate through the action of the *nitrifying* bacteria, in accordance with reactions (10-7) and (10-8).

Nitrates are suitable hydrogen acceptors in oxidations by many bacteria, which reduce the nitrate to nitrite. In addition to these organisms, a large number of bacteria are capable of reducing nitrate and nitrite to gaseous nitrogen. These organisms (*14*, Vol. 1, p. 433) are aerobes except for their ability to use nitrates and nitrites; they attack proteins vigorously and they are unable to ferment carbohydrates anaerobically. Most of these organisms attack both nitrites and nitrates. The reduction of nitrite may be represented as follows:

$$
R \cdot NH_2 + HNO_2 \longrightarrow R \cdot OH + N_2 + H_2O \quad\quad\quad (10\text{-}75)
$$

Amides and amines are known to react readily with nitrous acid, yielding gaseous nitrogen and substantial amounts of energy. For example, with urea,

$$
\overset{60}{NH_2 \cdot CO \cdot NH_2} + \overset{94}{2HNO_2} \longrightarrow \overset{56}{2N_2} + \overset{44}{CO_2} + \overset{54}{3H_2O} + 167.0\,kcal \quad (10\text{-}76)
$$

$$\text{(urea)} \quad\quad \text{(aq.)} \quad\quad \text{(gas)} \quad \text{(gas)} \quad \text{(liquid)}$$

$$H° = \quad -75.6 \quad\quad -57.0 \quad\quad\quad -94.4 \quad -205.2$$

With acetamide,

$$
\overset{59}{\underset{(\text{acetamide})}{CH_3 \cdot CO \cdot NH_2}} + \overset{47}{\underset{(\text{aq.})}{HNO_2}} \longrightarrow \overset{28}{\underset{(\text{gas})}{N_2}} + \overset{18}{\underset{(\text{liquid})}{H_2O}} + \overset{60}{\underset{(\text{acetic acid})}{CH_3COOH}} + 83.1 \, \text{kcal} \quad (10\text{-}77)
$$

$H° = \quad -74.9 \qquad -28.5 \qquad\qquad -68.4 \qquad -118.1$

Reactions similar to (10-77) are probably responsible for the phenomenon of *rising sludge* in final settling tanks of the activated sludge process. O'Shaugnessy and Hewitt (56) have shown that this phenomenon is caused by gaseous nitrogen in the absence of sufficient dissolved oxygen, and that it occurs only when nitrification has taken place in the aeration tanks. It is accompanied by a reduction in the nitrite concentration and a rise in pH, in one case from 6.9 to 8.3. No loss of ammonia occurred. Key and Etheridge (56, Pt. 11, p. 278, 1936) have observed the same phenomenon, and report that the gas responsible for the rising sludge is practically pure nitrogen. This would tend to confirm the assumption that urea is rapidly decomposed and is completely gone when the mixed liquor reaches the final settling tanks, for the reaction with urea would also produce carbon dioxide. Key and Etheridge report that the nitrogen produced is practically equivalent to $1\frac{1}{2}$ times the nitrogen in the nitrite destroyed, and approximately equivalent to the nitrogen in the nitrate.

The nitrogen gas sometimes reported in anaerobic digestion of sewage sludge is believed by some to be derived from dissolved atmospheric nitrogen (56). In the case of fresh sewage sludge containing nitrites and amides or amines, it might readily result from decompositions of the type in reactions (10-76) and (10-77). Nitrogen is usually noted in greatest amounts during the earlier stages of digestion.

10-16. BREAKDOWN OF FATS AND GREASE

Buswell and Neave (47, p. 29) found that the grease content of the dry sewage solids of Urbana averaged about 29 percent. About half the grease was present as soaps, one third as glycerides, and the remainder as mineral oils, unsaponifiable solids and water. The marked resistance to destruction of pure fats and soaps once led to the belief that they have strong germicidal powers. According to Buswell and Neave, in 1930: "Recent work has been shown that commercial soaps and lower fatty acids have little germicidal value over the ordinary range of hydrogen ion concentration. In nature, fats are resistant to bacterial attack because they tend to occur in dense aggregates which do not contain the other elements necessary for bacterial nutrition. In sewage sludge the finely divided nature of the grease and its distribution amongst fibrous and nitrogenous matter probably contribute largely to its degradation."

True fats are, in general, composed of only carbon, hydrogen, and oxygen. The true fats or lipids, as distinguished from the volatile or essential oils, are nonvolatile and are composed of esters of the fatty acids (mainly glyceryl esters). They are insoluble in water but are soluble in ether, chloroform and benzene. The simple lipids may be classified as follows:

1. Fats—esters of fatty acids with glycerol, solid at room temperature.
2. Oils—esters of fatty acids with glycerol, liquid at room temperature.
3. Waxes—esters of fatty acids with alcohols (commonly monohydroxy) other than glycerol.

Two groups of the simpler fatty acids that commonly occur in natural fats are presented in Table 10-6. Only a few representative members of each group, which are encountered most frequently, are presented (*52*, p. 767).

Esters are acids in which the hydrogen of the carboxyl group has been replaced by an alkyl radical. Fats are esters formed by the union of fatty acids with alcohols with the elimination of water; for example,

$$CH_2OH + HOOC \cdot (CH_2)_{16} \cdot CH_3 \underset{\text{saponification}}{\overset{\text{esterification}}{\rightleftharpoons}} CH_2 \cdot OOC \cdot (CH_2)_{16} \cdot CH_3 + H_2O$$

$$\begin{array}{ll} | & | \\ CHOH & CHOH \\ | & | \\ CH_2OH & CH_2OH \end{array}$$

(glycerol)   (stearic acid)   (fat)

$$(10\text{-}78)$$

This example is a monoglyceride of stearic acid, which apparently does not occur in nature. Glycerol is a trihydroxy alcohol, and may yield mono-, di-, and triglycerides. Only triglycerides occur in natural fats and oils and they are usually mixed glycerides, at least one acid being different from the other two acids in the ester. It will be noted with mixed glycerides that the position of the various acids on the glycerol molecule will affect the structural formula of the fat, although it may have no influence on its content of carbon, hydrogen, and oxygen. The presence of mono- or diglycerides in a fat or oil indicates that the material has undergone partial saponification. The presence of free fatty acid also indicates that partial hydrolysis has taken place.

Soap is made by boiling an emulsion of fat and sodium hydroxide to yield sodium salts of the fatty acids and glycerol. This saponification reaction is indicated as follows:

$$CH_2 \cdot OOC \cdot (CH_2)_{16} \cdot CH_3 + NaOH \xrightarrow{\text{saponification}} CH_2 \cdot OH + NaOOC \cdot (CH_2)_{16} \cdot CH_3$$

$$\begin{array}{ll} | & | \\ CHOH & CHOH \\ | & | \\ CH_2OH & CH_2OH \end{array}$$

            (stearin)   (glycerol)   (sodium stearate)

$$(10\text{-}79)$$

Table 10-6
Some Fatty Acids Found in Natural Fats

Name	n	Formula	Occurrence
		Group A. Saturated fatty acids, acetic acid series, $C_nH_{2n+1}COOH$ (only those in which n is an odd number are commonly found in natural fats)	
Acetic	1	$CH_3 \cdot COOH$	Vinegar
Butyric	3	$CH_3 \cdot (CH_2)_2 \cdot COOH$	Milk fat
Caproic	5	$CH_3 \cdot (CH_2)_4 \cdot COOH$	Butter
Caprylic	7	$CH_3 \cdot (CH_2)_6 \cdot COOH$	Butter
Capric	9	$CH_3 \cdot (CH_2)_8 \cdot COOH$	Butter
Lauric	11	$CH_3 \cdot (CH_2)_{10} \cdot COOH$	Laurel oil, spermaceti
Myristic	13	$CH_3 \cdot (CH_2)_{12} \cdot COOH$	Nutmeg butter
Palmitic	15	$CH_3 \cdot (CH_2)_{14} \cdot COOH$	Animal and vegetable fats
Stearic	17	$CH_3 \cdot (CH_2)_{16} \cdot COOH$	Animal and vegetable fats
Arachidic	19	$CH_3 \cdot (CH_2)_{18} \cdot COOH$	Peanut oil

Name	n	Formula	Occurrence
		Group B. Unsaturated fatty acids	
		1. Oleic acid series, $C_nH_{2n-1}COOH$	
Oleic	17	$CH_3 \cdot (CH_2)_7 \cdot CH = CH \cdot (CH_2)_7 \cdot COOH$	Animal and vegetable fats
Erucic	21	$C_{22}H_{42}O_2$	Rapeseed oil
Cetoleic	21	$C_{22}H_{42}O_2$	Fish oils
		2. Linoleic acid series, $C_nH_{2n-3}COOH$	
Linoleic	17	$C_{18}H_{32}O_2$	Linseed and cottonseed oil
		3. Linolenic acid series, $C_nH_{2n-5}COOH$	
Linolenic	17	$C_{18}H_{30}O_2$	Linseed oil
		4. Series $C_nH_{2n-7}COOH$	
Clupanodonic	17	$C_{18}H_{28}O_2$	Japanese sardine oil
Arachidonic	19	$C_{20}H_{32}O_2$	Lecithin, cephalin

The fats usually used in soap manufacture consist principally of stearin, olein, and palmitin. Potash may be used instead of soda for soap manufacture; sodium soaps are hard soaps and potassium soaps are soft. Potassium and sodium soaps are soluble in water. When soap is added to hard water, the sodium or potassium is replaced by calcium, magnesium, iron, or manganese, and the resulting soaps are insoluble.

The residual soap in wastewater is readily precipitated by reaction with multivalent metallic ions in the water and is thus removable by settling and filtration. The principal advantage to the housewife of synthetic detergents over soap is that the former do not form precipitates with multivalent metallic ions and thus do not stain fabrics and produce films on dishes and plumbing fixtures. Although this characteristic of synthetic detergents is beneficial to the user, it is very damaging to our watercourses because it eliminates the main process by which the detergent might be removed from wastewater.

The fat constituent of soaps is subject to bacterial decomposition both aerobically and anaerobically, as shown next, and the energy yield from aerobic decomposition is very high. This provides another mechanism by which fats and soaps may be removed from wastewater and watercourses. On the other hand, the principal constituent of most of the original synthetic detergents, ABS, was very resistant to bacterial decomposition. The gradual substitution, in the late 1960s, of biodegradable detergents containing LAS constituted a significant step forward in water-pollution control.

The aerobic decomposition of the fatty acids and glycerol yields CO_2 and water. Numerous bacteria are known to accomplish the breakdown of fatty acids aerobically. Typical reactions for complete breakdown are as follows:

$$\underset{60}{CH_3 \cdot COOH} + \underset{64}{2O_2} \longrightarrow \underset{88}{2CO_2} + \underset{36}{2H_2O} + 199.9 \, kcal \qquad (10\text{-}80)$$

$$\quad \text{(acetic acid)} \quad \text{(aq.)} \qquad \text{(gas)} \quad \text{(liquid)}$$

$$H° = \quad -118.1 \qquad -7.6 \qquad -188.8 \quad -136.8$$

$$\underset{88}{CH_3 \cdot (CH_2)_2 \cdot COOH} + \underset{160}{5O_2} \longrightarrow \underset{176}{4CO_2} + \underset{72}{4H_2O} + 501.0 \, kcal \quad (10\text{-}81)$$

$$\qquad \text{(butyric acid)} \qquad \text{(aq.)} \qquad \text{(gas)} \quad \text{(liquid)}$$

$$H° = \qquad -131.2 \qquad\qquad -19.0 \qquad -377.6 \quad -273.6$$

$$\underset{284}{C_{18}H_{36}O_2} + \underset{832}{26O_2} \longrightarrow \underset{792}{18CO_2} + \underset{324}{18H_2O} + 2{,}598.6 \, kcal \quad (10\text{-}82)$$

$$\text{(stearic acid)} \quad \text{(aq.)} \qquad\quad \text{(gas)} \quad \text{(liquid)}$$

$$H° = \; -233.0 \qquad -98.8 \qquad -1{,}699.2 \quad -1{,}231.2$$

$$\underset{92}{C_3H_5(OH)_3} + \underset{112}{\tfrac{7}{2}O_2} \longrightarrow \underset{132}{3CO_2} + \underset{72}{4H_2O} + 382.8 \, kcal \qquad (10\text{-}83)$$

$$\text{(aq. glycerol)} \quad \text{(aq.)} \longrightarrow \text{(gas)} \quad \text{(liquid)}$$

$$H° = \quad -160.7 \qquad -13.3 \qquad -283.2 \quad -273.6$$

The energy yield from these reactions increases from 3.33 kcal/g of substrate for acetic acid to more than 9 kcal/g for the higher fatty acids. The oxygen requirement per part per million of substrate also increases with the number of carbon atoms from a low of 1.07 ppm for acetic acid to about 3 ppm for the higher fatty acids.

In studies of the anaerobic decomposition of formic, acetic, propionic, n-butyric, n-valeric, and stearic acids, Buswell and Neave (47, p. 67) found that the breakdown proceeded in accordance with the following general reaction:

$$C_nH_{2n}O_2 + \frac{n-2}{2} H_2O = \frac{n+2}{4} CO_2 + \frac{3n-2}{4} CH_4 \qquad (10\text{-}84)$$

This formula, which applies to saturated fatty acids only, checks with the more general formula (10-37) for all compounds of C, H, and O. Tarvin and Buswell (53, p. 51) found that these same acids, when inoculated with flora from sewage sludge, were decomposed from 85 to substantially 100 percent, as measured by gas yield. The results checked with Eq. (10-84). Stearic acid decomposed more slowly than the other fatty acids.

Equation (10-84) indicates that all the fatty acids higher than acetic absorb water in anaerobiosis and, therefore, yield more gas than the weight of the acid decomposed. The ratio of the weights of gas produced to acid decomposed is 1.12 for propionic acid and increases with the number of carbon atoms to 1.51 for stearic acid. The percentage of CH_4 in the gaseous products also increases with the number of carbon atoms in the acid, being 26.6 percent by weight (50 percent by volume) for acetic and 48.5 percent by weight (72.2 percent by volume) for stearic acid.

It has been previously stated that the reactions in the anaerobiosis of fatty acids are endothermic in terms of the heats of reaction. This point has been demonstrated for acetic acid in Eq. (10-43) and for formic acid in Eq. (10-44). The following reactions demonstrate the same thesis for butyric and palmitic acids:

$$\overset{176}{2CH_3 \cdot (CH_2)_2 \cdot COOH} + \overset{36}{2H_2O} \longrightarrow \overset{132}{3CO_2} + \overset{80}{5CH_4} - 25.0\,\text{kcal} \quad (10\text{-}85)$$

$$\text{(butyric acid)} \quad \text{(liquid)} \quad \text{(gas)} \quad \text{(gas)}$$
$$H° = \quad -262.4 \quad -136.8 \quad -283.2 \quad -91.0$$

$$\overset{512}{2CH_3 \cdot (CH_2)_{14} \cdot COOH} + \overset{252}{14H_2O} \longrightarrow \overset{396}{9CO_2} + \overset{368}{23CH_4} - 139.4\,\text{kcal} \quad (10\text{-}86)$$

$$\text{(palmitic acid)} \quad \text{(liquid)} \quad \text{(gas)} \quad \text{(gas)}$$
$$H° = \quad -450.0 \quad -957.6 \quad -849.6 \quad -418.6$$

REFERENCES

1. See R. A. Gortner, *Outlines of Biochemistry*, New York, John Wiley & Sons, Inc., 1938, pp. 232, 918.
2. C. G. Anderson, *An Introduction to Bacteriological Chemistry*, Edinburgh, E. and S. Livingston Limited, 1938, p. 40.
3. *Water Works*, p. 24 (Jan. 1927).
4. L. Arnold, *Am. J. Public Health*, *28*, 839 (1938).
5. B. C. J. G. Knight, *Bacterial Nutrition*, London, British Medical Research Council, 1936.
6. M. Stephenson, *Bacterial Metabolism*, Essex, England, Longman Group Ltd., 1930, p. 101.
7. A. Szent-Györgyi, *Oxidation, Fermentation, Vitamins, Health and Disease*, Baltimore, The Williams & Wilkins Company, 1939.
8. H. Heukelekian, *Sewage Works J.*, *2*, 219 (1930).
9. W. J. Penfold and D. Norris, *J. Hyg.*, *12*, 527 (1912).
10. R. F. Weston and W. W. Eckenfelder, *Sewage Ind. Wastes*, *27*, 802 (July 1955).
11. H. A. Barker, *Arch. Mikrobiol.*, *7*, 420 (1936).
12. H. Heukelekian and B. Heinemann, *Sewage Works J.*, *11*, 426 (1939).
13. C. T. Butterfield, *Public Health Rept.*, *671*, May 17, 1935.
14. R. E. Buchanan and E. I. Fulmer, *Physiology and Biochemistry of Bacteria*, Vol. 1, Baltimore, The Williams & Wilkins Company, 1928, p. 16.
15. C. T. Butterfield, *Public Health Rept.*, *2865*, Nov. 22, 1929.
16. H. Heukelekian and B. Heinemann, *Sewage Works J.*, *11*, 436 (1939).
17. C. T. Butterfield and E. Wattie, *Sewage Works J.*, *10*, 815 (1938).
18. *Public Health Rept.*, *46*, Feb. 20, 1931.
19. S. R. Hoover, L. Jasewicz, and N. Porges, *Sewage Ind. Wastes*, *25*, 1163 (1953).
20. A. M. Buswell, *Sewage Works J.*, *3*, 362 (1931).
21. T. D. Brock, *Principles of Microbial Ecology*, Englewood Cliffs, N.J., Prentice-Hall, Inc., 1966, p. 103.
22. W. W. Eckenfelder and D. J. O'Connor, *Biological Waste Treatment*, Elmsford, N.Y., Pergamon Press, Inc., 1961, pp. 14–15.
23. See L. Metcalf and H. P. Eddy, *American Sewerage Practice*, Vol. III, New York, McGraw-Hill Book Company, 1935, p. 94.
24. S. L. Neave and A. M. Buswell, *J. Amer. Water Works Assoc.*, 388 (1927); *Illinois State Water Surv. Bull. 30*, 1930; P. H. Mitchell, *General Physiology*, New York, McGraw-Hill Book Company, 1923, p. 635; A. P. Mathews, *Physiological Chemistry*, New York, Wm. Wood & Co., 1930, p. 454.
25. M. Bodansky, *Introduction to Physiological Chemistry*, New York, John Wiley & Sons, Inc., 1930, p. 183.
26. A. P. Mathews, *Physiological Chemistry*, New York, Wm. Wood & Co., 1930, p. 455.
27. H. W. Streeter, *Sewage Works J.*, *7*, 251 (1935); E. B. Phelps, *Modern Sewage Disposal*, Federation of Sewage Works Associations, 1938, p. 158.
28. E. J. Theriault, *Public Health Bull.*, *173*, Washington, D.C., U.S. Public Health Service, 1927, p. 132.
29. *Standard Methods for the Examination of Water and Wastewater*, 13th ed., New York, American Public Health Association, Inc., 1971.
30. E. B. Phelps, *Stream Sanitation*, New York, John Wiley & Sons, Inc., 1944, p. 64.
31. H. W. Streeter and E. B. Phelps, *Public Health Bull.*, *146*, Washington, D.C., U.S. Public Health Service, 1925.
32. C. N. Sawyer, P. Callejas, M. Moore, and A. Tom, *Sewage Ind. Wastes*, *22*, 26 (1950).

33. See L. J. Gillespie, *Physical Chemistry*, New York, McGraw-Hill Book Company, 1931, p. 268.

34. G. M. Fair and E. W. Moore, *Sewage Works J.*, *4*, 589 (1932).

35. E. W. Moore, *Sewage Works J.*, *13*, 561 (1941).

36. C. T. Butterfield and E. Wattie, *Public Health Rept.*, *53*, Oct. 28, 1938.

37. C. T. Butterfield, C. C. Ruchhoft, and P. D. McNamee, *Sewage Works J.*, *9*, 173 (1937).

38. C. C. Ruchhoft, P. D. McNamee, and C. T. Butterfield, *Sewage Works J.*, *10*, 661 (1938).

39. C. N. Sawyer and M. S. Nichols, *Sewage Works J.*, *11*, 462 (1939).

40. S. Grant, E. Hurwitz, and F. W. Mohlman, *Sewage Works J.*, *2*, 228 (1930).

41. C. N. Sawyer and G. A. Rohlich, *Sewage Works J.*, *11*, 946 (1939).

42. C. N. Sawyer, *Sewage Works J.*, *12*, 3 (1940).

43. M. T. Garrett and C. N. Sawyer, "Kinetics of Removal of Soluble BOD by Activated Sludge," *Proc. Seventh Ind. Waste Conf.*, *Purdue Univ.*, p. 51, 1952.

44. I. Gellman and H. Heukelekian, *Sewage Ind. Wastes*, *25*, 1196 (1953).

45. E. N. Helmers, J. D. Frame, A. E. Greenberg, and C. N. Sawyer, *Sewage Ind. Wastes*, *23*, 884 (1951).

46. L. Metcalf and H. P. Eddy, *American Sewerage Practice*, Vol. III, New York, McGraw-Hill Book Company, 1935.

47. A. M. Buswell and S. L. Neave, *Illinois State Water Sur. Bull. 30*, 1930.

48. A. M. Buswell and C. S. Boruff, *Sewage Works J.*, *4*, 454 (1932).

49. G. M. Fair and E. W. Moore, *Sewage Works J.*, *4*, 428 (1932).

50. *Sewage Works J.*, *5*, 36 (1933).

51. C. N. Sawyer and H. K. Roy, *Sewage Ind. Wastes*, *27*, 1356 (1955).

52. R. A. Gortner, *Outlines of Biochemistry*, New York, John Wiley & Sons, Inc., 1938, p. 721.

53. *Illinois State Water Surv. Bull. 32.*

54. *Science*, 559 (June 24, 1938).

55. *Illinois State Water Surv. Bull. 32*, pp. 43–45.

56. *Inst. Sewage Purif., J. Proc.*, Pt. 1, p. 221 (1935).

CHAPTER 11

The Oxygen Balance in Polluted Waters

Natural surface waters and tidal estuaries contain organic impurities under decomposition by bacteria and other microorganisms. The bacteria utilize the oxygen dissolved in the water in their feeding processes, and the dissolved oxygen is in turn replenished from the atmosphere. Where green plant life, such as green and blue-green algae, is plentiful in the water, oxygen is also supplied during daylight as a by-product of the photosynthesis of carbohydrates from carbon dioxide and bicarbonates.

If the concentration of decomposing organic matter is excessive, which may be the case if the water is polluted with sewage or organic industrial wastes, the dissolved oxygen concentration may be lowered to a level that is harmful to fish and other aquatic life, or it may be lowered to a concentration near zero, at which anaerobic decomposition takes place with the production of gaseous by-products and black water. The gaseous by-products are carbon dioxide, methane, and, to a lesser extent, hydrogen sulfide and mercaptans. The sulfide compounds cause foul odors and react with metals in the water to form black precipitates, resulting in black water. In many cases where sulfide odors are present in the atmosphere above the water surface, the sulfides may react with paints on buildings, boats, and structures to discolor or blacken the paint.

Many polluted waters are analyzed routinely by water-pollution control agencies during the warmer and drier months when oxygen levels are lowest and the rate of bacterial decomposition is greatest. Dissolved oxygen, BOD, and other determinations are made on samples collected from selected sampling stations at selected time intervals throughout the critical months. Such analyses are useful in measuring the extent and nature of pollution and in locating the points of minimum oxygen concentration, but they are usually not an adequate basis for oxygen-balance studies to estimate the allowable pollution loads and to plan means for pollution abatement.

The most important purpose of oxygen-balance studies is to estimate the allowable pollution loads at existing or potential points of pollution. The difference between the allowable and the existing or potential pollution load represents the removal that must be accomplished at each polluting point. To estimate the allowable pollution loads, parameters and rate constants must be used that differ from those which pertain to existing conditions. For example, the oxygen demand of bottom deposits would be reduced after treatment to reflect the reduced amount of the suspended matter forming such deposits; the removal of BOD by settling to the bottom of the receiving water would be reduced after treatment of the polluting wastes; the deoxygenation rate constants would be less for biologically treated wastes than for raw wastes. Moreover, the rate of supply of dissolved oxygen must be estimated for the selected design flow regime, which is usually not the same as the flow and temperature conditions occurring at the times of routine sampling.

Routine receiving-water analyses may be used in oxygen-balance studies, but for planning abatement works they should be supplemented with additional analyses to estimate first-stage oxygen demands, coefficients of deoxygenation and reaeration, and photosynthetic oxygen production. Surveys should also be made to estimate the quantity and strength of present and future wastes at each point of pollution, with BOD analyses of raw and treated wastes to estimate the rate constants, the first-stage oxygen demand, and the changes in these parameters with temperature. Analysis should also be made of the oxygen demand of substantial bottom deposits to determine what disposition is required of such deposits. If the times of flow are very long, such that nitrification may be an important factor, the second-stage oxygen demand and rate constants should be estimated.

In oxygen-balance studies to estimate allowable pollution loads, oxygen furnished through photosynthesis by green plants should not be relied upon, because there is no assurance that it will be available during critical low-flow periods. Studies should, therefore, be made of existing receiving-water conditions to estimate how much, if any, of the oxygen supply is from green plants. The rate of supply of dissolved oxygen from atmospheric reaeration may be estimated for existing conditions from BOD and dissolved oxygen analyses, provided corrections are made for oxygen supplied through photosynthesis, and due account is taken of the settling out of decomposable organic matter and the BOD added from bottom deposits.

The methods developed here for computing the oxygen balance are all based upon the use of simple rate formulations analogous to unimolecular chemical reactions. It is assumed that the rate of deoxygenation is proportional to the BOD remaining to be satisfied, that the rate of atmospheric reaeration is proportional to the degree of undersaturation of dissolved oxygen, that the rate of settling out of decomposable organic matter is proportional to the BOD remaining to be satisfied, and that the rates of addition of BOD from the bottom deposits and of dissolved oxygen from photosynthesis are constant. The errors introduced by these simplifying assumptions will be of minor importance if the oxygen-balance equations are applied to relatively short reaches of the receiving waters in which the rate constants will change very little.

To obtain practical and useful answers in oxygen-balance studies, it is essential to assume *steady-state* conditions. It is assumed that all flows, temperatures, BOD loads, and rate constants at each point remain constant with time. It is further assumed that the concentration of BOD and dissolved oxygen is uniform over the cross section at any point in the path of flow. For very large receiving waters, where it is impractical to distribute the pollution load uniformly over the entire width of the stream, a fractional width may be used within which most of the pollution is expected to remain. Since in most cases there will be variations in pollution loads, flows, and temperatures throughout the 24 hr, and from day to day, and there will be slugs of pollution and imperfect mixing with the receiving waters, the assumption of steady-state conditions will introduce far greater errors than will inaccuracies in the determination of rate constants.

Oxygen-balance computations may be made for unsteady conditions, such as hourly variations of BOD load or cyclic variations in dissolved oxygen content caused by tides. The method of approach has been demonstrated by Li (*1*). Li's method of computation is based on the assumptions that the cyclic variations are approximately the same, cycle after cycle, and that the period is sufficiently long so that longitudinal mixing of the receiving waters will not excessively damp the peaks and valleys of dissolved oxygen levels at a critical downstream point.

11-1. SELECTION OF DESIGN FLOW CONDITIONS

The selection of the design flow and water temperature to be used in the oxygen-balance studies requires mature consideration of all factors involved if the required objective is to be met at reasonable cost of pollution abatement works. If the receiving waters dry up and the climate is hot, as is the case in many arid desert regions, no oxygen-balance studies can be made and complete treatment of wastes may be required to minimize nuisance. Oxygen-balance studies are of use only where there is always a flow in the receiving waters.

The design flow should be a drought flow sufficiently small in magnitude that excessive damage will not occur with flows of smaller magnitude. It should also be a flow that will not be expected too frequently, on the average. On the other hand, if a larger design flow is selected, substantial savings may be possible in the cost of waste-treatment works without substantial increase in the probability and frequency of damage to the users of the receiving waters.

Some water-pollution control agencies suggest design flows equal to the average for the driest 7-day period of record; others suggest flows that range from a flow which will be exceeded about 98 percent of the time to a flow which will be exceeded about 90 percent of the time. Flow-duration curves should be studied in the selection of design flows, and each design flow should be selected to fit the particular case. With well-regulated streams where the flow does not vary greatly, much higher design flows may be adopted. Consideration should be given in pollution abatement studies to the feasibility and cost of low-flow augmentation as a means of reducing the cost of waste treatment. There are small streams with mill ponds where the flow may be zero during weekends because of storage behind mill dams. In such cases the right to release water continuously for dilution purposes may have to be purchased from mill owners.

The water temperature to be adopted for the oxygen-balance studies should be the highest temperature likely to occur during the period of low water. The higher the temperature, the greater is the rate of biochemical deoxygenation, and the less is the amount of oxygen available. If the water temperature adopted for the computation of allowable pollution loads is lower than may be expected during critical periods, the pollution abatement objective may not be attained.

Where large quantities of exhaust cooling waters at high temperature are discharged into the receiving waters, the temperature of the receiving waters may be raised well above that which would be expected under natural climatic conditions. This form of pollution, known as *thermal pollution*, is of growing economic importance on industrial streams. As of 1968, about 70 percent of industrial thermal pollution was attributed to the electric power industry. By 1980 perhaps one fifth of the total freshwater runoff in the United States will be used by the power industry for cooling purposes, and the release of heated water to streams may amount to something on the order of 100 trillion gallons per year (2). A few states have set arbitrary limits on maximum temperature in streams (e.g., 93°F) and maximum increase over ambient stream temperature (e.g., 5°F). The long-term effects of heated discharges have yet to be studied adequately; consequently, the criteria for controlling thermal pollution are as yet poorly developed. Thermal pollution is of threefold economic importance: reducing the value of the stream water for cooling purposes; increasing the difficulty and cost of organic pollution abatement; and damaging fish and other aquatic life. Consideration will have to be given in some cases to reduction in thermal pollution as a legitimate part of a pollution abatement project.

The design flow conditions to be used for tidal estuaries, lakes, or oceans should be those which will produce the worst conditions at critical points or in critical areas. Usually the worst conditions are during warm dry weather when the flow in the tributary streams is a minimum. This is not always the case, however. For example, in the case of heavy pollution of a tributary stream some distance upstream from an estuary, during low flows most of the BOD may be exerted before the pollution reaches the tidal flats, whereas during high flows most of the BOD may be discharged to the estuary for satisfaction. The stream flow and tides may bring in sufficient dissolved oxygen for oxidation of the organic matter mixed with the estuary waters, but some of the BOD seeps into the sand flats on ebb tides. If there is too much such BOD for the dissolved oxygen in the pore water, the sulfates in the seawater in the pores will be reduced to sulfides, and obnoxious odors will emanate from the exposed tidal flats.

A special condition exists in sewage-polluted streams that are also heavily laden with acid mine wastes. High acidity associated with such drainage may delay BOD exertion to points far downstream of the organic pollution source, an occurrence typical in many rivers, such as the Monongahela.

11-2. SATURATION CONCENTRATION OF DISSOLVED OXYGEN

The saturation concentration of dissolved oxygen in distilled water at various temperatures in contact with air containing 20.95 percent oxygen by volume on the dry basis and at sea level (760 Torr) is shown in Table 11-1 and Figure 11-1 for both dry and saturated air. The upper values in Table 11-1 and the values in Figure 11-1 were computed from the solubilities in water in contact with 1 atm of pure oxygen gas, as shown in Table 5-2, and were modified by the vapor pressures from Table 2-3 to obtain the solubilities in contact with air saturated with water vapor. The lower values in Table 11-1 are the results of experiments by Churchill, Elmore, and Buckingham (3, 4) of the Tennessee Valley Authority with air of unknown composition adjusted to 760 Torr barometric pressure.

Many experiments have been made to determine the solubility of oxygen in distilled and brackish waters, using both pure oxygen and air of unknown composition. When oxygen was used, the solubility of atmospheric oxygen was computed from the fact that dry air contains oxygen in amounts ranging from 20.9 to 21 percent. Since the determinations with oxygen and air have checked reasonably well, it has been assumed that the values used for the oxygen content of the air were correct. In some cases, air analyses were made to determine the carbon dioxide (usually assumed at 0.03 percent by volume), but determinations of the oxygen content appear not to have been made in most cases. According

Table 11-1
Solubility of Oxygen in Freshwater at 760 Torr

	Temperature (°C)							
	0	5	10	15	20	25	30	40
O$_2$ sat. at 1 atm of oxygen (ppm)[a]	69.8	61.2	54.3	48.7	44.3	40.4	37.2	32.9
Vapor pressure[b] (atm)	0.006	0.009	0.012	0.0168	0.0231	0.0313	0.0419	0.0728
O$_2$ sat. at 1 atm of dry air[c] (ppm)	14.64	12.84	11.40	10.23	9.30	8.47	7.80	6.90
O$_2$ sat. at 1 atm of saturated air (ppm)	14.56	12.72	11.27	10.05	9.08	8.20	7.47	6.40
O$_2$ sat. at 1 atm of air (TVA)[d] (ppm)	14.65	12.79	11.27	10.03	9.02	8.18	7.44	—

[a] From Table 5-2.
[b] From Table 2-3.
[c] Assumed to contain 20.95 percent oxygen by volume.
[d] From experiments with air of unknown composition adjusted to 760 Torr (*3*).

to Schlesinger (*5*), "the composition of air is not constant, even though the variation from time to time and from place to place is not great." Schlesinger states that the average oxygen content of dry air at sea level is 20.99 percent by volume, and that in the upper portions of the atmosphere the lighter gases are present in larger proportion. The oxygen content might, therefore, decrease at higher elevations.

The difference between 20.9 and 21.0 percent in the assumed oxygen content of dry air at sea level represents a difference in its solubility in water of 0.48 percent. This is from 10 to 30 percent of the difference in solubility of oxygen from dry and saturated air, and is far greater than any difference that might be expected from excess concentrations of carbon dioxide. If in some locations, such as high altitudes, dry air contains only 20 percent oxygen, the solubility at 760 Torr will be about 4.5 percent lower than the values in Table 11-1 and Figure 11-1. It is desirable, therefore, to make periodic analyses of dry air to determine the actual average content of oxygen at selected locations where oxygen balance studies are required.

The oxygen solubilities at 760 Torr barometric pressure must be adjusted to the average barometric pressure that prevails at the location where oxygen-

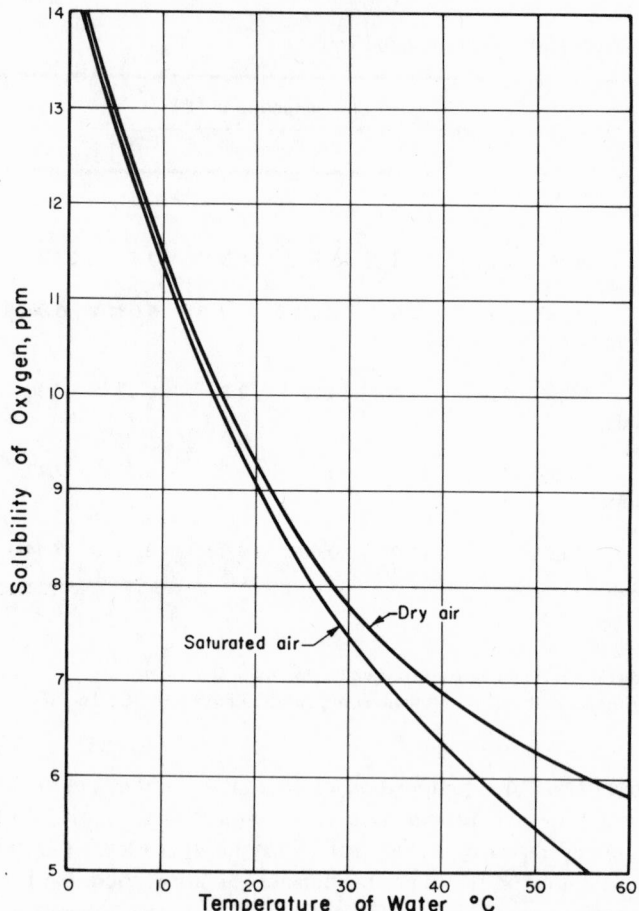

Figure 11-1
Solubility of oxygen in distilled water in contact with air at 760 Torr containing 20.95 percent oxygen.

balance studies are needed. Figure 11-2 shows the average barometric pressure for various elevations above sea level. It will be noted that for an elevation of only 500 ft above sea level, the barometric pressure, and hence the oxygen solubility, is decreased by about 2 percent. Therefore, in oxygen-balance studies made on watercourses more than about 100 ft above sea level, it is desirable to adjust the oxygen solubility for barometric pressure.

The solubility of dissolved oxygen is also less in brackish and seawater than in freshwater, as shown by Figure 5-1, which indicates that as the salinity increases the solubility of oxygen decreases, being only about 82 percent as great for sea-

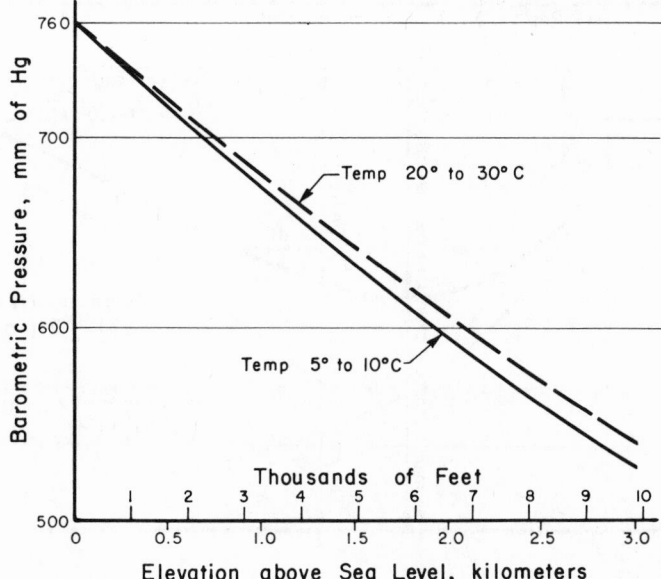

Figure 11-2
Average barometric pressure at various elevations above sea level.

water as for freshwater. In oxygen-balance studies of tidal estuaries, the solubility of oxygen should be adjusted for the average salt content in each reach.

11-3. DISSOLVED OXYGEN SAG CURVE

Oxygen-balance studies of a polluted stream usually result in one or more dissolved oxygen profiles along the course of the stream. Such a profile is referred to as a *dissolved oxygen sag curve* and is illustrated in Figure 11-3. The objective of pollution abatement is to remove sufficient pollution so that the minimum oxygen concentration at the bottom of the sag is not less than some predetermined standard, 4 or 5 ppm for propagation of fish and 1 to 2 ppm to prevent odor nuisance.

Figure 11-3 is a simple oxygen sag curve for a nontidal watercourse with only one pollution load. The initial dissolved oxygen concentration drops suddenly at the point of pollution, if there is an *immediate oxygen demand* associated with the wastes or if the wastes are voluminous as compared to stream flow and contain little dissolved oxygen. The initial BOD load, L_0, is the net demand after mixing with the receiving waters and after satisfaction of the immediate demand. The BOD of the receiving waters just upstream from the point of pollution must also be taken into account in estimating L_0. The difference be-

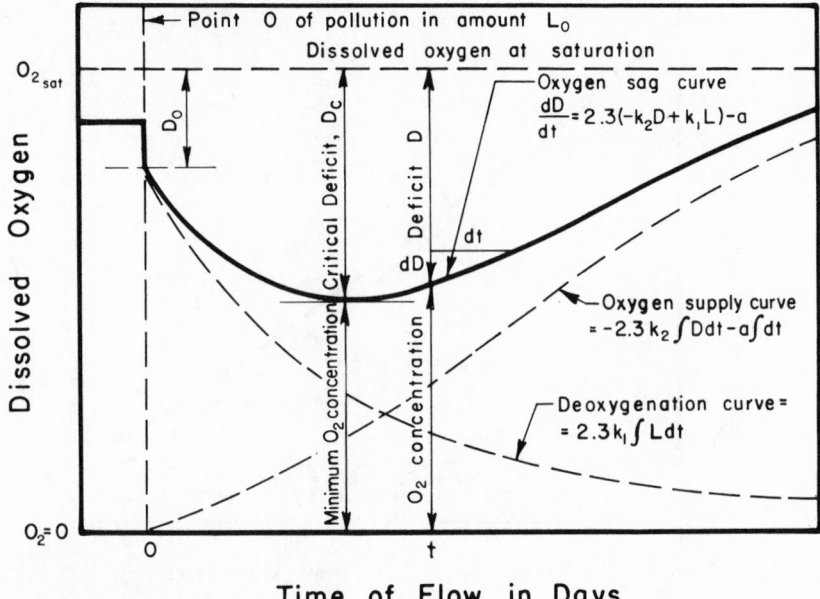

Figure 11-3
Simple oxygen sag curve.

tween the saturation concentration and the actual concentration at any station is known as the *oxygen deficit*, D. At any station, the rate of change of the oxygen deficit with time of flow downstream, which defines the shape of the oxygen sag curve, is as follows:

$$\frac{dD}{dt} = 2.3 \, (-k_2 D + k_1 L) - a \tag{11-1}$$

where D is the oxygen deficit in parts per million, L is the first-stage oxygen demand in parts per million, t is the time of flow in days, k_2 is the atmospheric reaeration constant in days^{-1}, k_1 is the deoxygenation constant in days^{-1}, and a is the oxygen production in the euphotic zone by photosynthesis in ppm/day. Equation (11-1) with a equal to zero was originally formulated by Streeter and Phelps (*6*).

It is evident from Figure 11-3 and Eq. (11-1) that the dissolved oxygen concentration at any station is the algebraic sum of the initial concentration at an upstream station, the oxygen added from the atmosphere and from photosynthesis between stations, and the amount of oxygen used in the bacterial decomposition of the polluting organic matter in the water. The critical deficit

D_c is reached at the bottom of the sag when the slope dD/dt is zero and the rate of supply of oxygen is equal to the rate of deoxygenation. The critical deficit is, therefore,

$$D_c = \frac{2.3\,k_1 L_c - a}{2.3\,k_2} \qquad (11\text{-}2)$$

where L_c is the first-stage demand at the bottom of the sag. Conversely, if k_1, k_2, and a are known, the allowable first-stage demand L_c for any selected critical deficit D_c may be estimated as follows:

$$L_c = \frac{2.3\,k_2 D_c + a}{2.3\,k_1} \qquad (11\text{-}3)$$

It should be noted that Eqs. (11-2) and (11-3) apply to the station at the bottom of the oxygen sag if D_c is less than the saturation concentration. Oxygen-balance computations are usually started from a point just upstream from the first point of pollution, and proceed downstream with trial-and-error solutions for each reach. It is possible, however, to start oxygen balance computations at the bottom of the sag, even though the location of this point is unknown with respect to pollution loads, and to work upstream to determine the allowable pollution loads. Equation (11-3) may be used to compute L_c whether or not there is settling out of BOD, addition of BOD from the bottom deposits, or both; however, these factors must be taken into account to compute upstream values of L and D.

In a grossly polluted stream, the dissolved oxygen concentration may be reduced to zero at the bottom of the sag as shown in Figure 11-4. It should be noted that, even though the concentration is zero in the reach $t_2 - t_1$, oxygen is nevertheless being supplied by reaeration from the atmosphere (and possibly by photosynthesis), and the rate of supply is a maximum, because D is equal to the saturation concentration. Therefore, the rate of deoxygenation in the reach $t_2 - t_1$ is limited by, and equal to, the rate of oxygen supply.

The rate of deoxygenation will decrease from the point of pollution to point t_1, at which point there will be a discontinuity as the rate is controlled by the oxygen available rather than by the BOD. At point t_2, the rate of deoxygenation again becomes proportional to the remaining value of L; and downstream from point t_2, dissolved oxygen is supplied faster than it is utilized by the bacteria. The value of L at point t_2 may be computed from Eq. (11-3). The time interval $t_2 - t_1$ may be computed by noting that the oxygen used in the reach is equal to the amount of dissolved oxygen supplied.

In Eq. (11-1), L is an independent variable and D is a dependent variable whose value depends on L. The value of L at one station may be computed

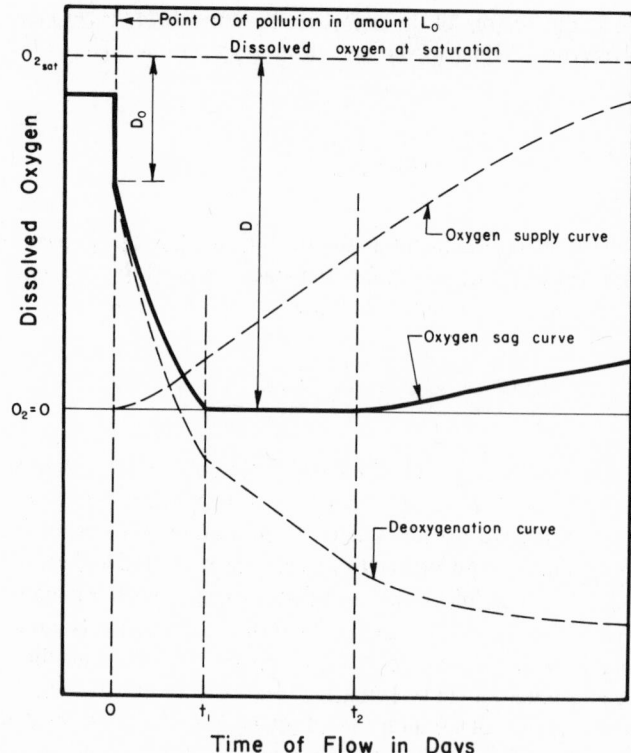

Figure 11-4
Oxygen sag curve with oxygen concentration reduced to zero.

from the value at another station by means of the following:

$$\frac{dL}{dt} = -2.3\,(k_1 + k_3)L + p \qquad (11\text{-}4)$$

where $-2.3k_1L$ is the deoxygenation rate in ppm/day, $-2.3k_3L$ is the rate of settling out of BOD to the bottom deposits in ppm/day, and p is the rate of addition of BOD to the overlying water from the bottom deposits in ppm/day. The L curve represented by Eq. (11-4) is shown in Figure 11-5. The deoxygenation curve is also shown. Note that the two curves are identical only when there is no settling out of BOD and no addition of BOD from the bottom deposits. It is the deoxygenation, and not the reduction in L, that must be added to the oxygen accumulation to compute the oxygen sag curve.

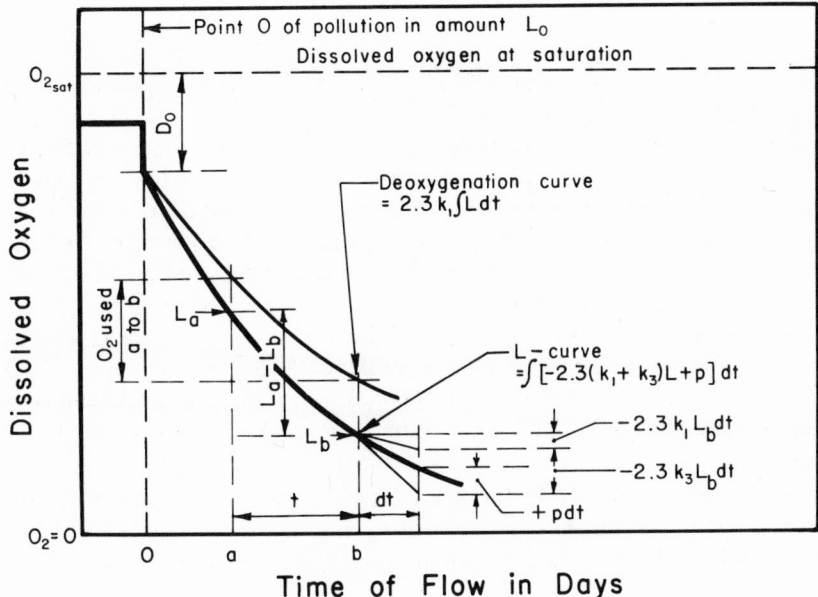

Figure 11-5
Deoxygenation curve and L curve.

The integral of Eq. (11-4) between stations a and b is as follows:

$$L_b = \left(L_a - \frac{p}{2.3\,(k_1 + k_3)}\right)10^{-(k_1+k_3)t} + \frac{p}{2.3\,(k_1 + k_3)} \tag{11-5}$$

If BOD is added to the overlying water from the bottom deposits, but settling out of BOD is negligible, Eq. (11-5) becomes

$$L_b = \left(L_a - \frac{p}{2.3\,k_1}\right)10^{-k_1 t} + \frac{p}{2.3\,k_1} \tag{11-5a}$$

If there is settling out of BOD, but a negligible addition to the overlying water from the bottom deposits, Eq. (11-5) becomes

$$L_b = L_a \cdot 10^{-(k_1+k_3)t} \tag{11-5b}$$

If the BOD settled out and the BOD added from the bottom deposits are both negligible, Eq. (11-5) becomes

$$L_b = L_a \cdot 10^{-k_1 t} \tag{11-5c}$$

Equation (11-5c) is the original form of the deoxygenation equation, as developed by Streeter and Phelps (6).

BOD dark-bottle tests of river-water samples collected at stations a and b and one or more intermediate stations at time intervals about equal to the time of flow, t, should reveal the approximate magnitude of k_3 and p under the conditions existing at the time of sampling. The tests should be made at the temperature of the river water and on samples that are truly representative of the river water at each station. Moreover, the length of the river reach from a to b should be short enough so that k_1 has the same value throughout. The BOD tests on all samples should, therefore, result in approximately the same k_1 value. If this is found to be the case, Eqs. (11-5) to (11-5c) should be used in trial-and-error computations to estimate the approximate values of k_3 and p from the bottle test values of L.

The differential Eq. (11-1) for the oxygen sag may be integrated for the general case, if the value of L in terms of L_a as given by Eq. (11-5) is first substituted in Eq. (11-1). The general form of the oxygen sag equation is as follows:

$$D_b = \frac{k_1}{k_2 - k_1 - k_3} \left[L_a - \frac{p}{2.3(k_1 + k_3)} \right] \left[10^{-(k_1 + k_3)t} - 10^{-k_2 t} \right]$$

$$+ \frac{k_1}{k_2} \left[\frac{p}{2.3(k_1 + k_3)} - \frac{a}{2.3 k_1} \right] (1 - 10^{-k_2 t}) + D_a \cdot 10^{-k_2 t} \tag{11-6}$$

If BOD is added to the overlying water from the bottom deposits, but settling out of BOD is negligible, Eq. (11-6) becomes

$$D_b = \frac{k_1}{k_2 - k_1} \left(L_a - \frac{p}{2.3 k_1} \right) (10^{-k_1 t} - 10^{-k_2 t})$$

$$+ \frac{1}{2.3 k_2} (p - a)(1 - 10^{-k_2 t}) + D_a \cdot 10^{-k_2 t} \tag{11-6a}$$

If there is settling out of BOD, but a negligible addition to the overlying water from the bottom deposits, Eq. (11-6) becomes

$$D_b = \frac{k_1}{k_2 - k_1 - k_3} L_a [10^{-(k_1 + k_3)t} - 10^{-k_2 t}] - \frac{a}{2.3 k_2} (1 - 10^{-k_2 t}) + D_a \cdot 10^{-k_2 t} \tag{11-6b}$$

If there is a negligible production of oxygen by photosynthesis, a is zero in Eqs. (11-6), (11-6a), and (11-6b). If k_3, p, and a are all negligible, Eq. (11-6) becomes

$$D_b = \frac{k_1}{k_2 - k_1} L_a \, (10^{-k_1 t} - 10^{-k_2 t}) + D_a \cdot 10^{-k_2 t} \qquad (11\text{-}6c)$$

Equation (11-6c) is the original integrated form of the oxygen sag curve presented by Streeter and Phelps (6).

If BOD bottle tests are made to evaluate L, k_1, k_3, p, and a, and if dissolved oxygen determinations are made to find the average value of D at each station, the reaeration constant k_2 may then be estimated for each reach by trial-and-error solutions of the oxygen sag equation that applies, Eqs. (11-6) to (11-6c). The light- and dark-bottle oxygen technique, which will be described, may be used to estimate a and p and to check the value of k_2. The value of k_2 will depend upon the water temperature and the flow conditions obtaining at the time of sampling. The reaeration constant k_2 is a function only of the hydraulic parameters of the stream and the liquid film coefficient K_L, and it may be estimated independently as shown hereinafter. The values of k_1, k_2, k_3, p, and a, estimated as shown previously for a particular existing condition in a polluted stream, may be used as guides for the selection of new parameters to apply to the critical design flow in determining allowable pollution loads. The equations may be used again, with the selected parameters, for trial-and-error solutions of the allowable pollution loads for a selected minimum dissolved oxygen content at the bottom of the sag.

11-4. BOTTOM DEPOSITS

Sludge deposits in watercourses polluted with organic wastes may exert heavy oxygen demands on the overlying waters if the conditions favor the accumulation of sludge over relatively long periods of time. This is particularly true in lakes, ponds, reservoirs, estuaries, and sluggish river reaches where scouring velocities are seldom attained even during flood stage. In fact, high river stage may result in greater accumulation of bottom deposits in sluggish waters by scouring the sludge previously deposited during dry weather in upstream reaches and redepositing it in sluggish waters.

In early studies by Streeter (7), it was assumed that the accumulated BOD in sludge deposits is oxidized aerobically at a rate somewhat lower than the rate in the overlying waters. Under these conditions, Streeter argued, the BOD of the sludge would increase until the daily oxygen demand of the sludge equaled the daily rate of deposit of BOD; thereafter, the oxygen demand in the overlying waters would be the same as if no BOD had settled out and no oxygen demand were exerted from the bottom deposits. Streeter estimated that, with a deoxygenation rate constant for the sludge of 0.03 to 0.05, equilibrium would be established in 30 to 40 days.

The steady-state conditions assumed by Streeter do not occur in an actual stream. Most sludge deposits, where they are a problem, are the result of scour of other reaches during heavy runoff and cold weather, whereas the oxygen demand of these deposits upon the overlying waters becomes critical only during low flows and when the water is warm. Moreover, the oxygen demand on the overlying waters of the BOD settled out in 1 day is totally different from the BOD, because the deposits are decomposed anaerobically, and the oxygen demand is exerted by anaerobic decomposition products that diffuse upward through the pore water of the sludge to the overlying water. It is, therefore, essential in oxygen-balance studies to consider the settling out of BOD and the oxygen demand of bottom deposits upon the overlying waters. It is also necessary to evaluate the oxygen produced in photosynthesis by the bottom slimes and attached plants in waters where the light penetrates to the bottom, which will be described.

Experimental studies were made by Fair et al. (8) on the oxygen demand of sludge deposits on the overlying waters for periods up to 450 days. The sludges were placed on the bottoms of carboys through which water containing dissolved oxygen was passed. The detention period of the water ranged from 4 hr to about 2 days, with drops in dissolved oxygen content ranging from 0.5 to 3.5 ppm. It was found that the oxygen demand of the deposits was independent of the concentration of dissolved oxygen in the overlying water with effluent concentrations ranging from 3.6 to 8 ppm. It was concluded that the rate of oxygen demand of the sludge is controlled not by the rate of diffusion of oxygen from the water into the deposit, but by the rate of diffusion and transport of oxidizable substances from the interior of the deposit to the overlying water. The substances that pass upward from the sludge may be expected to exert a chemical oxygen demand as well as a BOD. In the studies to determine the effect of the sludge depth, in which fresh sewage solids were employed, the recorded increase in ammonia and BOD in the effluent water was significantly large for about the first month. In the studies on the effect of temperature, in which a natural river mud was used, no measurable increase in BOD was found after the first day or two, but measurable increases in ammonia, nitrite, and nitrate were observed for about 2 months.

Fick's law of diffusion, Eq. (3-2), may be used to verify the conclusion of Fair et al.: that the oxygen demand of the bottom deposits is controlled by the diffusion upward, through the pores of the sediment, of oxidizable substances rather than the diffusion downward of dissolved oxygen. In their study of the composition of a number of different types of muds and pollutional deposits, Fair and co-workers found that the porosity (the pore water in which diffusion takes place) ranged from 60 to 90 percent. Using a porosity of 90 percent, a diffusion coefficient D_c of 2×10^{-5} cm^2/sec for dissolved oxygen, a concentration of 8 ppm at the interface and zero at 1 cm below the interface, the rate of

Table 11-2
Oxygen Demand of Bottom Deposits

Carboy No.	Mean Depth (cm)	Initial Volatile Solids (kg/m^2)	Initial Areal Oxygen Demand		
			L_{d_0} (g/m^2)	Initial Demand (g/m^2/day)	k_4
I	10.2	3.77	739	4.65	0.0027
II	4.75	1.38	426	3.09	0.0031
III	2.55	0.513	227	1.70	0.0032
IV	1.42	0.188	142	1.08	0.0033
V	1.42	0.188	134	1.02	0.0033

diffusion of oxygen into the sludge is computed as only about 0.12 g/m^2/day. As may be seen from Table 11-2, the measured initial areal demand is at least 8 times this rate, even for the sludge layer of minimum depth. On the other hand, using a value of 10^{-5} cm^2/sec for D_c, the diffusion upward of oxidizable substances at a rate of 1.7 g/m^2/day requires a concentration gradient of about 226 ppm/cm. From Table 11-2, carboy III, for example, L_d (the total areal BOD of the bottom deposits) is 227 g/m^2, which for a depth of 2.55 cm is 8,900 ppm. The required concentration gradient of 226 ppm/cm indicates that only about 7 percent of the available L_d needs to be dissolved in the pore water and subject fo diffusion.

The results of the studies by Fair et al. (8) on the oxygen demand of sludge deposits of various depths at temperatures of 20 to 25°C are presented in part in Table 11-2. The sludge studied was a mixture of sewage solids and inert materials with a 5-day, 20°C BOD of 331 g/kg of volatile solids.

The areal demand for dissolved oxygen of bottom deposits at a point in a stream, in the absence of settling out of BOD, is given by

$$Hp = -\frac{dL_d}{dt} = 2.3\,k_4 L_d \qquad (11\text{-}7)$$

where Hp is the areal demand in g/m^2/day, H is the depth of the stream in meters, p is the rate of addition of BOD to the overlying water in ppm/day, L_d is the total areal BOD of the bottom deposits in g/m^2 at any time t after decomposition has started, and k_4 is the areal demand rate constant.

Integration of Eq. (11-7) yields

$$L_d = L_{d_0} \cdot 10^{-k_4 t} \qquad (11\text{-}8)$$

where L_{d_0} is the initial areal BOD of the bottom deposits at the start of decomposition.

Table 11-2 indicates that the value of k_4 is approximately the same up to depths of about 4 cm, and decreases slightly at greater depths. The value of L_d is about in proportion to the depth of the deposit. It is evident, therefore, that the areal demand Hp increases almost in proportion to the depth of the deposit. A 1-m depth of fresh sludge, such as used by Fair and co-workers, might be expected to have an initial areal demand Hp of 40 to 50 g/m²/day. The corresponding value of p would be 40 to 50 ppm/day for a 1-m water depth and 4 to 5 ppm/day for a 10-m water depth.

Equation (11-8) shows that, as the bottom deposits decompose, L_d is reduced. Consequently, the oxygen demand on the overlying waters is also reduced. Table 11-3 shows the remaining fractions of the initial value of L_d after

Table 11-3
Fractions of Oxygen Demand of Bottom Deposits
Remaining After Various Times

t (days)	Summer $(10^{-0.0032t})$	Winter $(10^{-0.0016t})$
100	0.48	0.69
200	0.23	0.48
500	0.03	0.16
1,000	0.01	0.03

various times, computed by Eq. (11-8), using the value of k_4 from Table 11-2 for summer temperatures and a value half as large for winter temperatures.

11-5. ATMOSPHERIC REAERATION CONSTANT

The atmospheric reaeration constant k_2 may be estimated from the liquid film coefficient K_L as follows:

$$k_2 = \frac{K_L}{2.3H} \tag{11-9}$$

where H is the mean or hydraulic depth of the receiving waters (V/A_0, where V is the volume and A_0 is the quiet surface area). In oxygen balance studies, k_2 is usually expressed in days^{-1}. Therefore, the liquid film coefficient K_L

must be expressed in depth per day: feet per day if H is in feet, or meters per day if H is in meters.

To estimate K_L for the receiving waters from Eq. (3-7), the values of the dissolved oxygen diffusion coefficient D_c and the rate of renewal r of the liquid film must be known approximately (see Table 3-3). Dobbins (9) has found a fairly good correlation between r and the mean velocity gradient G of a flowing stream, as shown in Figure 11-6. The mean velocity gradient may be computed by means of Eq. (3-19) from the viscosity, μ, and the dissipation function W. For a flowing stream, the value of the dissipation function may be computed from the head loss as follows:

$$W = \frac{Q\gamma h_f}{V} = \frac{\gamma h_f}{t} \qquad (11\text{-}10)$$

where W is the work done per second per unit of volume, Q is the discharge per second, γ is the weight of water per unit of volume, and h_f is the head loss in the reach having a time of flow of t seconds and a volume V. In English units,

$$W = \frac{62.5 h_f}{t} \text{ ft-lb/sec/ft}^3 \qquad (11\text{-}10a)$$

The line on Figure 11-6 represents an empirical relation between r and G based primarily on measurements of k_2 by Krenkel and Orlob (10) in a laboratory experimental channel, and by Churchill, Elmore, and Buckingham (11) in rivers of the Tennessee valley. In the Krenkel–Orlob measurements, aeration by photosynthesis was excluded, and in the measurements by Churchill and associates, corrections were made to eliminate both the oxygen produced by photosynthesis and the oxygen used in the respiration of plankton.

Figure 11-6 also shows numerous points computed by Dobbins from experimental k_2 values on the Ohio River (12), in which settling out of BOD, oxygen demand of bottom deposits, and oxygen supplied by photosynthesis were neglected. The values of r computed from the k_2 values for most of these points are considerably greater than shown by the line for the corresponding velocity gradients, and thus indicate a considerable amount of oxygen supplied by photosynthesis. The same is true for the single point taken from the Clarion River data.

The reaeration coefficient k_2 increases with increase in water temperature. This effect appears to result primarily from changes in the coefficient of diffusion D_c with temperature, and is taken into account by using Eq. (3-7) for computing K_L. Equation (10-16), with a value of 1.0159 for θ, was first used by Streeter (13) to express the effect of temperature on the value of k_2. The Committee on Sanitary Engineering Research (14), ASCE, found considerable variation in the evaluation of θ by more recent experimenters, but reported a

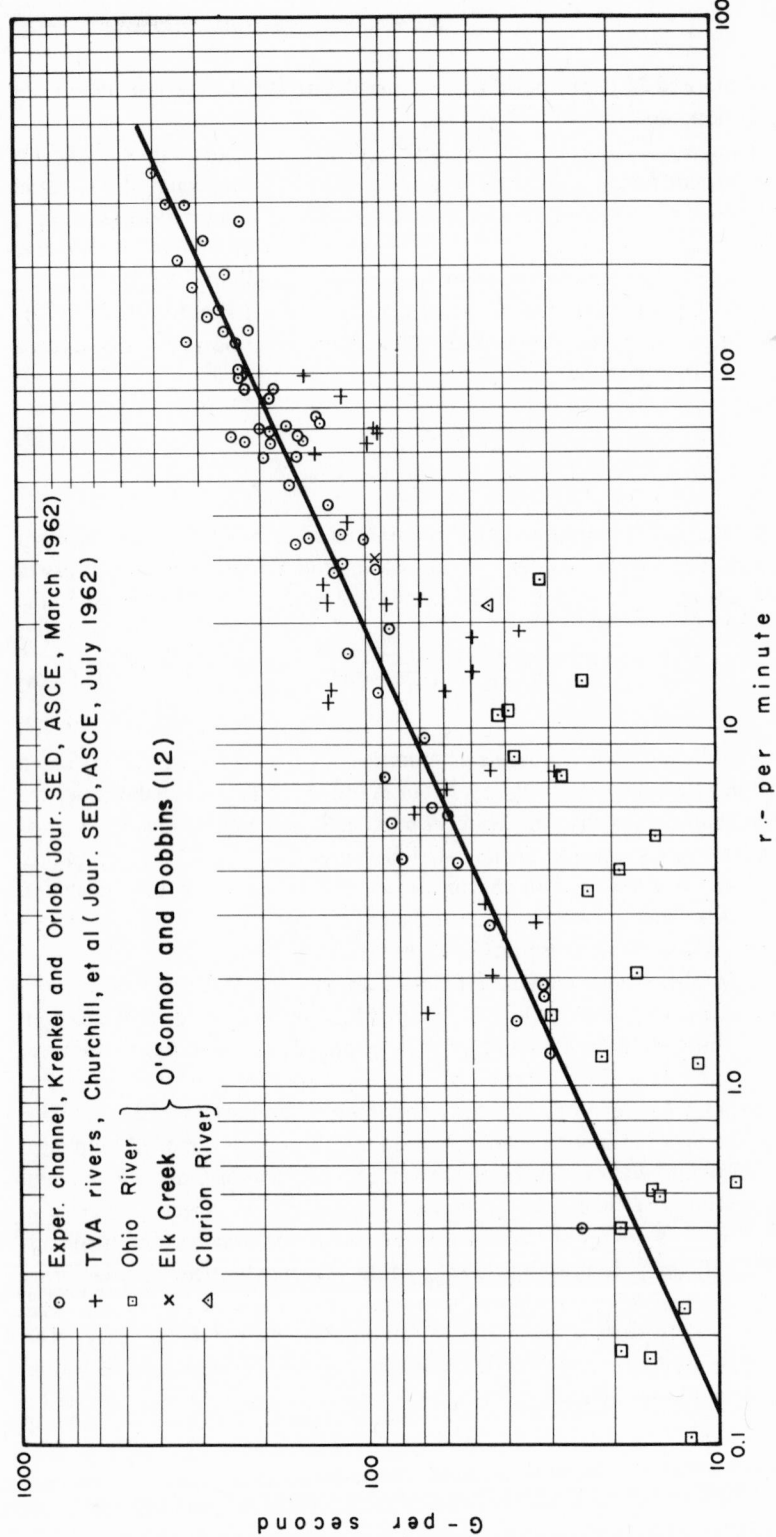

Figure 11-6
Relation between rate of surface renewal and mean velocity gradient.

value of 1.0241 as a result of its own carefully controlled laboratory experiments. This value indicates that the rate of reaeration should increase at the geometric rate of 2.41 percent/°C. A study of Figure 3-1 shows that D_c for dissolved oxygen increases about 4 percent/°C. Since from Eq. (3-7) K_L is proportional to $\sqrt{D_c r}\coth\sqrt{rL^2/D_c}$, k_2 should increase by more than 2 percent/°C, but the rate of increase should be greater for smaller values of r.

The use of Figure 11-6 to estimate the rate of surface renewal, r, from the mean velocity gradient, G, must be limited to streams where the velocity is high enough to stabilize the flow through the drag on the bottom. In sluggish and still waters, including tidal waters, the value of r is controlled by windblown surface currents and waves. The turbulence pattern produced by air currents is unsteady and unpredictable as the wind changes rapidly in direction and velocity, but its magnitude is sufficient even at very low wind velocities to account for a substantial rate of reaeration.

The drag of the wind on a water surface carries the surface water at a velocity ranging from about 1 to 5 percent of the wind velocity, depending on the size of the lake or estuary and the velocity of the wind. Experiments (15) on Lake Erie have shown that the velocity of the surface water was about 5 percent of the wind velocity. Experiments on Owasco Lake, N.Y., by Ackermann (in 15) showed that the ratio was about 3 percent for a wind velocity of about 5 miles/hr and about 1 percent for a wind velocity of 30 miles/hr. Experiments directed by Camp at Massabesic Lake, N.H., confirmed the relative velocities of surface currents and wind indicated above, and showed that wind-induced surface currents were accompanied by currents in the opposite direction, below the surface, and a stagnation zone (no velocity), which was depressed from a few inches below the surface at the start of the wind to several feet below after several hours of wind in the same direction. The velocity gradient at the water surface is, therefore, much higher at the start of a wind than later on.

According to O'Connor and Dobbins (12), the rate of surface renewal r is substantially equal to the velocity gradient at the water surface. With a 5-mile/hr breeze, a surface velocity at 3 percent of the wind velocity, and a stagnation zone at 3 ft below the surface, the renewal rate r is about 4/min. The corresponding value of K_L is 6.2 ft/day (1.9 g/m^2/day) at 20°C. The value of k_2 would be 0.27 for a 10-ft depth and 0.09 for a 30-ft depth. This example indicates that most of the atmospheric reaeration on ponds, lakes, estuaries, and sluggish streams is caused by wind-induced surface currents. It is not practicable to estimate k_2 for such waters from hydraulic and meteorological parameters. The reaeration coefficient k_2 may be estimated for such waters, however, along with k_1, L, and the dissolved oxygen produced in photosynthesis, by means of the light- and dark-bottle oxygen technique at several representative points in the receiving waters, as described next.

11-6. EVALUATION OF PARAMETERS BY OXYGEN ANALYSES

The light- and dark-bottle technique has been used primarily for estimating the production of oxygen in the euphotic zone by photosynthesis. In this method, samples of water are collected at preselected depths at each station. Portions of each sample are placed in bottle pairs, one of each pair in clear glass to be left exposed to natural light and the other covered with an opaque material to exclude light completely. Another portion of each sample is analyzed to determine the initial concentration of dissolved oxygen. Each pair of bottles is immediately suspended at the test depth from which the sample was collected and exposed for any desired period, from 24 hr to several days. At the end of the exposure period, the bottles are retrieved and analyzed for final dissolved oxygen concentration. The loss of oxygen from the dark bottle represents BOD. The change in oxygen content in the light bottle is the net result of both BOD and photosynthesis. The difference between the final dissolved oxygen concentrations represents the oxygen produced by photosynthesis during the exposure period. The value of a at each station and exposure depth is the oxygen produced divided by the exposure period in days.

To determine the average value of a for the entire water depth at the station, it is necessary to use several pairs of bottles at intervals of depth throughout the euphotic zone. At three stations in Baltimore Harbor, Hull (16) found the lower limit of photosynthesis to be from 10.3 to 12.8 ft (3.1 to 3.9 m) below the water surface. Bottle pairs were placed at 1, 5, 10, and 15 ft below the water surface. The hydraulic depth of the Harbor at midtide is about 4.46 m. The average production of oxygen by photosynthesis in Baltimore Harbor during five 4-day periods, August to October 1949, was about 2.7 $g/m^2/day$. The corresponding value of a at midtide is about 0.64 ppm/day. Similar but less precise experiments conducted during 1951 on San Diego Bay and reported by Nusbaum and Miller (17) showed a production of oxygen by photosynthesis ranging from about 0.5 $g/m^2/day$ for a mean depth of 2.1 m to 3.7 $g/m^2/day$ for a mean depth of 9.8 m. The corresponding values of a for San Diego Bay ranged from 0.25 to 0.38 ppm/day.

Hull (18) reports the results of an experimental study with 24-hr exposure periods on a 55-mile reach of the Delaware River Estuary conducted from March to December 1961, which showed that the mean depth of the euphotic zone ranged from about 1.5 to 2.7 m from May to November, and that the mean production of dissolved oxygen by photosynthesis varied from 0.46 $g/m^2/day$ at Torresdale in May to 6.93 $g/m^2/day$ at League Island in September. The corresponding values of the mean 1-day BOD were 0.28 and 1.57 ppm, respectively, and the mean water temperatures were 13.6 and 26°C. The photosynthetic production of oxygen increased with water temperature and BOD. In most of these experiments, bottle pairs were placed at depths of 0.5, 3, 6, and 9 ft below the water surface, since no production of oxygen was found at 10 ft or lower. Table 11-4 shows Hull's estimate of photosynthetic oxygen produc-

Table 11-4
Estimated Values of 1-Day Biochemical Oxygen Demand and a for Delaware River Estuary, Kaighn Point to Reedy Island, August–September 1961

(1)	(2)	(3)	(4)	(5)	(6)	(7)
		Mean Water Temperature	H	Photosynthetic Oxygen		1-day BOD
Reach	Month	($^\circ$C)	(m)	(g/m^2/day)	a (ppm/day)	(ppm)
Kaighn Point– League Island	September	27.0	13.0	5.97	0.46	1.52
League Island– Eddystone	September	25.8	9.8	6.30	0.64	1.90
Eddystone– Cherry Island	August	27.2	9.75	5.37	0.55	1.25
Cherry Island– New Castle	August	26.8	7.26	3.06	0.42	0.70
New Castle– Pea Patch Island	August	26.6	3.32	2.52	0.76	0.62
Pea Patch Island– Reedy Island	August	26.7	11.15	2.46	0.22	0.56

tion and BOD in the 44-mile reach of the Delaware River Estuary from Kaighn Point, Camden, to Reedy Island in August and September 1961, modified herein to show the hydraulic depths, H, at midtide and the values of a.

Under steady-state conditions at a point in the receiving waters, both D and L are constant (preferably averaged over a period of several days), and the rate of supply of oxygen is equal to the rate of deoxygenation as follows:

$$2.3 k_2 D + a = 2.3 k_1 L \qquad (11\text{-}11)$$

The value of k_2 may be estimated from light- and dark-bottle oxygen analyses by means of Eq. (11-11), if k_1, L, a, and D are determined. At least two points and preferably three should be determined on the BOD curve for reliable estimates of k_1 and L, which means that two or three dark bottles should be used at one or more submergence depths. The value of D should be computed as the average from dissolved oxygen determinations of the water near the surface made at frequent intervals throughout a period of several days, and a should be based on an exposure period of several consecutive 24-hr exposure periods.

Table 11-5
Relation of Deoxygenation Rate to 1-Day
Biochemical Oxygen Demand

k_1	$2.3k_1L$
0.13	1.15 × 1-day BOD
0.10	1.12 × 1-day BOD
0.07	1.074 × 1-day BOD
0.04	1.01 × 1-day BOD

The data obtained by Hull for the Delaware River Estuary may be used for approximate estimates of the value of K_L and the reaeration coefficient k_2. The rate of deoxygenation, $2.3k_1L$, may be estimated from the 1-day BOD by assuming reasonable values of k_1. Table 11-5 shows the relation between $2.3k_1L$ and the 1-day BOD computed from the BOD Eqs. (10-14), (10-15a), and (10-15b) for various values of k_1. Column 2 of Table 11-6 shows values of

Table 11-6
Approximate Values of k_2 for Delaware River Estuary, August–September 1961

(1) Reach	(2) k_1 (assumed)	(3) $2.3k_1L$ (ppm/day)	(4) O_2 concen- tration (% sat.)	(5) D (ppm)	(6) K_L (g/m^2/day)	(7) k_2 (days^{-1})
Kaighn Point– League Island	0.10	1.75	15	6.0	2.8	0.003
League Island– Eddystone	0.13	2.18	15	6.0	2.5	0.112
Eddystone– Cherry Island	0.10	1.40	25	5.2[a]	1.6	0.071
Cherry Island– New Castle	0.07	0.75	55	3.2[b]	0.75	0.045
New Castle– Pea Patch Island	0.07	0.67	65	2.5[c]	—	—
Pea Patch Island– Reedy Island	0.04	0.57	75	1.7	2.3	0.09

[a] For K_L of 2.3, D should be about 3.8.
[b] For K_L of 2.3, D should be about 2.0.
[c] Since rate of supply by photosynthesis exceeds rate of deoxygenation, water should be supersaturated with dissolved oxygen.

k_1 assumed by Camp for the reaches of the Delaware River Estuary. Column 3 shows values of $2.3\,k_1 L$ computed from the 1-day BOD values of Table 11-4 and the ratios of Table 11-5. Column 4 shows the average of the dissolved oxygen concentrations at low-water slack and high-water slack on September 15, 1958, as reported by the Delaware State Water Pollution Commission and reproduced by Hull (*18*). Hull presented no data on the dissolved oxygen profile for August and September 1961. Column 5 shows the oxygen deficits, D, based on the figures in column 4 and on 7.0 ppm as the saturation concentration (7.8 ppm in freshwater at 27°C reduced to 7.0 ppm for 9,000 ppm chlorides). There is reason to believe that K_L should have been about the same throughout the estuary during the test period. If K_L was 2.3 from Eddystone to Reedy Island, D should have been about 3.8 for the third reach and 2.0 for the fourth reach. The photosynthetic production of oxygen in the fifth reach was greater than the deoxygenation rate, thus indicating supersaturation. If it is assumed that K_L was 2.3 from Eddystone to Reedy Island, the corresponding values of k_2 were 0.102, 0.138, 0.30, and 0.09 for the last four reaches.

The oxygen demand, p, of bottom deposits may be determined *in situ*, at a station in the polluted watercourse, by passing water which contains dissolved oxygen through a bell jar that has been placed with its bottom rim tightly on the deposit. The water to be passed through the bell jar should have the same temperature and mineral content as the natural waters overlying the deposits, but its BOD and plankton count should be negligible. The water initially trapped under the bell jar should be replaced with the test water in about 1 hour, after which a constant rate of passage of test water should be used so as to produce a decrease in dissolved oxygen content that can be measured with accuracy. The procedure proposed is similar to that used by Fair *et al.* (*8*) except that bell jars are to be used instead of carboys. Each run should last for at least 24 hr and preferably for several days. Continuous analysis of dissolved oxygen is desirable on both influent and effluent for precision in determining the oxygen demand. The oxygen demand Hp in g/m^2/day is the drop in dissolved oxygen in mg/liter, divided by the detention period in days, multiplied by the volume of the bell jar in cubic meters, and divided by the area of the deposit under the jar in square meters. The value of p in ppm/day is the oxygen demand in g/m^2/day divided by the temporal mean water depth H at the station in meters.

If the water depth at the station where p is to be determined is shallow enough for photosynthetic production of oxygen by bottom slimes and attached plants, the light and dark technique should be used with two bell jars, one opaque and the other of clear glass. The opaque jar tests should yield the oxygen demand of the bottom deposits, and the clear glass jar should include the compensating effects of photosynthesis. In oxygen-balance studies of shallow waters, it is desirable to include in the value of p the amount of photosynthetic production of dissolved oxygen from bottom slimes and attached plants that may reasonably be expected during the critical period.

11-7. DISSOLVED OXYGEN PROFILES WITH LONGITUDINAL MIXING

An approximate equation may be developed for the dissolved oxygen profile in tidal estuaries or sluggish streams or ponds where longitudinal mixing occurs as a result of tides or wind-induced currents. The following development is based on the assumption that both the BOD L and the oxygen concentration c do not vary greatly throughout the temporal mean depth H, and may be assumed constant at a station. Figure 11-7 shows the changes in concentration in the longitudinal direction in an element of unit width, depth H, and length dx.

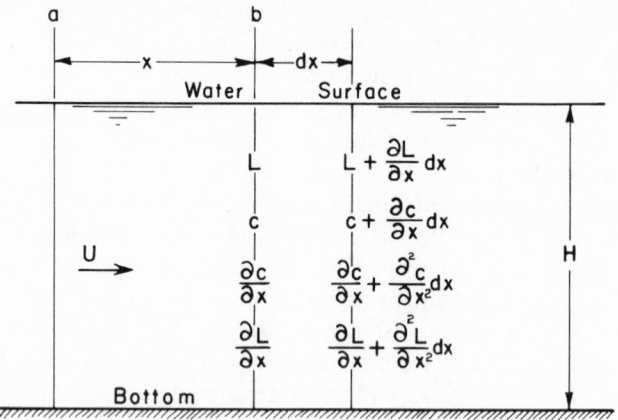

Figure 11-7
Concentration gradients through element.

The differential equation for the oxygen profile for the steady state is

$$\epsilon \frac{d^2 c}{dx^2} - U \frac{dc}{dx} + 2.3 k_2 (c_s - c) + a - 2.3 k_1 L = 0 \qquad (11\text{-}12)$$

where ϵ is the turbulent transport coefficient in the direction of x, U is the temporal mean velocity of the flowing stream, c_s is the saturation concentration of dissolved oxygen, and c is the concentration of dissolved oxygen at any station x distance from station a. Since L also changes with x, its value in terms of x must be substituted in Eq. (11-12) before it can be integrated.

The BOD profile in the direction of x for the steady state is given by the following differential equation:

$$\epsilon \frac{d^2 L}{dx^2} - U \frac{dL}{dx} - 2.3 (k_1 + k_3) L + p = 0 \qquad (11\text{-}13)$$

The integral of Eq. (11-13) between a and b for the boundary conditions, $L = L_a$ at $x = 0$ and L is a constant for $x = \infty$, is

$$L_b = \left[L_a - \frac{p}{2.3\,(k_1 + k_3)} \right] 10^{j_1 x} + \frac{p}{2.3\,(k_1 + k_3)} \tag{11-14}$$

in which

$$j_1 = 0.434 \left[\frac{U}{2\epsilon} - \sqrt{\frac{U^2}{4\epsilon^2} + \frac{2.3\,(k_1 + k_3)}{\epsilon}} \right] \tag{11-15}$$

The integration of Eq. (11-12) between a and b, with the value of L taken from Eq. (11-14) and with the boundary conditions such that $c = c_a$ at $x = 0$ and $c = c_s$ at $x = \infty$, leads to the following:

$$c_b = c_a \cdot 10^{j_2 x} + \left[c_s + \frac{a}{2.3\,k_2} - \frac{k_1 p}{2.3\,k_2(k_1 + k_3)} \right] (1 - 10^{j_2 x})$$

$$- \frac{k_1}{k_2 - k_1 - k_3} \left[L_a - \frac{p}{2.3\,(k_1 + k_3)} \right] (10^{j_1 x} - 10^{j_2 x}) \tag{11-16}$$

or

$$D_b = \frac{k_1}{k_2 - k_1 - k_3} \left[L_a - \frac{p}{2.3\,(k_1 + k_3)} \right] (10^{j_1 x} - 10^{j_2 x})$$

$$+ \frac{k_1}{k_2} \left[\frac{p}{2.3\,(k_1 + k_3)} - \frac{a}{2.3\,k_1} \right] (1 - 10^{j_2 x}) + D_a \cdot 10^{j_2 x} \tag{11-16a}$$

in which

$$j_2 = 0.434 \left[\left(\frac{U}{2\epsilon} - \sqrt{\frac{U^2}{4\epsilon^2} + \frac{2.3\,k_2}{\epsilon}} \right) \right] \tag{11-17}$$

The preceding theoretical development, Eq. (11-12) to (11-17), is similar to that of O'Connor (*19*), except that O'Connor assumes that a, k_3, and p are all zero. If U is insignificantly small compared to ϵ, it may be assumed to be zero, and

$$j_1 = -0.434 \sqrt{\frac{2.3\,(k_1 + k_3)}{\epsilon}} \tag{11-15a}$$

and

$$j_2 = -0.434 \sqrt{\frac{2.3 \, k_2}{\epsilon}}$$ (11-17a)

If ϵ is small compared to U so that it may be taken at zero, U/ϵ in Eqs. (11-15) and (11-17) becomes ∞, and j_1 and j_2 are indeterminate. In this case, Eqs. (11-12) and (11-13) may be integrated with the term including ϵ omitted. The resulting equations for the BOD and oxygen profiles will be identical with Eqs. (11-5) and (11-6), where t is x/U.

In applying these equations to oxygen-balance studies of a tidal estuary, water samples should be collected at the same slack period, either high water or low water. The parameters p, a, and $2.3 \, k_2$, which are determined initially in $g/m^2/day$, should be estimated by using the mean hydraulic depth at midtide. The design oxygen balance should be estimated at the slack period when the oxygen concentration is least, usually low water.

In freshwater, j_1 may be evaluated by means of Eq. (11-14) between two stations, x distance apart, after L_a, L_b, p, k_1, and k_3 have been estimated from dissolved oxygen determinations, provided L_a and L_b have significantly different values. If U is significantly large, ϵ may be estimated from j_1 by trial-and-error solutions of Eq. (11-15). If U is negligible in magnitude, ϵ may be estimated from j_1 by trial-and-error solutions of Eq. (11-15a). If L_a and L_b are approximately equal, j_1 cannot be evaluated by means of Eq. (11-14). This may be the case where p is large compared to L, and it is usually the case in tidal estuaries where ϵ is large.

If Eq. (11-14) cannot be used to evaluate j_1 and ϵ, a tracer (20) may be used. A known mass of the tracer should be released instantaneously (or as quickly as practicable) in uniform concentration over the cross section, at a selected station of the watercourse or a hydraulic model thereof. The time rate of change in concentration, c, at any station x distance from the point of release is given by the following differential equation:

$$\frac{dc}{dt} = \epsilon \frac{d^2c}{dx^2} - U \frac{dc}{dx} - 2.3 \, k_5 c$$ (11-18)

where $2.3 \, k_5 c$ is the rate of decay of the tracer. The solution (21) of Eq. (11-18) for the instantaneous release of a mass M, with the boundary condition that c is 0 when t is ∞, is

$$c = \frac{M}{2A\sqrt{\pi \epsilon t}} \, 10 \left\{ \exp \left[-0.434 \frac{(x - Ut)^2}{4 \epsilon t} - k_5 t \right] \right\}$$ (11-19)

where A is the cross-sectional area of the channel, assumed to be constant throughout the distance x. The rate of decay of the tracer must be determined prior to the use of Eq. (11-19) for the evaluation of ϵ. After various times t following the release of the tracer, the average concentration should be measured at each station distance x from the point of release. Trial-and-error solutions of Eq. (11-19) for each station should yield an approximate value of ϵ for the test conditions. If the velocity is negligible for the conditions of the test, U is 0 in Eq. (11-19).

The turbulent transport coefficient, ϵ, for an estuary may be evaluated for a particular stream flow from the steady-state concentration profile of salinity or chlorides, provided the velocity U is not negligible and provided vertical stratification of chlorides is not pronounced. The differential equation for the salinity or chloride concentration s, with x measured in the downstream direction, is

$$\epsilon \frac{d^2s}{dx^2} - U \frac{ds}{dx} = 0 \tag{11-20}$$

The solution of this equation is

$$s = s_0 \cdot 10^{-0.434\,(U/\epsilon)x} \tag{11-21}$$

where s_0 is the salinity or chloride concentration at or near the downstream end and x is measured upstream.

In most estuaries where the freshwater flow is large, there is a tendency for the freshwater to stratify and flow over the top of the heavier salt water and for the heavier salt water to move shoreward on the incoming tide as a wedge under the lighter freshwater. Since seawater is about 2.5 percent heavier than fresh water, these density currents are very pronounced, and violent vertical mixing is required to minimize vertical stratification of chlorides. Equation (11-20) does not account for stratification of salinity. For this reason the horizontal salinity gradient is of limited usefulness in computing the value of ϵ.

It does not follow that, because the salinity is stratified vertically, the same degree of stratification is to be expected of dissolved oxygen or BOD. These substances occur in such low concentrations that the density of the solution is unaffected. Hence, they are diffused vertically with a much lower degree of turbulent mixing than is required for salt water. Any conservative property of a wastewater, such as a stable dissolved chemical ion, that is present in the estuary in measurable concentrations may be used to evaluate ϵ by means of Eq. (11-21).

Values of ϵ have been estimated by O'Connor (*19*) for the Delaware River Estuary at 6.7 miles²/day on August 23, 1956, with U at 0.58 mile/day; at 10.8 miles²/day on November 13, 1956, with U at 1.27; for the James River in 1951, at about 10 with U at about 1.2. O'Connor (*21*) has estimated ϵ for the

Hudson River at New York and the Upper Bay at about 30 miles^2/day with U about 0.73 mile/day in the river and negligible in the Upper Bay. Estimates by Selleck and Pearson (*20*) of the value of ϵ for the Dumbarton Bridge–Coyote Creek region of San Francisco Bay in the summer of 1959 ranged from about 1.8 to 3.6 miles^2/day.

It should be noted that where longitudinal mixing is substantial, as in a tidal estuary, the mean or steady-state value of L in the receiving waters at a point of major pollution cannot be computed directly from the river and wastewater flows and pollution loads, as in the case of a river without substantial longitudinal mixing. The same is true of the dissolved oxygen concentration at the point of pollution. The preceding equations are not, therefore, sufficient means for estimating the allowable pollution loads at specific points of pollution. They may be used for studying existing BOD and dissolved oxygen profiles for polluted waters and for computing the changes to be expected in these profiles by changes in the values of parameters. Hypothetical dissolved oxygen profiles may then be computed on the basis of reduced values of L until a design profile is derived that will meet the minimum dissolved oxygen requirements for the receiving waters. The sum of the allowable pollution loads from all sources of pollution is approximately equal to the ratio of the design value of L to the existing value at any station times the sum of the existing pollution loads.

A materials balance of BOD and dissolved oxygen for the entire body of water that is subjected to longitudinal mixing will also be helpful in estimating the allowable pollution loads. The total daily discharge of BOD into the body of water from all sources minus that which is settled out and oxidized equals the total daily BOD discharged from the body of water. The total daily supply of dissolved oxygen to the body of water from all sources minus that which is used for oxidation of BOD equals the total daily discharge of dissolved oxygen from the body of water.

11-8. AERATION AT WEIRS

An extensive series of experiments (*22, 23*) have been conducted on English rivers by the Water Pollution Research Laboratory, Stevenage, Herts., to determine the effect of fall over weirs on the oxygen concentration of water. The weir systems studied were classified as follows:

Group I: Free weirs where the water has an unbroken fall from the sill to a pool below, or onto the bed of the stream or a masonry apron.

Group II: Step weirs at which the water is broken up before it reaches the downstream level.

Group III: Weirs preceded or followed by a short slope whose fall has been included in the total fall.

Group IV: Mills where the falling water is used as a source of power.

Group V: Sloping passages or turbulent runs with a gradient of about 1 in 10.

Group VI: Sluices where the water is delivered from an opening whose upper edge is below the upstream water level.

These experiments were conducted at rates of discharge ranging from about 1 to 500 mgd (imperial gallons), at falls up to about 10 ft and at temperatures (*24*) from 0 to 40°C. The results were expressed in terms of the ratio of the oxygen deficit by means of the following empirical equation (*23*):

$$\frac{D_b}{D_a} = 1 + 0.11\,ab\,(1 + 0.046\,T)h \tag{11-22}$$

where D_a and D_b are the oxygen deficits in parts per million above and below the weir, T is the water temperature in degrees Celsius, h is the height of the fall in feet, and a and b are parameters that depend on the quality of the water and the form of the weir, respectively. The value of a is about 1.25 in clean river water, about 1.0 in polluted river water, and about 0.8 in sewage effluents. The value of b is about 1.0 for a free weir and averages about 1.3 for the step weirs examined in the original work.

Many of the experiments (*22*) were conducted with the water supersaturated with oxygen by photosynthesis, in which case the fall resulted in a loss of oxygen to the atmosphere. The oxygen transfer was found to be the same whether absorbed from, or lost to, the atmosphere. Equation (11-22) holds for both cases, except that the deficits have negative values where the water is supersaturated.

Equation (11-22) takes no account of the depth of flow over the weir or of the rate of discharge. These parameters did not appear to affect the results. The oxygen transfer occurring on slopes was found to be less than 20 percent, on the average, of the transfer with weirs of the same fall. Some of the experiments (*22*) showed that nearly all the oxygen transfer takes place in the turbulent splash or jump below the fall and that very little occurs during the fall. The oxygen transfer through water wheels at mills was found to be about half that of a free weir at the same fall. The effect of the depth of the pool below the weirs was not evaluated.

It is evident from the English studies that most of the oxygen transfer accompanies the turbulent dissipation of kinetic energy at the bottom of the fall. For efficient transfer the high velocity must be dissipated at the water surface, usually as a hydraulic jump. This is the principle of the highly efficient turbine aerators now being used in the activated sludge process of wastewater treatment. A deeply submerged jet will not produce enough turbulence at the water surface for efficient aeration.

REFERENCES

1. Li, Wen-Hsiung, "Unsteady Dissolved Oxygen Sag in a Stream," *J. San. Eng. Div., Amer. Soc. Civil Engrs., 88,* 75 (May 1962).

2. *Chemical Engineering,* Mar. 25, 1968.

3. M. A. Churchill, H. L. Elmore, and R. A. Buckingham, "The Prediction of Stream Reaeration Rates," Chattanooga, Tenn., Tennessee Valley Authority, July 1961.

4. See also "Solubility of Atmospheric Oxygen in Water," *J. San. Eng. Div., Amer. Soc. Civil Engrs., 86,* 41 (July 1960).

5. H. I. Schlesinger, *General Chemistry,* Essex, England, Longman Group Ltd., 1938.

6. H. W. Streeter and E. B. Phelps, *Public Health Service Bull. 146,* Washington, D.C., U.S. Public Health Service, 1925.

7. H. W. Streeter, *Modern Sewage Disposal,* Federation of Sewage Works Associations, p. 198, 1938.

8. G. M. Fair, E. W. Moore, and H. W. Thomas, Jr., "The Natural Purification of River Muds and Pollutional Sediments," *Sewage Works J., 13,* 270 (1941).

9. W. E. Dobbins, private communication, Sept. 1962.

10. P. A. Krenkel and G. T. Orlob, "Turbulent Diffusion and the Reaeration Coefficient," *J. San. Eng. Div., Amer. Soc. Civil Engrs.,* 53 (Mar. 1962).

11. M. A. Churchill, H. L. Elmore, and R. A. Buckingham, "The Prediction of Stream Reaeration Rates," *J. San. Eng. Div., Amer. Soc. Civil Engrs.,* 1 (July 1962).

12. D. J. O'Connor and W. E. Dobbins, "The Mechanism of Reaeration in Natural Streams," *Trans. Amer. Soc. Civil Engrs., 123,* 641 (1958).

13. H. W. Streeter, *Public Health Rept. 41,* 247 (1926).

14. Committee on Sanitary Engineering Research, "Effect of Water Temperature on Stream Reaeration," *J. San. Eng. Div., Amer. Soc. Civil Engrs.,* 59 (1961).

15. G. C. Whipple, G. M. Fair, and M. C. Whipple, *Microscopy of Drinking Water,* New York, John Wiley & Sons, Inc., 1927, p. 158.

16. C. H. J. Hull, "Algae and Organic Waste Assimilation in Tidal Estuaries," *Proc. Maryland–Delaware Water and Sewage Assoc.,* 1961, p. 37.

17. I. Nusbaum and H. E. Miller, "The Oxygen Resources of San Diego Bay," *Sewage Ind. Wastes, 24,* 1512 (1952).

18. C. H. J. Hull, "Photosynthetic Oxygenation of a Polluted Estuary," Rep. XIII, Low-Flow Augmentation Project, The Johns Hopkins University, Jan. 1962.

19. D. J. O'Connor, "Oxygen Balance of an Estuary," *J. San. Eng. Div., Amer. Soc. Civil Engrs.,* 35 (May 1960).

20. See R. E. Selleck and E. A. Pearson, "Tracer Studies and Pollution Analyses of Estuaries," Publ. 23, State Water Pollution Control Board, Sacramento, Calif., 1961.

21. D. J. O'Connor, "Organic Pollution of New York Harbor—Theoretical Considerations," *J. Water Pollution Control Federation,* p. 905 (Sept. 1962).

22. A. L. H. Gameson, "Weirs and the Aeration of Rivers," *J. Inst. Water Engrs. 11,* 477 (Oct. 1957).

23. M. J. Barrett, A. L. H. Gameson, and C. G. Ogden, "Aeration Studies at Four Weir Systems," *Water and Water Eng., 64,* 407 (1960).

24. A. L. H. Gameson, K. G. Vandyke, and C. G. Ogden, "The Effect of Temperature on Aeration at Weirs," *Water and Water Eng., 62,* 489 (1958).

CHAPTER 12

Bacterial Decline Rates
in Polluted Waters

12-1. COLIFORMS AND ENTERIC PATHOGENS

As indicated in Section 6-1, the use of the coliform count as a standard of safety
of waters to be used for drinking, bathing, and recreation is based on the assumption that the numbers of infectious organisms of all types, if present at the source
of pollution, have been reduced at least in proportion to the reduction in the
count of coliform bacteria. Kehr and Butterfield, (1) in a survey of the literature
(1943), show that this is essentially true for the enteric pathogens, *Salmonella
typhosa* and *S. schottmuelleri*. The rate of decrease in concentration of these
organisms appears to be about the same as for the coliforms, whether the decrease is brought about by treatment of the sewage or water, including chlorination, or by natural decline in the watercourse. Thus the ratio of the pathogen
count to the coliform count, where these pathogens are known to be present in
fresh sewage, should remain about the same even after both counts are greatly
reduced. This may be the case with many other pathogenic bacteria, but the
supporting evidence is not available.

As shown in Chapter 6, the pathogenic viruses that have so far been studied
have about the same relative survival as coliforms in freshwater, except that
higher chlorine doses and longer contact times are required for the viruses. In
salt water, however, the coliforms die off much more rapidly than the viruses. It
has also been indicated in Chapter 6 that the cysts of the amoebic dysentery
protozoan and some nematodes require heavy chlorine doses. Doubtless there

337

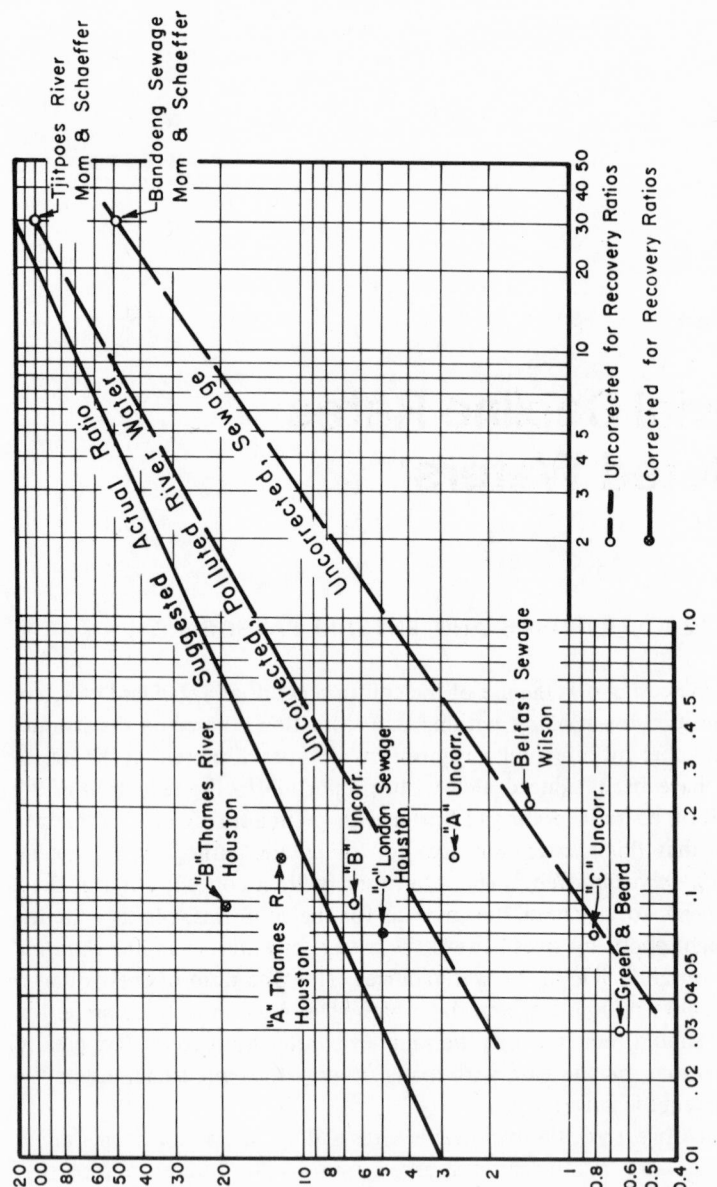

Figure 12-1

Eberthella typhosa per million coliforms for varying typhoid fever morbidity rates. [After Kehr and Butterfield (1).]

are many other exceptions to the general rule that will come to light with further research work. The coliform count is nevertheless the best index to the safety of a water and to the effectiveness of treatment and dispersion of sewage into receiving waters.

Figure 12-1 shows the relation between the count of *E. typhosa* per million coliforms in sewage and polluted river water and typhoid fever morbidity. The data used were sparse and somewhat crude. The upper line is the suggested ratio as indicated by the available data for the absolute numbers of *E. typhosa* in the absence of epidemic conditions; the two lower lines indicate the ratios that might be expected, uncorrected for recovery losses, in polluted river waters and sewage. The ratios in Figure 12-1 are the approximate minimum ratios to be expected, in general, in sewage and waters polluted by sewage from fairly large populations. In the absence of cases of typhoid fever and of typhoid carriers, there would presumably be no *E. typhosa* in the sewage, a condition that may be expected in small towns. Similarly, much higher ratios may be expected in some small towns where a single carrier can excrete proportionately much larger numbers of *E. typhosa*.

The rate of decrease in count of bacteria in natural waters is conveniently expressed as follows:

$$\frac{N_t}{N_0} = 10^{-k_b t} \tag{12-1}$$

where N_t is the count after time t days, N_0 is the initial count, and k_b is the bacterial decrease rate constant, the value of which depends largely on the temperature of the water. Figure 12-2 shows the effect of water temperature on the values of k_b for the decrease of coliforms and *E. typhosa* in polluted rivers, as estimated by Kehr and Butterfield (1). The data used by Kehr and Butterfield for compiling Figure 12-2 included some data from stored samples, but allowances were made for the differences in rates of decline in stored samples and natural waters.

It has become customary to express the decline rate in terms of the time required for a 90 percent reduction in count. This time, t_{90}, may be computed from Eq. (12-1) as follows:

$$t_{90} = \frac{1}{k_b} \tag{12-2}$$

It may be noted from Figure 12-2 and Eq. (12-2) that t_{90} is about 2.0, 1.0, and 0.5 days, respectively, for water at 5, 15, and 25°C.

Much of the experimental data indicates an initial rise in bacterial count, followed by a decline whose rate progressively decreases. Equations (12-1) and (12-2) apply only to the initial decline phase after the peak count has been

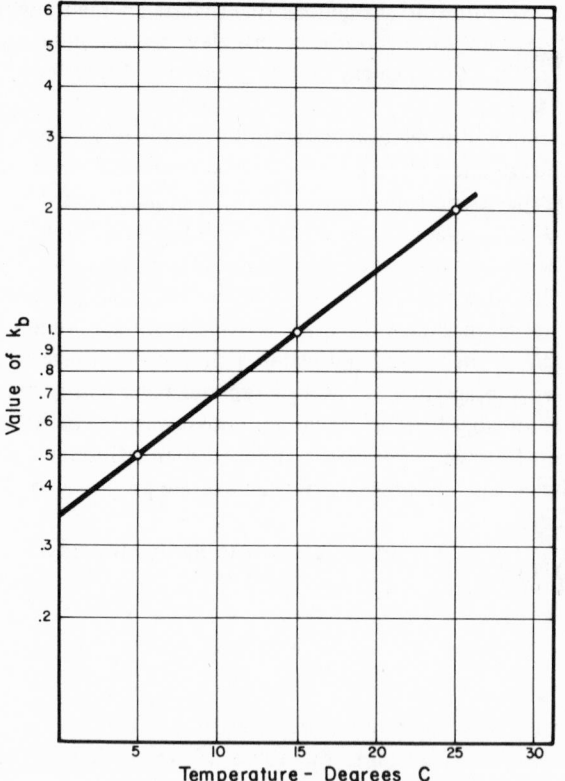

Figure 12-2
Effect of water temperature on values of k_b for death rate of coliforms and *E. typhosa* in polluted rivers. [After Kehr and Butterfield (*1*).]

reached. The bacterial changes with time have been studied extensively by Streeter (*2*) for freshwater and by Pearson (*3*) for salt water.

12-2. BACTERIAL CHANGES IN POLLUTED RIVERS

The studies by Streeter (*2*) were based on the intensive analysis, during 1914–1916, of the changes in bacterial content associated with the natural purification (*4*) of the 123-mile reach of the Ohio River from a point immediately below Cincinnati to one immediately above Louisville, and, also, on the observations by Butterfield (*5*) on changes in bacterial content of samples of Ohio River water collected during 1932 and stored in the laboratory under various conditions. The bacterial content of the stored samples showed in every case an initial rise much greater than that in the river; and though this increase was followed in both instances by an orderly decrease, the decrease was decidedly more rapid in the river under summer conditions than it was in the stored samples collected during the same period. During the winter period, however, the general trend of both the river and the stored sample curves throughout the decline phase appeared to be similar.

Figure 12-3 shows the progressive changes in bacterial count in samples of Ohio River (2) water collected in 1932 at station 475, immediately downstream from Cincinnati, and at station 598, immediately upstream from Louisville, and stored in the laboratory at temperatures of 10, 20, and 37°C. The stored samples were analyzed by means of gelatin plate counts incubated for 48 hr at 20°C. Also shown in Figure 12-3 is the average ratio of the *B. coli* index in the Ohio River from samples collected at various times of flow below the Cincinnati sewer outlets during April to November 1914–1916. The average ratios of the 48-hr gelatin count at 20°C and the 24-hr agar count at 37°C in the samples collected from the Ohio River during 1914–1916 were practically the same as for the *B. coli* index.

It will be noted from Figure 12-3 that the initial increase in bacterial numbers was much greater in the stored samples than in the river, and that the initial increase ratio in the stored samples of the relatively pure water from station 598 was much greater than the initial increase ratio in the stored samples of polluted water from station 475. Butterfield's studies (5) tend to show that, when any change is made in the environment of a portion of water which materially alters conditions, this disturbance of balance permits a temporary increase in the numbers of bacteria. The more extensive this interference with normal conditions, the greater the increase in bacterial numbers. Increases in numbers of bacteria have also been noted in polluted streams downstream from the confluence with a large branch. One of the most important conclusions to be drawn from Figure 12-3 is that the bacterial changes which take place in stored samples are not representative of the changes which take place in the river itself.

The value of t_{90} for the coliform decline in the Ohio River, from Figure 12-3, is about 3 days. This corresponds with a k_b value of about 0.33 at an average water temperature of 19.5°C. This is a very much slower rate than is indicated by Figure 12-2.

Streeter's studies show that the increased rate of decline in bacterial numbers in the Ohio River, during the summer periods, was accompanied by a substantial reduction in turbidity. He concluded that the reduction in turbidity was caused by sedimentation, and that a corresponding reduction in bacterial count might also be caused by settling. No reduction in turbidity was noted during high river stages in the winter, and the decline in bacterial count was roughly the same in the river and in stored samples of river water at low temperature.

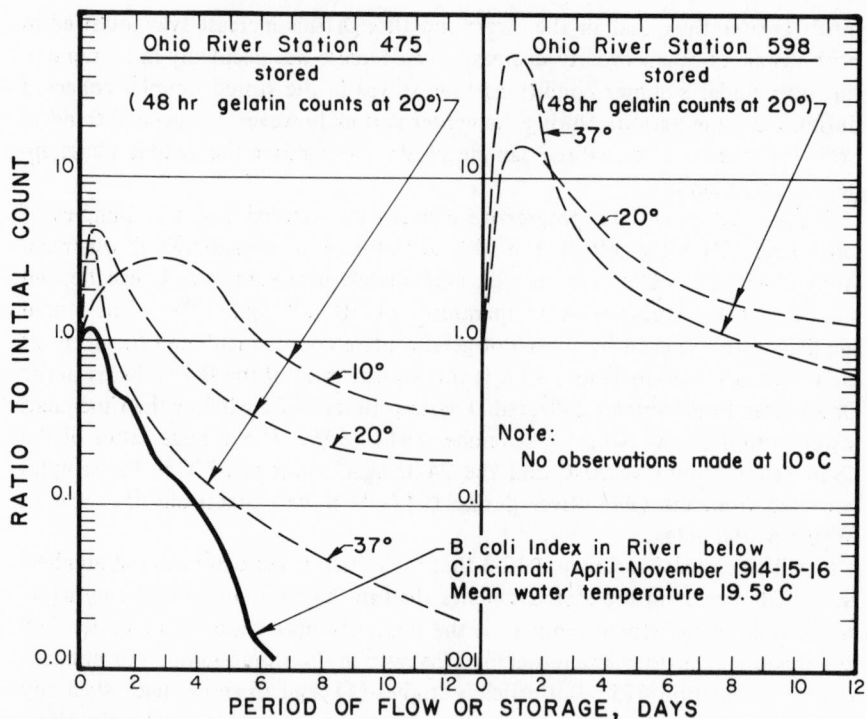

Figure 12-3

Progressive changes in bacterial content observed in Ohio River water. [After Streeter (2).]

12-3. BACTERIAL CHANGES IN SEAWATER

Pearson (*3*) has surveyed the literature on the viability of bacteria in seawater and has presented the results of experiments conducted from 1885 to 1954, inclusive. The results of some of these experiments are plotted in Figure 12-4. It is to be noted that all these experiments were conducted on samples stored in the laboratory, except those experiments by Beard and Meadowcroft on San Francisco Bay water, those by Zo Bell on Pacific Ocean water, some of those by Williams on Puget Sound water, and some of those by Nusbaum and Garver on San Diego Bay water. Beard and Meadowcroft (1935) placed their samples in cylindrical flasks suspended in natural bay water. The ends of the flasks were covered by semipermeable membranes prepared by soaking filter papers in a solution of pyroxylin in glacial acetic acid. Zo Bell (1936) used samples in semipermeable tubes suspended in the ocean. Williams (1950) enclosed sewage-seawater mixtures in cellophane dialysis tubing and suspended these samples in natural seawater. Nusbaum and Garver (1954) placed samples in dialysis tubing suspended in San Diego Bay.

These experiments with samples suspended in seawater were not representative of what takes place in seawater except for the temperature. The effects of mixing, dilution, light, and sedimentation were excluded. All the survival data for bacteria in seawater, as shown in Figure 12-4, are representative of results from stored samples. A comparison of the rates of decline in stored seawater samples, as shown in Figure 12-4, with the rates of decline in stored freshwater samples, as shown in Figure 12-3, indicates that the mortality rate is very much higher in seawater than in freshwater (6 to 8 times).

The determination of the rate of decline of bacterial concentrations in seawater from samples collected after various times of flow is accompanied by much larger sources of error than similar determinations in freshwater streams, because of the difficulty of estimating times of flow and because of the much greater effect of sedimentation in seawater. When sewage is discharged into seawater, it may rise to the surface, because it is lighter than seawater, and spread out in a *sewage field* or *slick*. The rising column and spreading field of sewage mixes with, and is diluted by, seawater so that at the edge of the field the dilution is so great that the mixture is indistinguishable to the eye from the seawater beyond. Where deep ocean outfalls are used (e.g., Hyperion), the discharged effluent rises to an isotherm and spreads out below the surface.

The usual method of evaluating the bacterial pollution of seawater that receives sewage is to make areal surveys at sampling stations located on a grid pattern or along radial lines. The surface current velocities must be measured or estimated to evaluate times of flow between sampling stations. Tracers are helpful in estimating times of flow. As the field of mixed sewage and storm water flows over the seawater, suspended matter heavier than seawater will settle out of the field into the seawater and on to the bottom, carrying bacteria with it. A survey (6) of contamination in the vicinity of the North Trunk sewer outfall, Seattle, Washington, in 1949, showed that, although the coliform MPN per 100 ml in samples collected 1.5 ft below the surface was nearly always higher than the MPN in samples collected at depths of 20 ft or greater (up to 100 ft), the total number of coliforms below the sewage field was about the same as, or greater than, the number in the sewage field. Sedimentation is thus shown to be a significant cause of the decline in numbers of bacteria in sewage discharged to seawater.

A survey of the literature on the marine disposal of wastes was presented in a progress report (7) of the Committee on Sewerage and Sewage Treatment, ASCE, 1961. The decline rate of coliforms at three southern California submarine outfalls was investigated in terms of the time required for a 90 percent reduction, t_{90}, by dilution, mortality, sedimentation, and by all three effects combined.

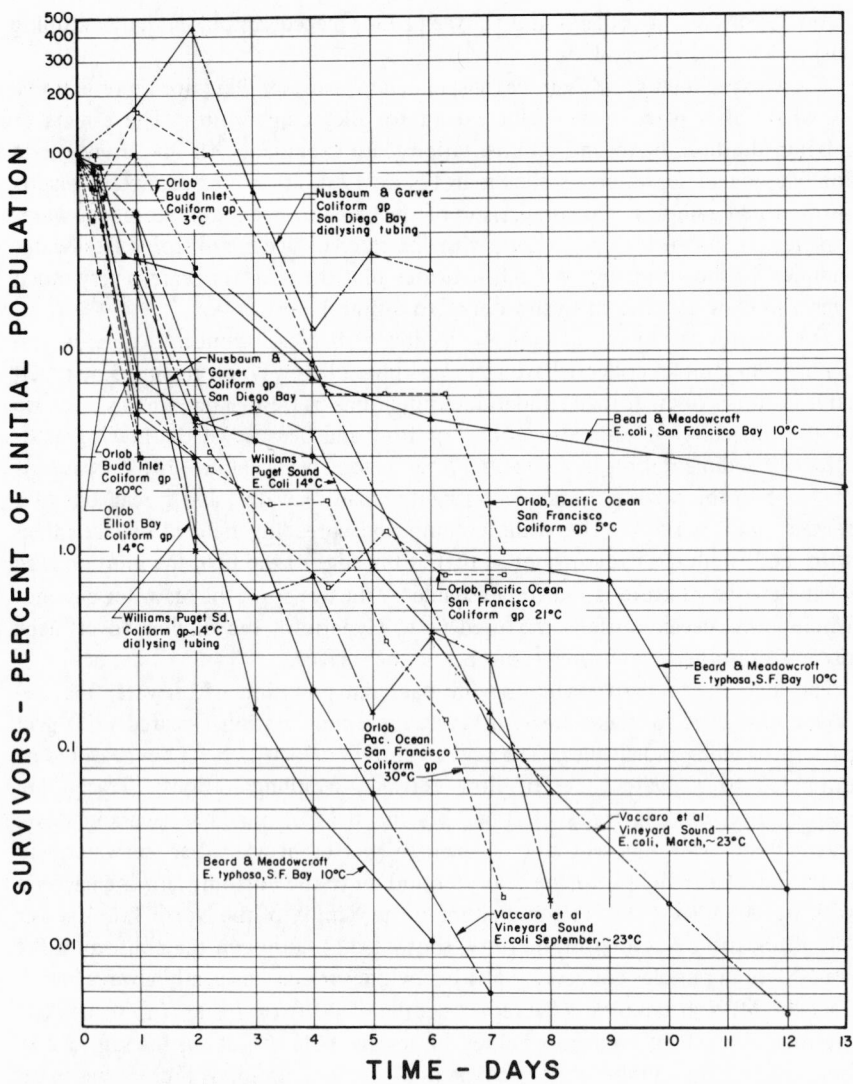

Figure 12-4
Survival of bacteria in raw seawater. [After Pearson (3).]

The results of this investigation in terms of k_b, computed by Camp from the reported values of t_{90} by means of Eq. (12-2), are presented in Table 12-1. The rate constant of 1.35 for mortality, which corresponds with a value of t_{90} of about 0.74 day agrees reasonably well with the initial rates of decline shown in Figure 12-4.

Table 12-1
Rates of Decrease of Coliforms in Seawater

| Plant | Treatment | Rate Constant k_b per Day | | | |
		Dilution	Mortality	Sedimentation	Combined
Hyperion	Secondary	1.2	1.35	1.15	3.7
Hyperion	Primary	1.2	1.35	4.55	7.1
Orange County	Primary	2.4–4.8	1.35	10–12	16.0

Table 12-1 shows the overwhelming importance of dilution and sedimentation, particularly sedimentation, as agents for reducing the concentration of bacteria in seawater into which sewage is discharged. The committee estimates the persistence of coliforms that have settled to the bottom as relatively long, with a probable t_{90} of 1 to 10 months, but it concludes that the probability of resuspension of these bacteria is small.

REFERENCES

1. R. W. Kehr and C. T. Butterfield, "Notes on the Relation Between Coliforms and Enteric Pathogens," *Public Health Rept.*, *58*, p. 589 (Apr. 9, 1943).
2. H. W. Streeter, "A Formulation of Bacterial Changes Occurring in Polluted Water," *Sewage Works J.*, *6*, 208 (1934).
3. E. A. Pearson, "An Investigation of the Efficacy of Submarine Outfall Disposal of Sewage and Sludge," Publ. 14, State Water Pollution Control Board, Sacramento, Calif., 1956.
4. "A Study of the Pollution and Natural Purification of the Ohio River: II. Report of Surveys and Laboratory Studies," *Public Health Service Bull. 143*, 1924.
5. C. T. Butterfield, "Observations on Changes in Numbers of Bacteria in Polluted Water," *Sewage Works J.*, *5*, 600 (1933).
6. *Washington Pollution Control Commission Tech. Bull. 2*, 1949.
7. *J. San. Eng. Div.*, Amer. Soc. Civil Engrs., *23* (Jan. 1961).

CHAPTER 13

Dispersal of Wastes into Receiving Waters

13-1. SHORELINE RELEASES

In engineering studies to determine the effect of wastewater treatment on the receiving stream, it has been the practice until quite recently to assume uniform mixing of the effluent with the receiving stream a short distance downstream from the outfall sewer. Little effort has been made to design outlet works in fresh waterways to achieve the degree of mixing that is required for most effective use of the watercourse. Consequently, most sewer outlets have been constructed to discharge at or near the water's edge, with more attention directed toward avoidance of obstruction to navigation, docking, and dredging operations than to dispersion in the receiving waters. As a result of this practice, local unsightly nuisances are common near sewer outlets, and the effluent may hug the shoreline of a large river and be plainly visible for miles. Moreover, the high bacterial content of the water along the shore may interfere with the satisfactory use of waterworks' intakes for many miles downstream.

Falk (1) has studied the dilution of shoreline releases of industrial wastewater into large streams with results as shown in Figure 13-1. Plant A is located on a stream 500 ft wide with a depth of about 10 ft at the time the measurements were made. The river flow was about 5,000 ft³/sec and the plant waste plus cooling water entered through a ditch at a flow of about 10 ft³/sec. The total available dilution was thus 500 to 1. Conductivity measurements were used to determine dilution. Plant B is located on a river about 1,700 ft wide and 13 ft

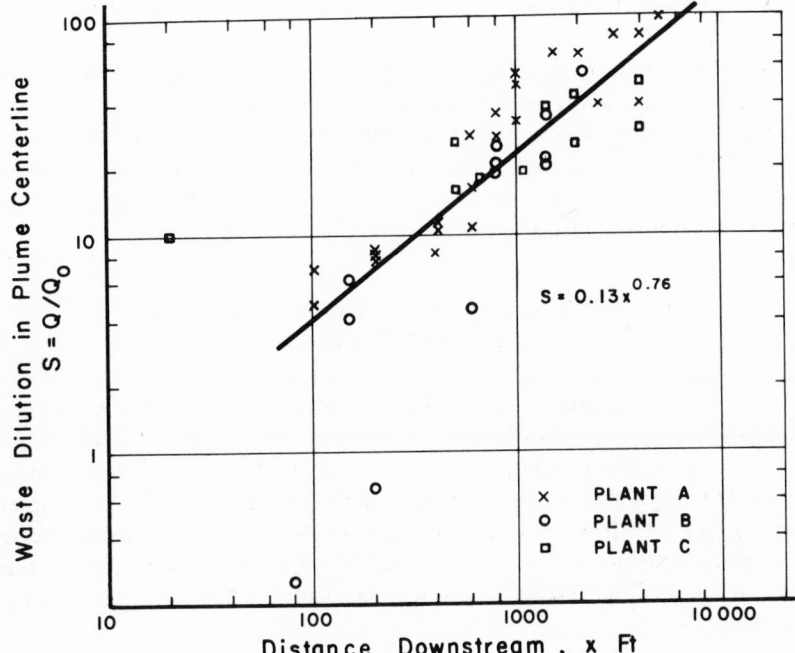

Figure 13-1
Dilution with shoreline release. [After Falk (1).]

deep. The river flow was about 94,000 ft^3/sec and wastes plus cooling water flow was about 4.5 ft^3/sec, yielding an available dilution of 21,000 to 1. Dilution was measured on the basis of pH changes. Plant C is located on the St. Lawrence River near Maitland, Ontario, where the flow is about 140,000 ft^3/sec. The available dilution was several thousand to one. Dilution was measured on the basis of pH and BOD determinations.

More attention has been given to dispersion facilities for disposal of waste-waters in seawater primarily because the sewage slick is plainly visible to the public and because of bacterial pollution of bathing beaches. Most of our knowledge on the design of dispersal works has developed as a result of studies on marine disposal along the west coast.

The most difficult problems of dispersal of wastewater effluents are associated with freshwater lakes. Freshwater lakes are used for water supply and for recreational purposes, both of which must be protected. Wind-induced currents on lakes are as variable in velocity and direction as they are on the ocean, but the advantages arising from differences in density of effluent and receiving water are not available for lakes.

The objective in the design of dispersal works is to locate the outlet so as to cause the least damage to, and interference with, other uses of the water and watercourse, and to effect a degree of dilution of the wastewater with relatively pure water so that, under nearly all conditions of expected use of the receiving waters and watercourse, the pollution will be kept within a :ceptable standards. The method of approach may be divided into two phases: (1) the initial mixing of the wastewater with the receiving water and (2) the further dilution of the mixture by turbulent diffusion in currents of the receiving waters.

13-2. INITIAL MIXING OF WASTEWATER WITH RECEIVING WATER

For effective initial mixing it is necessary to divide the discharge of wastewater among a sufficient number of submerged jets to produce the required initial dilution with the receiving water. The kinetic energy of each jet is dissipated by turbulent mixing with the receiving waters, as illustrated by Figure 13-2, adapted from a paper by Albertson et al. (2). The deceleration of the fluid in the jet occurs through simultaneous acceleration of the fluid surrounding the jet, so that the total rate of flow past successive sections of the jet increases with distance from the outlet.

The studies by Albertson and co-workers were limited to jets discharging into

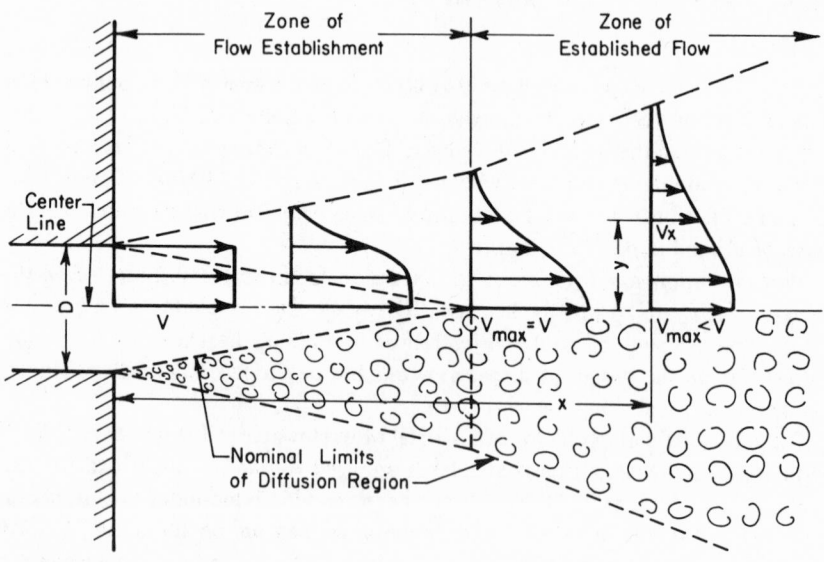

Figure 13-2
Jet diffusion. [After Albertson et al. (2).]

fluids of the same density. For convenience in the experimental work, air jets were discharged into still air. The results are applicable to wastewater discharges into receiving waters of the same density. Both circular orifices and slots were used in the studies. A zone of flow establishment was noted, as shown on Figure 13-2, which extended out beyond the orifice or slot a distance of about 6 times the diameter of the orifice or width of the slot. The zone of established flow was studied for a distance beyond the orifice of 64 times the diameter and beyond the slot of 2,300 times the width of the slot. The total rate of flow in the expanding jet was found to increase with the distance in the case of circular orifices and with the square root of the distance in the case of slots. Orifices are, therefore, much more efficient than slots.

Albertson and associates found that the dilution ratio for initial mixing with circular orifices, without gravity effects, is

$$S_0 = \frac{q}{q_0} = 0.32 \frac{L}{D} \tag{13-1}$$

where S_0, or q/q_0 is the ratio of the total flow to the orifice discharge rate at distance L from an orifice of diameter D (assuming no contraction). If there is contraction, D is the diameter of the jet at the *vena contracta*. If, for example, an initial dilution of 30 is required, L/D is 94. By comparison for a slot, L/B is 2,340 from a similar equation presented by Albertson et al., in which B is the width of the slot.

Albertson and co-workers found that the boundary of the expanding jet from a circular orifice diverged at a slope of approximately 1 to 5 from the center line. The diameter of the expanding jet D' is, therefore, about $0.4L$ or, from Eq. (13-1),

$$\frac{D'}{D} = 1.25 S_0 \tag{13-2}$$

The orifices should be spaced at intervals sufficiently large to avoid interference between expanding jets until the required initial dilution, S_0, is reached. The required minimum spacing is D', which may be estimated from Eq. (13-2). If in the preceding example S_0 is 30, D'/D is 37.5. The required orifice diameter D is, therefore, controlled by the required initial dilution ratio S_0 and the distance L within which this dilution must be achieved. In our example, if a 6-in. orifice is used, L is 47 ft and the spacing D' is 18.7 ft.

According to Pearson (3), Folsom and Ferguson, who worked with gasoline jets submerged in gasoline, found that the boundary of the expanding jet diverged at a slope of 1 to 4.31 and that the coefficient in Eq. (13-1) was 0.234 instead of 0.32. Tollmien, working with air in air, found that the boundary diverged at a slope of 1 to 3.92; Rice, working with freshwater in salt water, with differences in specific gravity of 0.01 to 0.035, found that the boundary diverged at a slope of 1 to 4.8. Colley and Harris, working with freshwater in salt water, with differences in specific gravity of 0.008 to 0.0795, found that the coefficient in Eq. (13-1) was about 0.33, although the dilution was 10 to 60 percent greater in their experiments than indicated by a coefficient of 0.33. Rawn and Palmer, in experiments with freshwater jets in seawater, found that the boundary of the expanding jet diverged at a slope of 1 over 6 to 8.

It is apparent from this discussion that differences in results were obtained by the various investigators, but there is no clear indication that dilution in fluids of like density is different from dilution in fluids of different densities. In the studies by Albertson et al. it was assumed that the jet induced movement of the diluting fluid by drawing it into the jet at right angles to the axis of the jet. The studies assumed a steady state, but took no account of currents in the diluting fluid or of the boundaries controlling the motion of the diluting fluid. Such limiting and controlling factors must be considered in dispersion of wastewater.

Figure 13-3 represents a mixing of wastewater with river water by means of

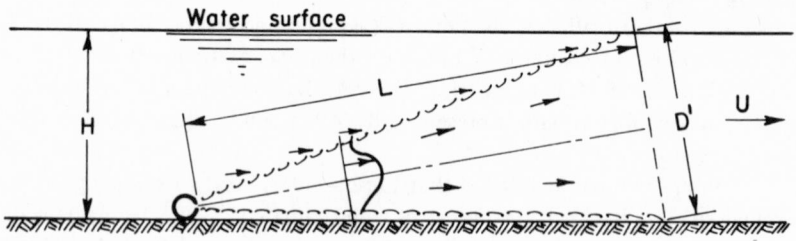

Figure 13-3
Jet mixing with river water.

jets. In this case the jet is shown directed downstream. After the dispersal system is in operation, D in Eqs. (13-1) and (13-2) is constant and q_0 varies independently with the rate of discharge of wastewater into the receiving stream. It is evident from Figure 13-3 that if the expanding jet is to occupy the full depth, only one value of q will satisfy Eqs. (13-1) and (13-2) for a particular value of q_0. D' is approximately equal to H, and the river velocity U is approximately $q/0.785HD'$. For example, if D is 0.5 ft, q_0 is 1.96 ft^3/sec, and H (or D') is 15 ft, from Eq. (13-2), S_0 is 24 and q is 47 ft^3/sec. U is, therefore, about 0.27 ft/sec, and from Eq. (13-1), L is 37.5. What happens to the dispersion pattern if the stream velocity is greater or less than that required to satisfy the conditions of Eqs. (13-1) and (13-2) and Fig. 13-3?

It is obvious that if the stream velocity is equal to, or greater than, the velocity of the wastewater as it issues from the orifice, the jet will induce no motion in the surrounding water. The wastewater will tend to remain in a narrow stream

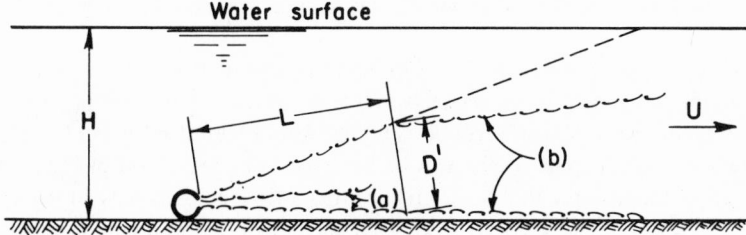

Figure 13-4
Jet mixing with river water of high velocity.

path parallel to the bottom, but will be subject to lateral mixing by the turbulence in the river water, as shown in Figure 13-4a. If the stream velocity is greater than the velocity required to satisfy the dilution illustrated by Figure 13-3 but less than the velocity of the wastewater as it issues from the orifice, the slope of the boundary of the expanding jet will be about 1 over 5 to a point beyond which the river velocity is greater than the mean velocity in the expanding jet. If in the preceding example the river velocity is 0.6 ft/sec, the diameter of the expanding jet D', computed from Eq. (13-2) is 6.65 ft at the point where the mean velocity in the jet is 0.6 ft/sec. The value of q is then 20.9 ft^3/sec and S_0 is 10.7. From Eq. (13-1), L is 16.7 ft. It is reasonable to suppose that the jet will become a part of the normal stream flow pattern downstream from this point and will be subject to lateral mixing by the turbulence in the river water, as shown in Figure 13-4b.

If the wastewater jet is directed upstream in river water, the mixing pattern will be approximately that shown in Figure 13-5. A zone of stagnation will develop at a point where the mean velocity in the expanding jet is approximately equal to the stream velocity in the reverse direction. The stream will flow around this zone as it does around a solid obstacle, but it will carry the flow of the expanding jet with it. The flow induced into the expanding jet is recirculated as a roller. This roller will contain some of the wastewater, but most of it will be

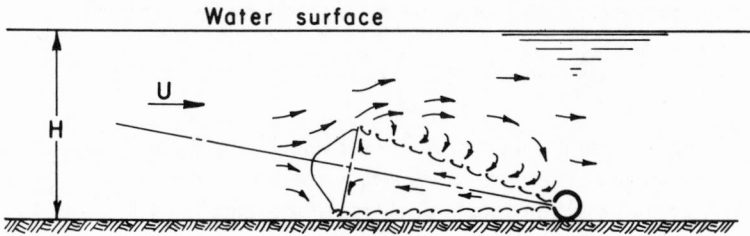

Figure 13-5
Jet directed upstream in river water.

exchanged into the river water by the turbulence accompanying the impact of the jet. This type of dispersion is probably more effective in mixing over a wide range of flow conditions than dispersion with downstream jets.

If the stream velocity U is nearly zero, as may be the case in a pond or lake with no wind, the jet pattern will be as indicated in Figure 13-3, except that there will be a stagnation zone at the end of the expanding jet and all of the induced flow will be recirculated in a roller. If the stream flow consists only of the wastewater, there will be no dilution water available, and all the water in the jet pattern and in the vicinity of the outlet will be wastewater.

For successful dispersion of wastewater into a lake, wind-induced currents are essential. The flow through the lake must, of course, be sufficient for adequate dilution of all the wastes discharged into the lake, but lake velocities associated with flow are usually too small to be significant factors in dispersion of wastewater. When the wind velocity is very low and when the lake is covered with sheet ice, wastewaters accumulate around the wastewater outlet. Wind-induced currents must be of sufficient volume and frequency during the months of no ice to afford the required dilution of the wastes discharged during the year. A continuous record of the velocity and direction of the wind for several years at a weather station near the outlet is desirable for reliable estimates of the volume and frequency of wind-induced currents. Float studies are also desirable for determining the vertical distribution of wind-induced current velocities for various wind velocities and durations.

Figure 13-6 shows jet mixing of wastewater with lake water, with jets directed

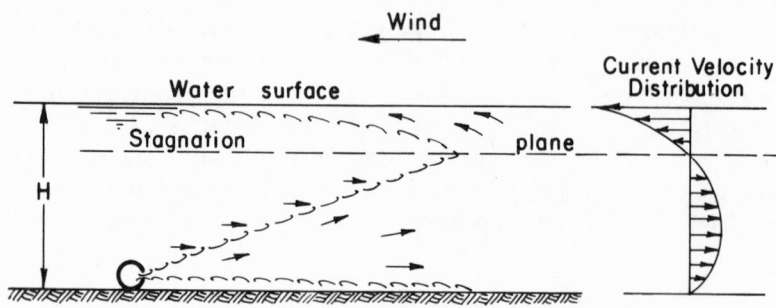

Figure 13-6
Jet mixing in lake with onshore wind-induced currents.

offshore and with winds onshore. A vertical velocity distribution curve is shown for the wind-induced currents. Under steady conditions, the rate of discharge of water shoreward equals the rate of discharge of the undertow lakeward. As indicated in Section 11-5, the wind velocity and direction must be sustained for considerable time before steady flow conditions are reached. At the start of a wind, the velocity gradient at the surface is high, the stagnation plane is just below the water surface and there is very little underflow in the opposite direction. As the wind continues, the onshore surface current is increased and the stagnation plane is lowered. As the water piles up on the shore, it develops the head necessary to sustain the underflow in the reverse direction. A study of Figure 13-6 will indicate that, if the depth and velocity of the undertow is sufficient, nearly all the wastewater may be carried offshore in the undertow, whereas if the depth and velocity are insufficient, most of the wastewater may be carried shoreward in the wind-induced surface current.

Figure 13-7 illustrates jet mixing of wastewater with lake water with jets di-

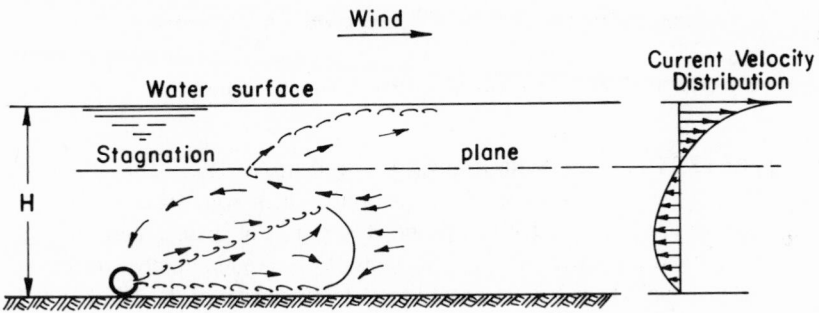

Figure 13-7
Jet mixing in lake with offshore wind-induced currents.

rected offshore and with winds offshore. In this case it is unlikely that the undertow will bring much of the wastewater shoreward.

In deep lakes in the temperate zone, the water is stratified according to density most of the year, except during the spring and fall *turnovers* when the water passes through its maximum density at about 4°C. In such lakes, there is a *stagnation zone* at the bottom and a *zone of circulation* at the top; these zones are separated by a *transition zone* or *thermocline* in which the temperature changes rapidly with depth. The vertical velocity distributions shown in Figures 13-6 and 13-7 for wind-induced currents will be confined to the zone of circulation in deep lakes, except during the periods of turnover when they may extend to the lake bottom.

When sewage is discharged as a horizontal jet into seawater, the expanding jet turns upward, as shown in Figure 13-8, because the mixture is lighter than seawater. An extensive set of experiments (388 in number) was conducted by Rawn and Palmer (4) around 1929 on horizontal freshwater jets in seawater. The experiments were conducted within a basin in Los Angeles Harbor, which was protected from horizontal currents. The equipment was mounted on a large raft. The nozzle diameter, D, ranged from $\frac{1}{4}$ in. ($\frac{1}{48}$ ft) to 2 in. ($\frac{1}{6}$ ft), and depths,

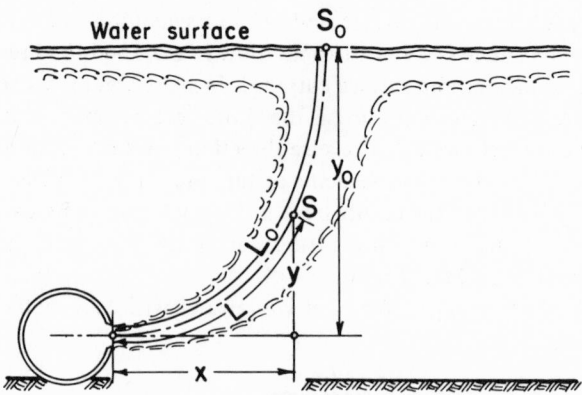

Figure 13-8
Single rising column of sewage in seawater. [After Rawn and Palmer (4).]

y_0, up to 13.25 ft. The values of y_0/D ranged up to 192. The results were presented in the form of empirical equations, which were not dimensionally homogeneous and were difficult to apply. The results of these experiments have been reexamined by Rawn et al. (5) in terms of the Froude number of the jet at the nozzle. Graphs have been prepared to show the values of the dilution ratios, S_0, as a function of y_0/D and the Froude number. Figures 13-9 and 13-10 have been adapted from the graphs by Rawn et al. The apparent acceleration of gravity, g', used for the Froude number, is the acceleration caused by the buoyant force and is related to the gravity constant g as follows:

$$g' = g \frac{\rho_1 - \rho}{\rho} \tag{13-3}$$

where ρ_1 is the density of the receiving water and ρ is the density of the wastewater. For sewage into seawater, g' is approximately $0.025g$.

Rawn (5) and associates have reported the actual values of S_0 obtained at the three outfalls of the Los Angeles County Sanitation District, all of which are equipped with jet diffusers. The dilutions were based on the maximum sewage concentrations at the ocean surface over the points of discharge, as determined by ammonia concentration measurements. Estimated values of the Froude number and y_0/D from data presented by Rawn et al. (5) have been plotted as three points on Figures 13-9 and 13-10. The points are marked 60, 72, and 90 in., which are the inside diameters of the three outfalls. The dashed curves have been interpolated between these points and the solid curves. It is possible to use these figures to estimate the dilutions that may be expected from horizontal jets submerged in seawater in depths up to more than 200 ft.

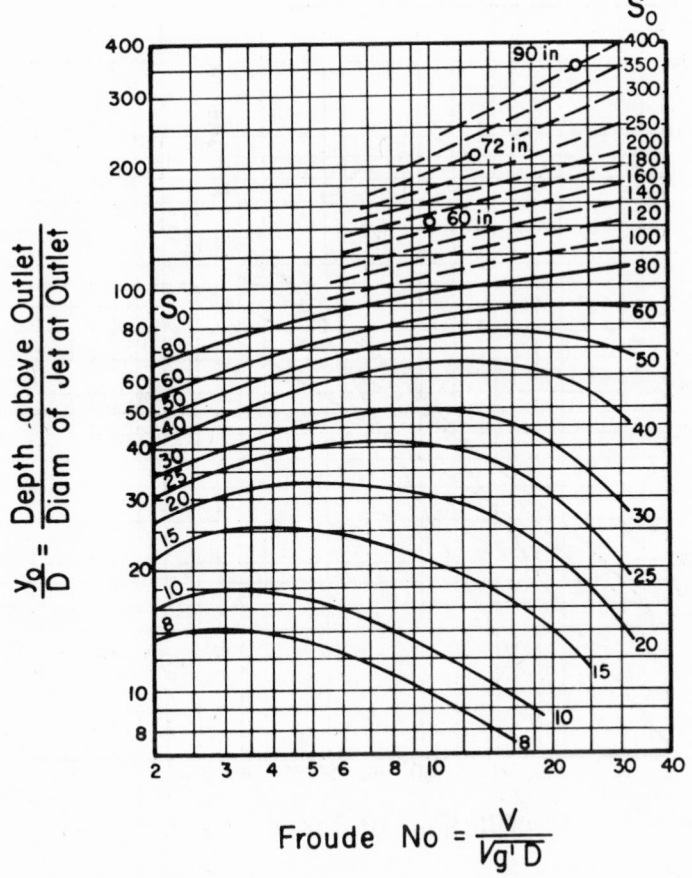

Figure 13-9
Dilution at top of rising column for horizontal discharge. Based on Rawn–Palmer (*4*) data.]

If the receiving seawater is stratified as a result of variations in temperature, salinity, or both, the rising column of mixed sewage and seawater may spread out horizontally below the surface of the sea, as shown in Figure 13-11, after Pearson (*6*). Figures 13-9 and 13-10 may be used to determine the approximate value of the dilution at the warm-water interface. If the density of the mixed sewage and cold seawater is greater than the density of the warm seawater at the top, the mixture will stay submerged. This relationship is represented, approximately, by

$$\frac{(S_0 - 1)\rho_c + \rho}{S_0} > \rho_w \qquad (13\text{-}4)$$

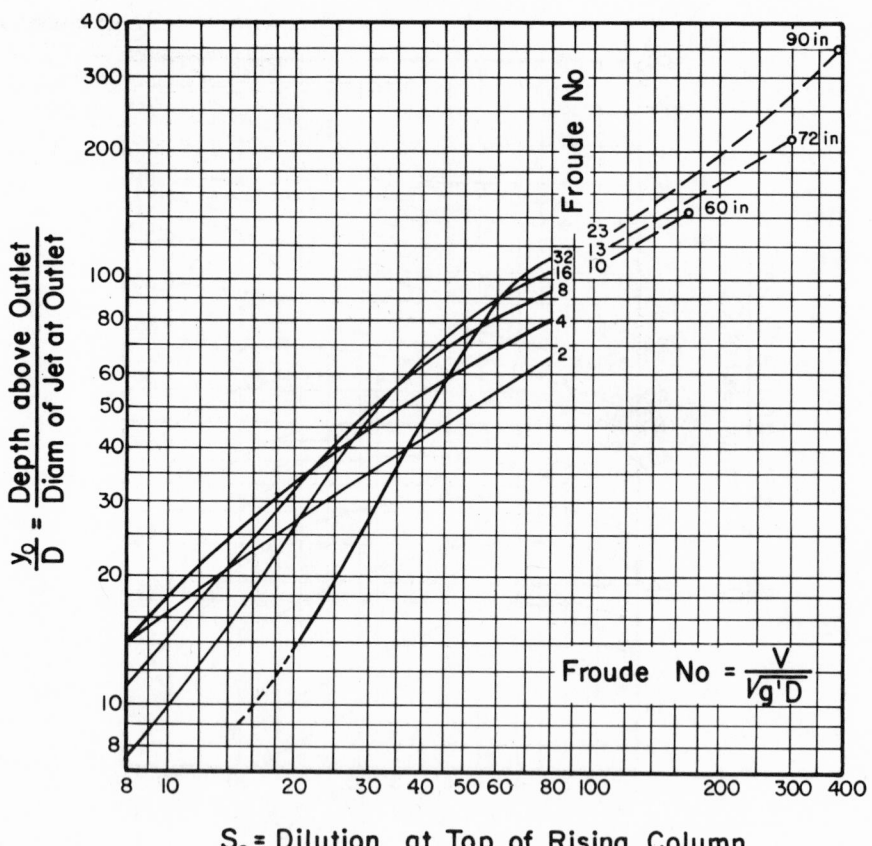

Figure 13-10

Dilution for constant Froude number for horizontal discharge. [Based on Rawn–Palmer (4) Data.]

where ρ_c is the density of the cold seawater, ρ_w is the density of the warm sea-water at the surface, and ρ is the density of the sewage. For example, with a temperature gradient as shown in Figure 13-11 throughout a depth of 100 ft and with the thermocline 70 ft from the bottom, ρ_c is about 1.0245 and ρ_w is about 1.0225. If the diameter of the jet at the orifice is 1.0 ft and the velocity is 5 ft/sec, y_0/D is 70 and the Froude number is about 5.6. From Figures 13-9 and 13-10, S_0 is about 55. With this value of S_0 and a value of ρ at 1.0, the value of the left side of the inequality (13-4) is about 1.0241. Since this is larger than 1.0225, the mixture will stay submerged. For all values of S_0 greater than about 12.3, from Eq. (13-4), the mixed sewage and cold seawater in the preceding example will remain submerged.

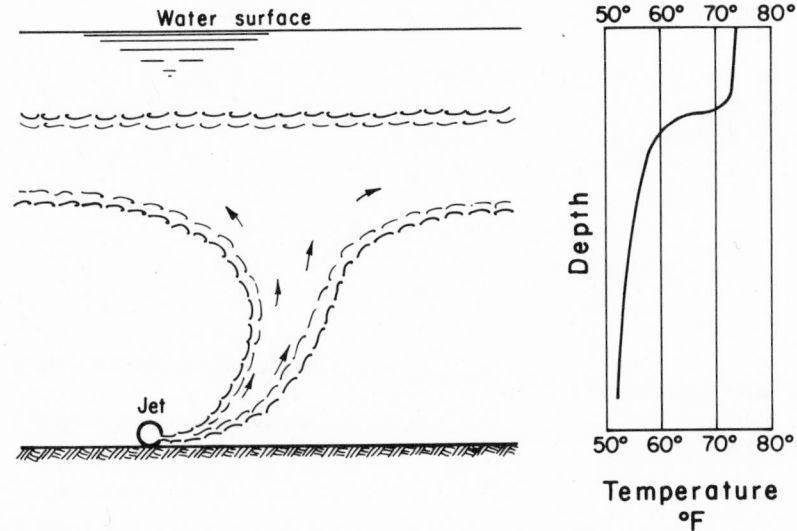

Figure 13-11
Submergence of sewage field by density stratification in ocean. [After Pearson (6).]

In the design of diffuser systems, the orifices should have the same diameter and the energy loss at each jet should be large as compared to the difference in energy available between the jets at each end of a diffuser leg, if good distribution of wastewater flow between orifices is to be obtained. If m is the desired ratio (7) of the rates of discharge from the two orifices, then

$$\frac{h_0}{h_0 - \Delta h} = \frac{1}{m^2} \qquad (13\text{-}5)$$

where h_0 is the velocity head for the upstream jet and Δh is the difference in energy head available. The decrease in energy head, Δh, is the loss of head in the pipe between the orifices from pipe friction and velocity changes, less the velocity head. If a diffuser leg in seawater is not laid level, Δh must include a further decrease in energy head, $\Delta y[(\rho_1 - \rho)/\rho]$, where Δy is the height of the last orifice above the orifice at the upstream end of the diffuser leg.

If the rates of discharge from the orifices are to vary by not more than 10 percent, m is 0.9 (or 1.1 if Δh is negative), and from Eq. (13-5), h_0 must be not less than 5.3 Δh (or 4.8 Δh if Δh is negative). If the rates are to vary by not more than 20 percent, m is 0.8 (or 1.2 if Δh is negative), and from Eq. (13-5), h_0 must be not less than 2.8 Δh (or 2.3 Δh if Δh is negative). If $\Delta y[(\rho_1 - \rho)/\rho]$ is zero, m will remain substantially constant for all rates of discharge into the diffuser, since both Δh and h_0 vary as the square of the discharge. If a diffuser leg discharging fresh wastewater into salt water is not laid level, $\Delta y[(\rho_1 - \rho)/\rho]$ is a constant for all rates of flow and Δh will not vary with the square of the discharge. For this case, m will not remain constant as the discharge changes.

The pressure head in a diffuser leg that discharges fresh wastewater into seawater is $h_0 + y_0[(\rho_1 - \rho)/\rho]$. For very deep diffusers, the hydraulic grade line in the diffuser will be several feet above the level of the sea. For a 200-ft depth, $y_0[(\rho_1 - \rho)/\rho]$ is about 5 ft.

The minimum velocity at the upstream end of a diffuser leg should be not less than 2 ft/sec, for which the velocity head is about 0.06 ft. If, for example, the loss of head in the diffuser pipe is 0.12 ft, Δh will be 0.06 ft for a level diffuser. For a variation in orifice discharge of not more than 20 percent, h_0 must be not less than 0.168, which corresponds with a jet velocity of 3.3 ft/sec. If the diffuser is in seawater and rises 5 ft between orifices, $\Delta y[(\rho_1 - \rho)/\rho]$ is about 0.125 ft and Δh is 0.185 ft. For a variation in orifice discharge of not more than 20 percent for this case, h_0 must be not less than 0.517, which corresponds with a jet velocity of 5.8 ft/sec. It is thus evident that, if the rate of discharge into a diffuser discharging into seawater is to vary widely between the minimum and peak, the diffuser leg must be level in order to reduce the required value of h_0. For example, with a 5 to 1 variation in flow, h_0 will vary 25-fold. With a level diffuser, the maximum value of h_0 would be 4.2 ft for the preceding example, whereas with the rising diffuser, the maximum value of h_0 would be 13.0 ft.

13-3. DIFFUSION OF WASTEWATER IN RECEIVING-WATER CURRENTS

If the velocity of the receiving-water current is U, it is evident that

$$UbH' = \text{approx. } QS_0 \tag{13-6}$$

where b is the width of the sewage field across the receiving-water current after initial mixing, H' is the thickness or depth of the sewage field, S_0 is the initial dilution, and Q is the rate of sewage discharge from the diffuser system.

In the case of jets directed downstream in a river where the velocity U is just sufficient to carry away the flow of the jets after expansion to the full depth, Figure 13-3, H' is equal to the full depth H, and full dilution is substantially accomplished in the initial mixing. If the river velocity U is somewhat greater, but not as great as the jet velocity, Figure 13-4(b), H' will be less than H, and further dilution must be obtained by diffusion accompanying the turbulence of the river. When the jet is directed upstream, Figure 13-5, better initial mixing will result over a wider range of flows.

In the case of jet mixing in a lake, it is necessary to estimate the dilution obtained with onshore currents. The jets should be directed offshore, as shown in

Figures 13-6 and 13-7. With onshore winds, Figure 13-6, H' is the depth of water above the stagnation plane, U is the mean velocity of the onshore current, and Q is the fraction of the sewage discharge estimated to be carried toward the shore. With offshore winds, Figure 13-7, H' is the depth of water below the stagnation plane, U is the mean velocity in the underflow, and Q is the fraction of the sewage discharge estimated to be carried toward the shore in the underflow.

In the case of ocean outfalls, the diffuser system is usually placed approximately parallel with the shore line, as shown in Figure 13-12 for San Diego (*8*). The diffusers are usually constructed at an angle of about 120° with the outfall sewer so that the width of the sewage field, *b* in Eq. (13-6), will be appreciable even with the currents parallel to the shore line. The jets, alternating in direction, are usually placed on both sides of a diffuser. The thickness of the sewage field, H', at the diffuser system may be computed by means of Eq. (13-6), provided the

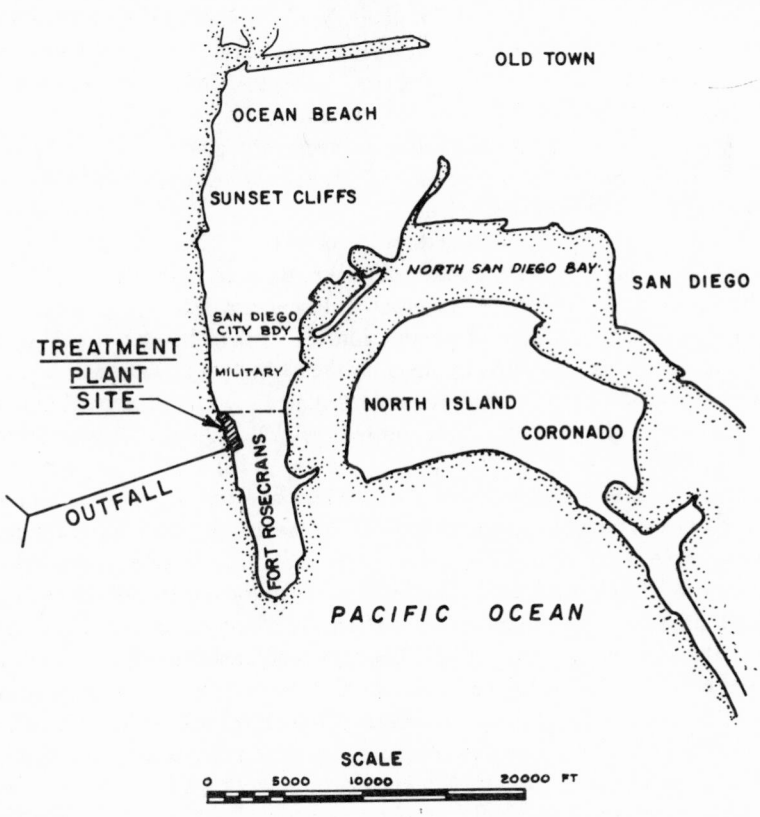

Figure 13-12
Location map for proposed ocean outfall for city of San Diego, California. [After Brooks (*8*).]

current velocity U is high enough to carry the diluted sewage away in a thickness H' not greater than one quarter to one third the ocean depth, y_0 of Figure 13-8. In the initial experimental work by Rawn and Palmer (4), H' was found to average about $\frac{1}{12}L$ (probably about $\frac{1}{8}y_0$); but it is doubtful whether steady-state conditions were approached in these experiments.

The critical condition for the design of an ocean outfall is for onshore currents, just as it is for lake outfalls. Vertical velocity distributions similar to those illustrated in Figures 13-6 and 13-7 are to be expected. The underflow current toward the shore with offshore winds illustrated by Figure 13-7 is known as *upwelling.* Upwelled waters (9) characteristically have lower temperatures and dissolved oxygen and higher nutrient concentrations than surface waters, except during long sustained freezing or near-freezing air temperatures. Flood tides may carry upwelled waters with low dissolved oxygen concentration onto beaches and into estuaries. Winds, wind-induced currents, and seawater temperatures have been extensively studied (10) along the southern California coast. The upper 6 in. of sea was found to move at a velocity about 2.5 percent that of the wind. In water deeper than 100 ft in Santa Monica Bay, the average current–wind factor was 2 percent at 15 ft and 1.4 percent at 50 ft. These measurements apparently neglected reverse flow below the stagnation plane for winds transverse to the shore line, and the current velocity changes brought about with sustained duration of winds were not reported.

In the disposal of sewage in a lake or in the ocean, the bacterial pollution of the shore is a major consideration. In a lake, BOD and suspended solids may also be a factor, but in deep-water ocean diffusers they are usually of minor importance. In most cases, the initial dilution, S_0, obtained at the diffusers is not sufficient, and advantage must be taken of the additional reduction to be had in the wind-induced currents. The required distance of the diffuser system from the shore (i.e., the length of the outfall) is determined by this additional reduction.

Figure 13-13 shows in plan a sewage field diffusing laterally in an ocean current of velocity U, as presented by Brooks (8). The same illustration may be used for a sewage field in a lake current. The width, Z, of the field expands from the initial width, b, at the diffuser system as a result of lateral diffusion at both edges of the field. The initial concentration, c_0, of an impurity is its concentration in the outfall divided by the dilution ratio S_0, resulting from the diffuser system. For example, if the coliform concentration in the outfall is 100 million per 100 ml and S_0 is 100, c_0 is 1 million per 100 ml. It will be noted from Figure 13-13 that the concentration, c, at any point in the sewage field is less than c_0, and that the concentration becomes less with increased distance from the diffuser system. For certain types of impurities, such as bacteria and radioactivity, there may be a further decrease in concentration with time of flow in the current resulting from die-off, decay, or settling out.

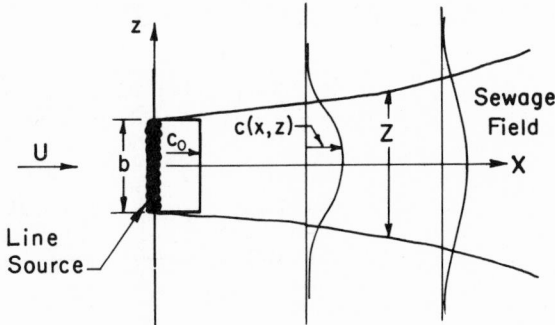

Ocean - Plan View

Figure 13-13
Schematic diagram of sewage field diffusing laterally in an ocean current (die-off not considered).

Brooks (8) has developed a solution for the concentration, c, based upon the following assumptions:

1. The turbulent diffusion law applies.
2. The turbulent transport coefficient ϵ is a function of b or of Z/b, and Z/b is a function of x and not of z.
3. Vertical mixing is negligible.
4. Longitudinal mixing in direction of current is negligible.
5. Flow is steady at velocity U.
6. The rate of disappearance of coliform is proportional to the concentration, or k_b of Eq. (12-1) is constant.

Lateral diffusion has been extensively measured in the ocean by means of the spreading of dye fields, chemicals, and radioactive wastes normal to the axis of the current. Pearson (3) has made a survey of the results of measurements recorded in the literature, which indicate that ϵ varies approximately with the $\frac{4}{3}$ power of the scale (presumably the width of the field or plume in the ocean current), a relation first proposed by Richardson. Figure 13-14 shows Pearson's plot of the results of the measurements, to which Brooks has added the determinations by Gunnerson (11) in Santa Monica Bay. The best-fit curve for Gunnerson's measurements has the formula $\epsilon = 0.005Z^{4/3}$. Two of Gunnerson's values were determined at a current velocity of 0.23 knot, and four at a velocity of 0.50 knot, which indicates that the current velocity is not a factor in the $\frac{4}{3}$ power relation.

Vertical mixing will take place, but it will be less in seawater than in a lake because of the stability caused by density stratification in the ocean. In either case, however, vertical mixing will be much more limited than lateral mixing be-

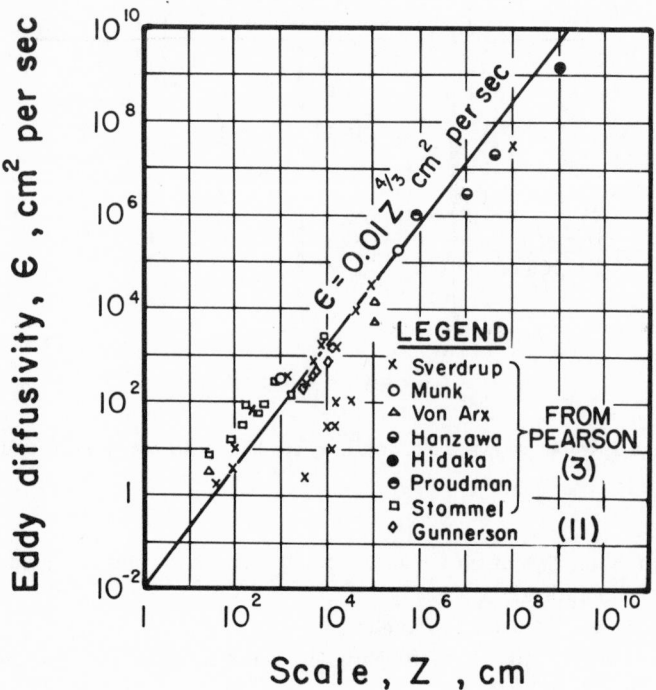

Figure 13-14

Measured values of eddy diffusivity in the horizontal direction in the ocean, showing increase with scale.

cause of the small ratio of receiving-water depth to the width of the sewage field. Longitudinal mixing may be neglected because of the relatively small concentration gradient along the axis of the current.

The basic differential equation for c under the steady-state conditions assumed by Brooks (8) is as follows:

$$\frac{\partial}{\partial z}\left(-\epsilon\frac{\partial c}{\partial z}\right) + U\frac{\partial c}{\partial x} + 2.3k_bc = 0 \qquad (13\text{-}7)$$

The equation was solved by Brooks by first working out the diffusion problem (k_b equal to zero) and then multiplying the result by the die-off factor 10^{-k_bt}, where t is the travel time x/U. The diffusion problem was solved for three cases: (1) constant value of ϵ, (2) linear increase in ϵ, and (3) the $\frac{4}{3}$ power relation. The initial value of the turbulent transport coefficient, ϵ_0, is a function of b and may be estimated from Figure 13-14 for $Z = b$. The width of the field Z for the three cases is shown in Table 13-1 and Figure 13-15.

Brooks believes that the solution involving the $\frac{4}{3}$ power relation is the most

Table 13-1
Width of Sewage Field Downstream from Diffuser System

Case I	$\dfrac{\epsilon}{\epsilon_0} = 1$	$\dfrac{Z}{b} = \left(1 + 2\beta\,\dfrac{x}{b}\right)^{1/2}$
Case II	$\dfrac{\epsilon}{\epsilon_0} = \dfrac{Z}{b}$	$\dfrac{Z}{b} = 1 + \beta\,\dfrac{x}{b}$
Case III	$\dfrac{\epsilon}{\epsilon_0} = \left(\dfrac{Z}{b}\right)^{4/3}$	$\dfrac{Z}{b} = \left(1 + \dfrac{2}{3}\,\beta\,\dfrac{x}{b}\right)^{3/2}$

$$\beta = \frac{12\epsilon_0}{Ub}$$

nearly correct. This solution also gives the greatest spread of the sewage field and, consequently, the greatest dilution. It will be noted that for a particular spread, Z/b (and corresponding dilution), $\beta(x/b)$ has a particular value. The distance, x, required for dilution is, therefore, directly proportional to the velocity U. To determine the required length of the outfall, current velocity–frequency studies are required so that a design velocity of acceptable frequency may be selected. This is particularly important where decrease in bacterial concentration is the prime objective, because the time, t, at the velocity, U, is also the time of die-off and settling out.

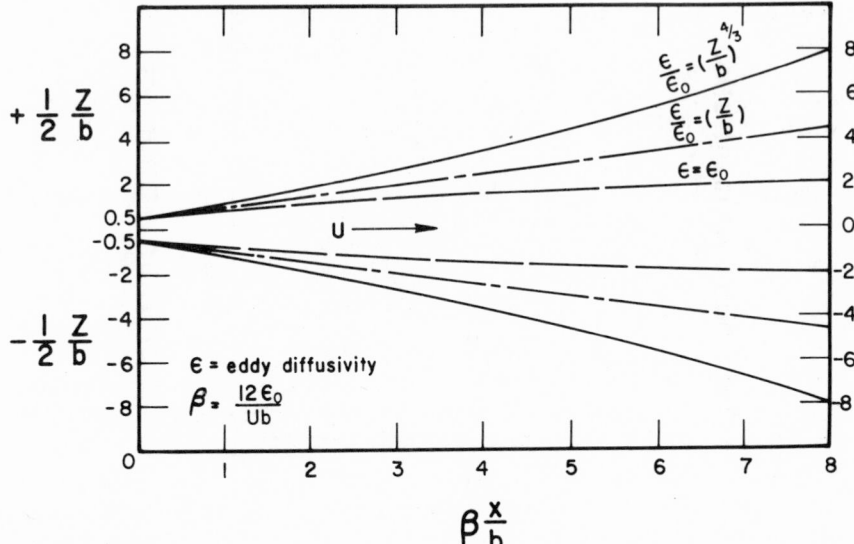

Figure 13-15
Lateral diffusion in a uniform current. [After Brooks (8).]

The maximum theoretical concentration will always occur on the axis of the sewage field, where z is zero in Figure 13-13. Brooks's solution of the differential equation for the concentration c along the axis (including the die-off factor $10^{-k_b t}$) is as follows for the three cases:

Case I: For $z = 0$, $c = c_0 \cdot 10^{-k_b t}$ erf $\sqrt{\dfrac{0.75}{\beta \dfrac{x}{b}}}$ (13-8)

Case II: For $z = 0$, $c = c_0 \cdot 10^{-k_b t}$ erf $\sqrt{\dfrac{1.5}{\left(1 + \beta \dfrac{x}{b}\right)^2 - 1}}$ (13-9)

Case III: For $z = 0$, $c = c_0 \cdot 10^{-k_b t}$ erf $\sqrt{\dfrac{1.5}{\left(1 + \dfrac{2}{3}\beta \dfrac{x}{b}\right)^3 - 1}}$ (13-10)

in which erf denotes the standard error function defined as

$$\text{erf } X = \frac{2}{\sqrt{\pi}} \int_0^X e^{-v^2} \, dv \qquad (13\text{-}11)$$

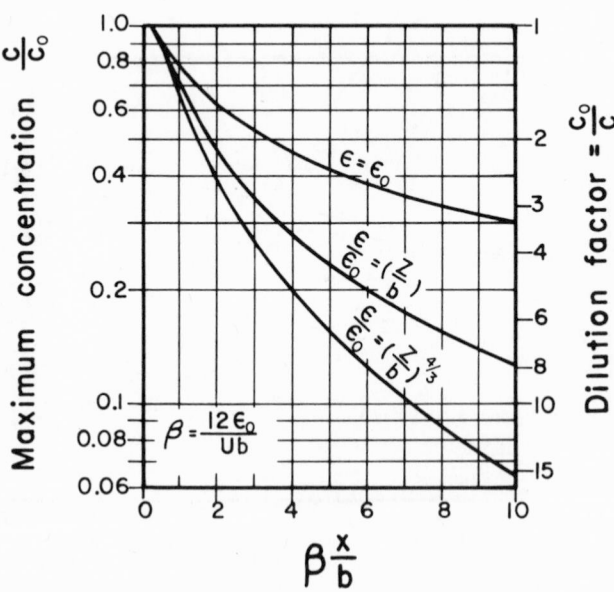

Figure 13-16
Decrease of concentration with distance along the axis of a sewage field (die-off not considered, i.e., $k_b = 0$). [After Brooks (8).]

The solution of Eqs. (13-8), (13-9), and (13-10), with k_b equal to zero, is shown in Figure 13-16 as presented by Brooks. The concentration as obtained from Figure 13-16 may be multiplied by $10^{-k_b t}$ for a final solution, including decay or bacterial die-off.

As an example of the application, consider an ocean outfall in which the coliform count is 100 million per 100 ml, which is reduced to c_0 of 1 million per 100 ml by an initial dilution ratio S_0 of 100. Assume further that b is 1,000 ft (3.05 $\times$ 10⁴ cm) and U is 0.4 ft/sec (12.2 cm/sec). From Figure 13-14, ϵ_0 is about 10^4 cm²/sec. Then β is $12\epsilon_0/Ub$, is $(12 \times 10^4)/(12.2 \times 3.05 \times 10^4)$, is 0.32. Now, if x, the distance to the shore, is 15,000 ft, $\beta(x/b)$ is 4.8, and the time of flow t is x/U, is 37,500 sec, or 0.434 day. From Figure 13-15, the width of the field Z at the shore is about 8,600 ft, using the $\frac{4}{3}$ power relation for ϵ. From Figure 13-16, the coliform count by dilution is about 1 to 6. If k_b is 5.0, $10^{-k_b t}$ is 0.0067. Then the estimated maximum concentration of coliforms at the shore is 1,000,000 $\times$ 1/6 $\times$ 0.0067, or 1,100 per 100 ml. It is evident from this example that unchlorinated sewage may be disposed of successfully with properly designed ocean diffuser systems and long ocean outfalls.

REFERENCES

1. L. L. Falk, "Some Modes of Waste Dilution in Receiving Waters," *Proc. 16th Ind. Waste Conf. Purdue Univ.*, 1962, p. 126.
2. M. L. Albertson, Y. B. Dai, R. A. Jensen, and H. Rouse, *Trans. Amer. Soc. Civil Engrs.*, *115*, 630 (1950).
3. E. A. Pearson, "An Investigation of the Efficacy of Submarine Outfall Disposal of Sewage and Sludge," Publ. 14, State Water Pollution Control Board, Sacramento, Calif., 1956.
4. A. M. Rawn and H. K. Palmer, "Predetermining the Extent of a Sewage Field in Sea Water," *Trans. Amer. Soc. Civil Engrs.*, *94*, 1036 (1930).
5. A. M. Rawn, F. R. Bowerman, and N. H. Brooks, "Diffusers for Disposal of Sewage in Sea Water," *Trans. Amer. Soc. Civil Engrs.*, *126*, 111, 344 (1961).
6. E. A. Pearson, "Submarine Waste Disposal Installations," Sixth Intern. Conf. Coastal Engineering, University of Florida, Dec. 1957.
7. C. V. Davis, *Handbook of Applied Hydraulics*, 2nd ed., New York, McGraw-Hill Book Company, 1952, p. 968. (3rd ed. is by Davis and K. E. Sorensen, 1969.)
8. N. H. Brooks, "Diffusion of Sewage Effluent in an Ocean Current," *Proceedings of the International Conference on Waste Disposal in the Marine Environment*, Elmsford, N.Y., Pergamon Press, Inc., 1959, p. 246.
9. *J. San. Eng. Div., Amer. Soc. Civil Engrs.*, *23* (Jan. 1961).
10. "Oceanographic Survey of the Continental Shelf Area of Southern California," Publ. 20, State Water Pollution Control Board, Sacramento, Calif., 1959.
11. C. G. Gunnerson, "Sewage Disposal in Santa Monica Bay, California," *J. San. Eng. Div.*, *Amer. Soc. Civil Engrs.*, *84* (Feb. 1958).

Water Quality Standards for the Interstate Waters of Massachusetts*

FRESHWATERS

Class A—Waters designated for use as public water supplies in accordance with Chapter III of the General Laws. Character uniformly excellent.

Standards of Quality

Item	Water Quality Criteria
1. Dissolved oxygen	Not less than 75 percent of saturation during at least 16 hr of any 24-hr period and not less than 5 mg/liter at any time.
2. Sludge deposits–solid refuse–floating solids–oils–grease–scum	None allowable.
3. Color and turbidity	None other than of natural origin.
4. Coliform bacteria per 100 ml	Not to exceed an average value of 50 during any monthly sampling period.
5. Taste and odor	None other than of natural origin.

*Condensed from the Water Quality Standards, Commonwealth of Massachusetts, Division of Water Pollution Control; as approved by the federal government on August 7, 1967. From a joint publication by the United States Environmental Protection Agency and the Massachusetts Water Resources Commission, Division of Water Pollution Control, December 1971.

Item	Water Quality Criteria
6. pH	As naturally occurs.
7. Allowable temperature increase	None other than of natural origin.
8. Chemical constituents	None in concentrations or combinations that would be harmful or offensive to humans, or harmful to animal, or aquatic life.
9. Radioactivity	None other than that occurring from natural phenomena.

Class B—Suitable for bathing and recreational purposes, including water contact sports. Acceptable for public water supply with appropriate treatment. Suitable for agricultural, and certain industrial, cooling and process uses; excellent fish and wildlife habitat; excellent esthetic value.

Standards of Quality

Item	Water Quality Criteria
1. Dissolved oxygen	Not less than 75 percent of saturation during at least 16 hr of any 24-hr period and not less than 5 mg/liter at any time.
2. Sludge deposits–solid refuse–floating solids–oils–grease–scum	None allowable.
3. Color and turbidity	None in concentrations that would impair any usages specifically assigned to this class.
4. Coliform bacteria per 100 ml	Not to exceed an average value of 1,000 during any monthly sampling period or 2,400 in more than 20 percent of samples examined during such period.
5. Taste and odor	None in concentrations that would impair any usages specifically assigned to this class and none that would cause taste and odor in edible fish.
6. pH	6.5–8.0.
7. Allowable temperature increase	None except where the increase will not exceed the recommended limit on the most sensitive receiving-water use and in no case exceed 83°F in warm-water fisheries, and 68°F in cold-water fisheries, or in any case raise the normal temperature of the receiving water more than 4°F.
8. Chemical constituents	None in concentrations or combinations that would be harmful or offensive to human, or harmful to animal or aquatic life, or any water use specifically assigned to this class.

Item	Water Quality Criteria
9. Radioactivity	None in concentrations or combinations that would be harmful to human, animal, or aquatic life for the appropriate water use. None in such concentrations which would result in radionuclide concentrations in aquatic life which exceed the recommended limits for consumption by humans.
10. Total phosphate	Not to exceed an average of 0.05 mg/liter as P during any monthly sampling period.
11. Ammonia	Not to exceed an average of 0.5 mg/liter as N during any monthly sampling period.
12. Phenols	Shall not exceed 0.001 mg/liter at any time.

Class C–Suitable for recreational boating; habitat for wildlife and common food and game fishes indigenous to the region; certain industrial cooling and process uses; under some conditions acceptable for public water supply with appropriate treatment. Suitable for irrigation of crops used for consumption after cooking. Good esthetic value.

Standards of Quality

Item	Water Quality Criteria
1. Dissolved oxygen	Not less than 5 mg/liter during at least 16 hr of any 24-hr period nor less than 3 mg/liter at any time. For seasonal cold-water fisheries at least 5 mg/liter must be maintained.
2. Sludge deposits–solid refuse–floating solids–oils–grease–scum	None allowable except those amounts that may result from the discharge from waste-treatment facilities that provide appropriate treatment.
3. Color and turbidity	None allowable in such concentrations that would impair any usages specifically assigned to this class.
4. Coliform bacteria	None in concentrations that would impair any usages specifically assigned to this class.
5. Taste and odor	None in concentrations that would impair any usages specifically assigned to this class, and none that would cause taste and odor to edible fish.
6. pH	6.0–8.5.
7. Allowable temperature increase	None except where the increase will not exceed the recommended limits on the most sensitive

Item	Water Quality Criteria
	receiving-water use and in no case exceed 83°F in warm-water fisheries and 68°F in cold-water fisheries, or in any case raise the normal temperature of the receiving water more than 4°F.
8. Chemical constituents	None in concentrations or combinations that would be harmful or offensive to human, or harmful to animal or aquatic life, or any water use specifically assigned to this class.
9. Radioactivity	None in concentrations or combinations that would be harmful to human, animal, or aquatic life for the appropriate water use. None in concentrations that would result in radionuclide concentrations in aquatic life which exceed the recommended limits for consumption by humans.
10. Total phosphate	Not to exceed an average of 0.05 mg/liter as P during any monthly sampling period.
11. Ammonia	Not to exceed an average of 1.0 mg/liter as N during any monthly sampling period.
12. Phenols	Not to exceed an average of 0.002 mg/liter at any time.

Class D—Suitable for esthetic enjoyment, power, navigation, and certain industrial cooling and process uses. Class D waters will be assigned only where a higher water-use class cannot be attained after all appropriate waste-treatment methods are utilized.

Standards of Quality

Item	Water Quality Criteria
1. Dissolved oxygen	Not less than 2 mg/liter at any time.
2. Sludge deposits–solid refuse–floating solids–oils–grease–scum	None allowable except those amounts that may result from the discharge from waste-treatment facilities that provide appropriate treatment.
3. Color and turbidity	None in concentrations that would impair any usages specifically assigned to this class.
4. Coliform bacteria	None in concentrations that would impair any usages specifically assigned to this class.
5. Taste and odor	None in concentrations that would impair any usages specifically assigned to this class.
6. pH	6.0–9.0.

Item	Water Quality Criteria
7. Allowable temperature increase	None except where the increase will not exceed the recommended limits on the most sensitive receiving-water use and in no case exceed 90° F.
8. Chemical constituents	None in concentrations or combinations that would be harmful to human, animal, or aquatic life for the designated water use.
9. Radioactivity	None in concentrations or combinations that would be harmful to human, animal, or aquatic life for the designated water use. None in concentrations that will result in radionuclide concentrations in aquatic life which exceed the recommended limits for consumption by humans.

Notes

1. All wastes shall receive appropriate waste treatment, which is defined as secondary treatment with disinfection or its industrial waste treatment equivalent except when a higher degree of treatment is required to meet the objectives of the water quality standards, all as determined by the Division of Water Pollution Control. Disinfection from October 1 to May 1 may be discontinued at the discretion of the Division of Water Pollution Control.

2. Appropriate water-supply treatment is as determined by the Massachusetts Department of Public Health.

3. These water quality standards do not apply to conditions brought about by natural causes.

4. Class B and C waters shall be substantially free of pollutants that will (a) unduly affect the composition of bottom fauna, (b) unduly affect the physical or chemical nature of the bottom, or (c) interfere with the spawning of fish or their eggs.

5. The average minimum consecutive seven-day flow to be expected once in 10 years shall be used in the interpretation of the standards except where noted.

6. The amount of disinfection required shall be equivalent to a free and combined chlorine residual of at least 1.0 mg/liter after 15 minutes contact time during peak hourly flow or maximum rate of pumpage.

COASTAL AND MARINE WATERS

Class SA—Suitable for any high-quality water use, including bathing and water contact sports. Suitable for approved shellfish areas.

Standards of Quality

Item	Water Quality Criteria
1. Dissolved oxygen	Not less than 6.5 mg/liter at any time.
2. Sludge deposits–solid refuse–floating solids–oil–grease–scum	None allowable.
3. Color and turbidity	None in such concentrations that will impair any usages specifically assigned to this class.
4. Coliform bacteria per 100 ml	Not to exceed a median value of 70 and not more than 10 percent of the samples shall ordinarily exceed 230 during any monthly sampling period.
5. Taste and odor	None allowable.
6. pH	6.8–8.5.
7. Allowable temperature increase	None except where the increase will not exceed the recommended limits on the most sensitive water use.
8. Chemical constituents	None in concentrations or combinations that would be harmful to human, animal, or aquatic life or would make the waters unsafe or unsuitable for fish or shellfish or their propagation, impair the palatability of same, or impair the waters for any other uses.
9. Radioactivity	None in concentrations or combinations that would be harmful to human, animal, or aquatic life for the designated water use. None in concentrations that would result in radionuclide concentrations in aquatic life which exceed the recommended limits for consumption by humans.
10. Total phosphate	Not to exceed an average of 0.07 mg/liter as P during any monthly sampling period.
11. Ammonia	Not to exceed an average of 0.2 mg/liter as N during any monthly sampling period.

Class SB–Suitable for bathing and recreational purposes, including water contact sports; industrial cooling; excellent fish habitat; good esthetic value; and suitable for certain shellfisheries with depuration (Restricted Shellfish Areas).

Standards of Quality

Item	Water Quality Criteria
1. Dissolved oxygen	Not less than 5.0 mg/liter at any time.
2. Sludge deposits–solid refuse–floating solids–oils–grease–scum	None allowable.
3. Color and turbidity	None in concentrations that would impair any usages specifically assigned to this class.

Item	Water Quality Criteria
4. Coliform bacteria per 100 ml	Not to exceed a median value of 700 and not more than 2,300 in more than 10 percent of the samples during any monthly sampling period.
5. Taste and odor	None in concentrations that would impair any usages specifically assigned to this class and none that would cause taste and odor in edible fish or shellfish.
6. pH	6.8–8.5.
7. Allowable temperature increase	None except where the increase will not exceed the recommended limits on the most sensitive water use.
8. Chemical constituents	None in concentrations or combinations that would be harmful to human, animal, or aquatic life or make the waters unsafe or unsuitable for fish or shellfish or their propagation, impair the palatability of same, or impair the water for any other usage.
9. Radioactivity	None in concentrations or combinations that would be harmful to human, animal, or aquatic life for the appropriate water use. None in concentrations that would result in radionuclide concentrations in aquatic life which exceed the recommended limits for consumption by humans.
10. Total phosphate	Not to exceed an average of 0.07 mg/liter as P during any monthly sampling period.
11. Ammonia	Not to exceed an average of 0.2 mg/liter as N during any monthly sampling period.

Class SC—Suitable for esthetic enjoyment; for recreational boating; habitat for wildlife and common food and game fishes indigenous to the region; industrial cooling and process uses.

Standards of Quality

Item	Water Quality Criteria
1. Dissolved oxygen	Not less than 5 mg/liter during at least 16 hr of any 24-hr period or less than 3 mg/liter at any time.
2. Sludge deposits–solid refuse–floating solids–oils–grease–scum	None except the amount that may result from discharge from a waste-treatment facility that provides appropriate treatment.
3. Color and turbidity	None in concentrations that would impair any usages specifically assigned to this class.

Item	Water Quality Criteria
4. Coliform bacteria	None in concentrations that would impair any usages specifically assigned to this class.
5. Taste and odor	None in concentrations that would impair any usages specifically assigned to this class and none that would cause taste and odor in edible fish or shellfish.
6. pH	6.5–8.5.
7. Allowable temperature increase	None except where the increase will not exceed the recommended limits on the most sensitive water use.
8. Chemical constituents	None in concentrations or combinations that would be harmful to human, animal, or aquatic life or which would make the waters unsafe or unsuitable for fish or shellfish or their propagation, impair the palatability of same, or impair the water for any other usage.
9. Radioactivity	None in concentrations that would be harmful to human, animal, or aquatic life for the designated water use. None in concentrations that would result in radionuclide concentrations in aquatic life which exceed the recommended limits for consumption by humans.
10. Total phosphate	Not to exceed an average of 0.07 mg/liter as P during any monthly sampling period.
11. Ammonia	Not to exceed an average of 1.0 mg/liter as N during any monthly sampling period.

Notes

1. Coastal and marine waters are those subject to the rise and fall of the tide.
2. Appropriate treatment is defined as the degree of treatment with disinfection required for the receiving waters to meet their assigned state or interstate classification and to meet the objectives of the water quality standards. Disinfection from October 1 to May 1 may be discontinued at the discretion of the Division of Water Pollution Control.
3. The water quality standards do not apply to conditions brought about by natural causes.
4. The waters shall be substantially free of pollutants that will (a) unduly affect the composition of bottom fauna, (b) unduly affect the physical or chemical nature of the bottom, or (c) interfere with the spawning of fish or their eggs.
5. The standards shall apply at all times in coastal and marine waters.
6. The amount of disinfection required shall be equivalent to a free and combined chlorine residual of at least 1.0 mg/liter after 15 minutes contact time during peak hourly flow or maximum rate of pumpage.

Index